Liquid State Chemical Physics

R. O. Watts

Computer Center, The Australian National University, Canberra, Australia

I. J. McGee

Department of Applied Mathematics, University of Waterloo, Waterloo, Ontario, Canada

A Wiley-Interscience Publication

John Wiley & Sons

New York · London · Sydney · Toronto

Library of Congress Cataloging in Publication Data

Watts, Robert Oliver.
Liquid state chemical physics.

"A Wiley-Interscience publication."
Includes bibliographical references and index.
1. Liquids. 2. Fluids. 3. Molecular dynamics.
4. Statistical mechanics. I. McGee, Ian James, 1937– II. Title.

QD541.W33 530.4′2 76-21793
ISBN 0-471-91240-9

Printed in the United States of America

10 9 8 7 6 5 4 3 2 1

Preface

During the past two decades the field of fluid state chemical physics has seen remarkable advances both in experimental procedures and in theoretical understanding. Prior to 1960 there was a distinct difference between the problems studied by experimentalists and those studied by theoreticians. Thus, the former were very much concerned with substances that were liquid at room temperature while the latter were still investigating the statistical mechanics of simple models such as hard spheres. In recent years significant improvements in experimental techniques and in computer technology have enabled the two approaches to converge so that both experimentalists and theoreticians can work together on the same problem. In this book the theory fundamental to advances in both of the two fields is developed in order that the interested reader may reach a reasonable understanding of the chemical physics of fluids. The book is written for use at the Honors year undergraduate and postgraduate levels and is intended to form the basis of a one-semester course in the subject. In addition we believe it will be a useful supplement to research groups which have been largely experimental.

In the book we have separated the fundamental theory needed to understand the relationship between intermolecular forces and the properties of fluids and applications of that theory to real systems. Following an introductory chapter an account is given of recent advances in the determination of intermolecular forces, including an account of the quantum scattering cross-sections and the vibrational levels of van der Waals' dimers. The background to the theory of dense fluids is given in a chapter on equilibrium statistical mechanics and it is in this chapter that advances in the theory of liquids are reviewed. Chapter 4 develops the theory of time dependent phenomena in dense fluids and examines such topics as nonequilibrium thermodynamics, linear response theory, and memory function formalism.

The remaining chapters consider the application of the theory developed earlier to real systems and compare the predicted bulk properties with experimental results. Chapter 5 gives an outline of the theory needed to extract structural information from X-ray and neutron diffraction data and goes on to examine the ability of statistical mechanical theories to predict this data. Following this, the application of these methods to the prediction of thermodynamic data and transport coefficients is discussed.

In Chapter 7 we demonstrate that an accurate description of the interaction potentials of the inert gases can be obtained using experimental data together with the theory developed earlier. This chapter also investigates the importance of many body interactions in the inert gases, discusses the influence of quantum effects, and describes one approach to calculating the equilibrium properties of liquid helium.

The next two chapters are used to investigate the theory of molecular fluids. Chapter 8 pays particular attention to simple molecular fluids, such as nitrogen and carbon monoxide, and also gives an account of more complicated systems such as liquid benzene. Chapter 9 is used to place theoretical studies of water and ionic solutions into the same framework as that used to discuss simpler liquids. Perhaps one of the more important features of this chapter is a section comparing the classical Debye-Hückel theory, based on the Poisson-Boltzmann equation, with the corresponding statistical mechanical treatment. This should remove some of the "mystique" associated with water and ionic solutions and show that they can be considered in the same general terms as are other fluids. The final chapter gives an account of the theory of liquid metals and molten salts, including an outline of pseudopotential theory. Once again, it is shown that the theoretical methods developed in the first few chapters can be applied to very complex fluids.

The general theme of the book is that modern experimental and theoretical methods enable the chemical physicist to correlate a wide range of bulk properties of fluids in terms of the intermolecular forces. Furthermore, it shows that theories which have generally been considered as being in the realm of the theoretical physicists can now be adapted to include substances of interest to the chemist. The time has come when such theories should be incorporated into the traditional physical chemistry teaching curriculum, and the work discussed in this book offers one method by which this can be done.

The authors are indebted to several people without whose help this book would never have been completed. In particular we mention Professor I. G. Ross, who allowed his undergraduate students to be "guinea pigs" for substantial sections of the book, Dr. D. J. Evans, who commented extensively on the manuscript, and Mrs. Jo Mahon and Mrs. Cheryl Riddell, who typed early drafts of the manuscript. We are particular indebted to Mr. Z. J. Derlacki, who drew most of the diagrams, and to Miss Fran Rocke, who typed the final manuscript.

R. O. WATTS
I. J. MCGEE

Canberra, Australia
Waterloo, Canada
March 1976

Contents

Units of Measurement

In this book we have followed the practice common in low temperature liquid state theory and reported energies in kelvin, distances in angstroms and densities as molecules per cubic angstrom. The only exception to this rule occurs in Chapter 9, where energies are given in kcal mole^{-1}. To enable the reader to transfer readily between kelvin and other measures of the energy, we give the following conversion factors:

$$\begin{aligned} 1\ \mathrm{K} &\equiv 8.6173\times 10^{-5}\ \mathrm{eV} \\ &\equiv 1.38066\times 10^{-23}\ \mathrm{J} \\ &\equiv 1.9872\times 10^{-3}\ \mathrm{kcal\ mole^{-1}} \\ &\equiv 0.69503\ \mathrm{cm^{-1}} \\ &\equiv 3.1671\times 10^{-6}\ \mathrm{au} \end{aligned}$$

Liquid State Chemical Physics

One

The Physics of Fluids

Modern theories of the relation between intermolecular forces and the properties of fluids can probably be traced to two fundamental studies, one experimental and the other theoretical, reported about 100 years ago. In 1869 Andrews [1] presented a set of measurements of the equation of state of carbon dioxide in which he showed that it was possible to pass from the vapor to the liquid state without there being a noticeable change in phase. This process can be carried out by suitable changes in the thermodynamic state of the system. Thus, if a gas is heated at constant volume until it is above some *critical temperature*, T_c, compressed, and then cooled to the starting temperature it is found that the resulting fluid has all the properties of a liquid. We can see how this is done from Fig. 1.1, where the so-called Andrews isotherms are shown schematically. Starting in the gas phase, at A, where the temperature is less than T_c, the fluid is heated at constant volume to point B, isothermally compressed to point C, and then cooled at constant volume to reach point D in the liquid region.

The second advance was primarily theoretical, and was reported by van der Waals [2] in 1873. His approach was to develop an explicit description of the equation of state so that the pressure of the system was given by the equation

$$p = f(V, T) \tag{1.1}$$

where V is the volume of the fluid and T its temperature. The van der Waals theory gave an expression for $f(V, T)$ that depended on two parameters, a and b, to be determined from experimental data. The theory is rather simple to describe and does not tell us very much about the fundamental structure of dense fluids. Nevertheless, it represents probably the first successful description of the macroscopic properties of a

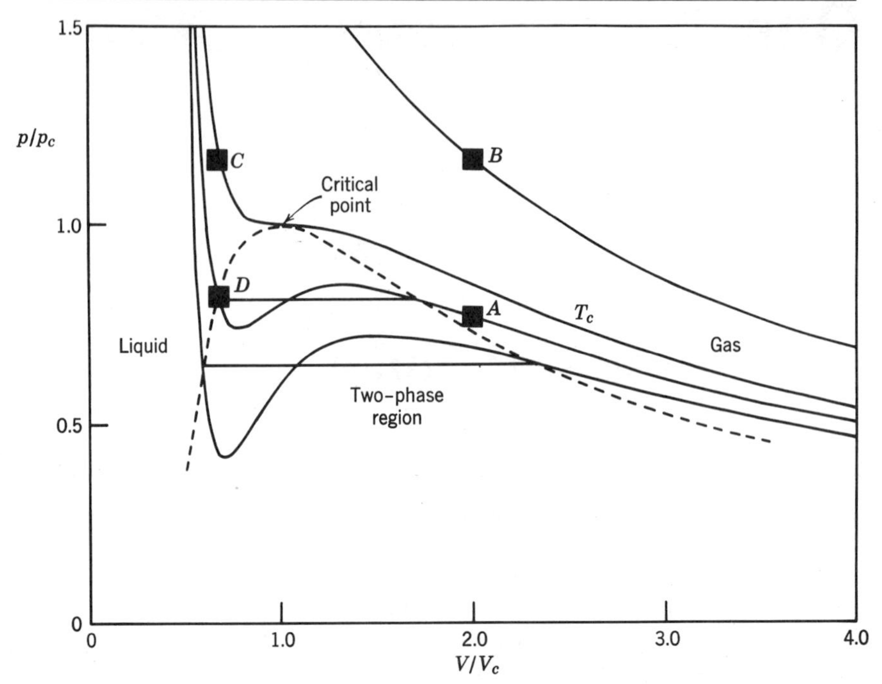

Figure 1.1 Andrews' isotherms calculated from the van der Waals equation.

fluid in terms of molecular properties and for that reason warrants further study.

1.1 The van der Waals Equation

Although the Andrews isotherms showed for the first time that it was possible to pass continuously between the gaseous and liquid phases, a proper understanding of the process took much longer to develop. The van der Waals theory, which was the first step toward an explanation of this aspect of fluid behavior, contains two important concepts. First, it attempts to develop a theory of macroscopic phenomena in terms of the properties of molecules, and second, it gives a formal description of the gas-liquid phase change. To see these properties we have to examine Fig. 1.1 again.

The most important qualitative features in Fig. 1.1 are associated with the *critical point.* From the figure we see that the critical temperature is associated with a critical volume, V_c, and a critical pressure, p_c. At the point (T_c, V_c, p_c) the slope of the function represented by eq. 1.1 is zero, so that

$$\left(\frac{\partial p}{\partial V}\right)_{T_c, V_c, p_c} = 0$$

Below the critical temperature, if a gas is compressed isothermally there is a sudden change in the character of the fluid as the *coexistence curve* is reached, and compressing the gas further takes the fluid into the liquid state. As is well known, the phase change occurs when the chemical potentials of the two phases are equal [3]. The importance of van der Waals' work was that it gave a mathematical description of these properties of a fluid. He proposed that the form of $f(V, T)$ in eq. 1.1 should be

$$\left(p + \frac{N^2 a}{V^2}\right)(V - Nb) = NkT \tag{1.2}$$

where a and b are parameters to be determined, N is the number of particles in the system, and k is Boltzmann's constant.

From the point of view of describing the properties of a fluid in terms of the properties of its constituent particles, parameters a and b are of particular interest. Van der Waals argued that two properties of molecules give rise to the gas-liquid phase change, namely, the strongly repulsive forces found when the molecules are close together and the weaker attractive forces that act when the molecules are further apart. If the fluid is at a relatively high temperature the most important property of a molecule is that it occupies a finite volume. Thus, even if the fluid is strongly compressed, at high temperatures the repulsive forces acting between the molecules imply that the volume of the system cannot be reduced below some limiting volume Nb. However, van der Waals also recognized that the attractive forces between the molecules are important at lower temperatures. He argued that these attractive forces would increase the pressure acting on the molecules and hence the term $N^2 a/V^2$ was added to p. Consequently, he modified Boyle's law, $pV = NkT$, by increasing the pressure of the system and decreasing the available volume. The result of these changes is eq. 1.2, and it is important to realize that this equation recognizes both attractive and repulsive intermolecular interactions.

The van der Waals equation also gives a good qualitative description of the Andrews isotherms. In particular it predicts the existence of a critical point, and this can be found by using the fact that at this point there is a point of inflection on the p-V isotherm

$$\frac{\partial p}{\partial V} = 0; \qquad \frac{\partial^2 p}{\partial V^2} = 0$$

Using these results it is found that

$$p_c = \frac{a}{27b^2}; \qquad T_c = \frac{8a}{27bk}; \qquad V_c = 3Nb \tag{1.3}$$

At temperatures less than T_c the p-V isotherms predicted by the van der Waals equation rise steadily from zero, at large V, cross the coexistence curve, and then reach a maximum. The isotherms then fall to some minimum value and rise rapidly, crossing the liquid side of the coexistence curve before going to very high pressures at small volumes. Figure 1.1, in fact, gives a description of the van der Waals equation for various temperatures. For clarity the parameters a and b were chosen so that $p_c = 1$, $T_c = 1$, and $V_c = 1$.

There are two main differences between the van der Waals equation and the experimentally determined Andrews isotherms. The first is the behavior of the fluid within the region bounded by the coexistence curve. Experimentally, much of this region is inaccessible although the mechanically unstable superheated liquid can be thought of as representing the fluid state in part of this region. The van der Waals equation is continuous through this region and has to be supplemented with thermodynamic calculations before it predicts the discontinuous gas-liquid transition. In addition, the negative value of the compressibility, $(\partial p/\partial V)_T$, in the two-phase region is not found experimentally. The second major difference between the behavior of real fluids and the van der Waals equation, which cannot be seen from the figure, is the qualitatively different behaviors of the experimental and theoretical critical isotherms.

Briefly, it is found that the van der Waals equation gives an incorrect mathematical description of the way in which the fluid approaches the critical point. A good example is the way in which the densities of the gas and liquid tend to the critical density as the two fluid phases approach this point along the coexistence curve. The van der Waals equation predicts

$$\rho_l(T) - \rho_g(T) = B(T_c - T)^\beta \tag{1.4}$$

where the index $\beta = 0.5$, ρ_i is the number density of phase i, and B is some constant. Experimentally, it is found that $\beta = 0.3 - 0.37$ [3]. This difference between the van der Waals theory and the experiment is rather important and has significant consequences for theories of the critical point. It can be shown that if the Helmholtz free energy, A, is an analytic function of V and T everywhere, then a Taylor expansion of A in powers of $(V - V_c)$ and $(T - T_c)$ gives $\beta = 0.5$ [3]. Consequently, the experimental measurements suggest that the free energy of a fluid is not an analytic function of V and T in the neighborhood of the critical point. It is now well established that real fluids do not follow the van der Waals equation near the critical point, and a careful study of the region has been made both experimentally [4] and theoretically [5]. Unfortunately, a complete description of the critical point behavior of various systems is outside the scope of this book and the reader is referred to other sources [3–5] for a detailed discussion.

1.2 Molecular Properties from Bulk Data

For our purposes the most important concept embodied in the van der Waals theory is the suggestion that by estimating parameters a and b from experimental measurements the intermolecular potential can be determined from the bulk properties of a material. This idea has existed for a very long time and can be traced back to the work of Newton if not earlier [6]. Briefly, the idea is to use measurements of certain properties of a system to determine the forces acting between the particles in that system, and then to use these forces to predict other phenomena. In a review published in 1970 Brush [6] concluded that such a program was not feasible. Nevertheless, within a year the argon-argon potential had been so accurately established, primarily from bulk data [7], that it was possible to predict certain properties of the solid more accurately than had been measured at that time [8, 9]. It is of interest to examine the background to the argon work as it is the prime motivation for writing this book.

Until about 1960, the argon-argon interaction had been modeled using simple functional forms containing no more than two or three parameters determined from bulk data. Although such models had limited success in that they were able to correlate a restricted range of bulk properties, it was not possible to find one set of parameters that would correlate all the bulk properties of a given substance. Apparently, the interaction potential between a group of atoms or molecules cannot be described by a simple analytic form. In an attempt to overcome this problem Guggenheim and McGlashan [10] introduced more flexibility into the modeling procedure, primarily by using three different functional forms to describe the interaction at various particle separations. This procedure was continued [11] until Barker and Pompe [12] introduced in 1968 a single functional form that accurately described the argon-argon interaction. Their model was very flexible and contained a relatively large number of adjustable parameters. In addition, the model had the correct behavior at long distances, agreeing with accurate quantum mechanical calculations in this region [13]. By this time, it was also apparent that even in the simple inert gases, interactions between many particles could not be written as a sum of pairwise additive terms [14]. The Barker-Pompe model explicitly allowed for three-body interactions in calculations of the bulk properties of argon. During the next few years the Barker-Pompe potential was refined first by Barker and Bobetic [15] using a range of solid state data and then by Barker, Fisher, and Watts [7] who used liquid state data, until it gave excellent agreement with practically all the bulk properties of argon.

There are two very important lessons to be learned from the development of the Barker model. First, it is not possible to derive an accurate description of the interaction potential between two particles unless a very wide range of experimental data is available. In particular, it is important to use data from the gas, liquid, and solid phases and to ensure that the data used are accurate. There were several occasions during the development of the argon potential when errors were detected in experimental properties, the best known being the problem with gas viscosities [12, 16]. Second, it is important that the model make use of all the available theoretical information. For example, if it is known from other arguments that the potential has a particular functional form at certain distances, then the model must also have that functional form. Similarly, if any estimates of atomic or molecular properties exist, such as values of electrical polarizabilities or permanent electrostatic moments, these quantities must also be included in the model. If this is done, it is possible to obtain accurate descriptions of the microscopic behavior of atoms and molecules from bulk data.

We have given some priority to the Barker model in this section primarily because it represents in a real sense the culmination of an idea first expressed by Newton. That is, it has given the first highly accurate interaction potential that has been derived primarily on the basis of experimental measurements of bulk data. However, it is important to note that this process could not have been done without a corresponding improvement in the theoretical methods available for calculating bulk data from a knowledge of the interaction potential. During the past 20 years there has been a significant improvement in theory. In particular, we can single out the computer simulation methods developed for both equilibrium [17] and nonequilibrium [18] calculations, the perturbation theories used to compute thermodynamic properties of complex systems from the properties of simple model fluids [19], and the various integral equation approximations [20]. More recently, there has been an improvement in the understanding of transport processes in dense fluids [21], and this has also given greater insight into the fluid state.

The remaining chapters of this book develop the various aspects of fluid state theory needed to describe the behavior of bulk systems in terms of intermolecular interactions. Chapter 2 gives a condensed account of the theory of intermolecular forces and discusses the recent advances in collision phenomena that has made their direct measurement possible. Following this, Chapters 3 and 4 give an account of the equilibrium and nonequilibrium statistical mechanics needed to compute bulk properties of fluids. The structure and thermodynamic properties of simple fluids are discussed in the next two chapters, and some emphasis is given to the utility of various approximate theories of liquids. Chapter 7 gives a full account of the current status of inert gas interactions, including a discussion of the theory needed to compute the equilibrium properties of liquid helium. Following this, we discuss the application of the theory developed in Chapters 2 through 4 to more complex fluids. Thus Chapter 8 gives an account of the theory of fluids composed of nonspherical molecules, Chapter 9 gives an account of recent studies of water and ionic solutions, and Chapter 10 discusses liquid metals and molten salts. The result is, we hope, a reasonably comprehensive account of the current state of the theory of dense fluids.

References

1. T. Andrews, 1869, *Phil. Trans. Roy. Soc.*, **159,** 575.
2. J. D. van der Waals, 1873, doctoral dissertation, Leiden.
3. J. S. Rowlinson, 1969, *Nature* (*London*), **224,** 541; J. M. H. Levelt-Sengers, 1973, *Physica*, **73,** 73; J. S. Rowlinson, 1972, *Ber. Bunsenges. Phys. Chem.*, **76,** 281.
4. H. L. Lorentzen, 1965, in *Statistical Mechanics of Equilibrium and Non-equilibrium*, J. Meixner, Ed., p. 262, North Holland, Amsterdam; H. W. Habgood and W. G. Schneider, 1954, *Can. J. Chem.*, **32,** 98; P. R. Roach, 1968, *Phys. Rev.*, **170,** 213.
5. B. Widom, 1965, *J. Chem. Phys.*, **34,** 3898; R. B. Griffiths, 1967, *Phys. Rev.*, **158,** 176; L. P. Kadanoff, W. Götze, D. Hamblen, R. Hecht, E. A. S. Lewis, V. V. Palciauskas, M. Rayl, J. Swift, D. Aspnes, and J. Kane, 1967, *Rev. Mod. Phys.*, **39,** 395.
6. S. G. Brush, 1970, *Arch. Rational Mech. Anal.*, **39,** 1.
7. J. A. Barker, R. A. Fisher, and R. O. Watts, 1971, *Mol. Phys.*, **21,** 657.
8. R. A. Fisher and R. O. Watts, 1972, *Mol. Phys.*, **23,** 1051.
9. S. Gewurtz, H. Kiefte, D. Landheer, R. A. McLaren, and B. P. Stoicheff, 1972, *Phys. Rev. Lett.*, **29,** 1454.
10. E. A. Guggenheim and M. L. McGlashan, 1960, *Proc. Roy. Soc.* (*London*), *Ser. A*, **255,** 456.
11. M. L. McGlashan, 1965, *Discuss. Faraday Soc.* **40,** 59; A. E. Sherwood and J. M. Prausnitz, 1964, *J. Chem. Phys.*, **41,** 429; J. C. Rossi and F. Danon, 1965, *Discuss. Faraday Soc.*, **40,** 97; R. J. Munn and F. J. Smith, 1965, *J. Chem. Phys.*, **43,** 3998; D. D. Konowalow and S. Carrá, 1965, *Phys. Fluids*, **8,** 1585; J. H. Dymond and B. J. Alder, 1969, *J. Chem. Phys.*, **51,** 309.
12. J. A. Barker and A. Pompe, 1968, *Aust. J. Chem.*, **21,** 1683.
13. P. J. Leonard, 1968, Masters Thesis, University of Melbourne; A. E. Kingston, 1964, *Phys. Rev.*, **135,** A1018; A. Dalgarno and A. E. Kingston, 1961, *Proc. Phys. Soc.* (*London*), **78,** 607; R. J. Bell and A. E. Kingston, 1966, *ibid.*, **88,** 901.

14. W. Götze and H. Schmidt, 1966, *Z. Phys.*, **192,** 409; G. C. Chell and I. J. Zucker, 1968, *J. Phys., C. (London)*, **1,** 35, 1505; A. Hüller, W. Götze, and H. Schmidt, 1970, *Z. Phys.*, **231,** 173.
15. M. V. Bobetic and J. A. Barker, 1970, *Phys. Rev.*, **B2,** 4169.
16. In their paper, Barker and Pompe [12] suggested that the then current gas viscosities were too low, particularly at high temperatures. They came to this conclusion after making many attempts to find a single potential function that fitted both the argon viscosities and all other data. In a note added in proof, Barker and Pompe drew attention to new measurements of the viscosities that were in agreement with their predictions.
17. N. Metropolis, A. W. Rosenbluth, M. N. Rosenbluth, A. H. Teller, and E. Teller, 1953, *J. Chem. Phys.*, **21,** 1087.
18. A. Rahman, 1964, *Phys. Rev.*, **136,** *A*405; B. J. Alder and T. E. Wainwright, 1960, *J. Chem. Phys.*, **33,** 1439; *ibid.*, 1959, **31,** 459.
19. W. R. Smith, 1973, "Perturbation Theory in the Classical Statistical Mechanics of Fluids," Chapter 2 in *Specialist Periodical Reports of the Chemical Society: Statistical Mechanics*, K. Singer, Ed., Vol. 1, Chemical Society, London.
20. R. O. Watts, 1973, "Integral Equation Approximations," Chapter 1 in *Specialist Periodical Reports of the Chemical Society: Statistical Mechanics*, K. Singer, Ed., Vol. 1, Chemical Society, London.
21. H. Mori, 1965, *Progr. Theor. Phys. (Kyoto)*, **33,** 423; *ibid.*, 1965, **34,** 399; R. W. Zwanzig, 1960, in *Lectures in Theoretical Physics*, Vol. 3, p. 106, W. E. Britten, B. W. Downs, and J. Downs, Eds., Interscience, New York; R. Kubo, 1958, in *Lectures in Theoretical Physics*, Vol. 1, p. 120, W. E. Britten, L. G. Dunham, Eds., Interscience, New York.

Two

Intermolecular Forces and Scattering Phenomena

2.1 Quantum Mechanics of Intermolecular Potentials

When two particles are held at a fixed distance apart there exists a potential energy of interaction which, taking the total energy of the particles at infinite separation to be zero, may be either positive or negative. In principle it is possible to calculate this potential energy using quantum mechanics [1, 2] but in practice for most systems of interest the accuracy of such calculations is severely limited. Consequently, it is common to assume some model of the interaction containing adjustable parameters that are determined by fitting the model to various properties of the system. Before these empirical and semiempirical models are discussed the quantum mechanical problem is examined briefly.

In quantum mechanics the state of a system at equilibrium is described by a wave function, ψ_T, that is a function of the positions of all the particles in the system and is specified by a set of quantum numbers. We assume throughout this chapter that all the atoms and molecules in the system are in their ground state—their quantum numbers have their lowest allowed values. The wave functions are obtained as solutions to the Schrödinger equation

$$H_T\psi_T = E_T\psi_T \tag{2.1}$$

where H_T is the total Hamiltonian operator and E_T, the energy eigenvalue, is the total energy of the system in the particular quantum state being considered.

The Hamiltonian operator, H_T, can be written as the sum of two parts

$$H_T = H_N + H_e \tag{2.2}$$

where the nuclear Hamiltonian H_N is the sum of the nuclear kinetic energy operators and the electronic Hamiltonian H_e is the sum of the electron kinetic energy operators together with the Coulombic interactions. The latter interactions include nucleus-electron, electron-electron, and nucleus-nucleus terms. To simplify the problem we *assume* that the electrons are moving much more quickly than the nuclei and write the total wave function of the system as a product of two terms

$$\psi_T(\mathbf{R}_i, \mathbf{r}_j) = \psi_N(\mathbf{R}_i)\psi_e(\mathbf{r}_j \mid \mathbf{R}_i) \tag{2.3}$$

where $\mathbf{R}_i$ are vectors giving the positions of the nuclei and $\mathbf{r}_j$ are vectors giving the positions of the electrons in some arbitrary reference frame. This assumption is known as the Born-Oppenheimer approximation [3] and asserts that the nuclear positions are independent of the electron coordinates, that is, there is no coupling between the electronic and nuclear motions.

When the electronic Hamiltonian H_e used in eq. 2.2 is applied to an electronic wave function ψ_e the energy eigenvalues depend on the positions of the nuclei

$$H_e\psi_e = E(\mathbf{R}_i)\psi_e \tag{2.4}$$

If we apply H_T, written as in eq. 2.2, to the approximate wave function given in eq. 2.3 we obtain

$$(H_N + H_e)\psi_N\psi_e = E_T\psi_N\psi_e$$

which leads to

$$[H_N + E(\mathbf{R}_i)]\psi_N = E_T\psi_N \tag{2.5}$$

That is, H_N is the kinetic energy operator and $E(\mathbf{R}_i)$ is the potential energy term for the *nuclear problem.* It is usual to redefine the energies $E(\mathbf{R}_i)$ and E_T as follows

$$E(\mathbf{R}_i) = \Phi(\mathbf{R}_i) + \sum_k E_k \tag{2.6}$$

$$E_T = \Delta_T + \sum_k E_k \tag{2.7}$$

where E_k is the energy of the kth particle in the system when it is isolated from the other molecules but is otherwise in the same electronic state. Equations 2.6 and 2.7 define the total potential energy of the system $\Phi(\mathbf{R}_i)$ in such a way that when all the particles are infinitely far apart both the potential energy of the system and the total energy Δ_T are zero.

Having defined the potential energy in this way we see that two infinitely separated groups of molecules do not interact. This leads directly to the assumption that the total potential energy can be written as a sum of simpler interactions

$$\Phi(\mathbf{R}_i) = \sum_{i<j} \phi^{(2)}(\mathbf{R}_i, \mathbf{R}_j) + \sum_{i<j<k} \phi^{(3)}(\mathbf{R}_i, \mathbf{R}_j, \mathbf{R}_k) + \cdots \tag{2.8}$$

where $\phi^{(2)}(\mathbf{R}_i, \mathbf{R}_j)$ is a two-body potential function depending only on the positions of nuclei in particles i and j and $\phi^{(3)}(\mathbf{R}_i, \mathbf{R}_j, \mathbf{R}_k)$ is a three-body potential function depending only on the positions of nuclei in particles i, j, and k. The two-body function can be obtained by solving eq. 2.5 written for a system containing two particles only; to calculate $\phi^{(3)}$ eq. 2.5 is solved for a three-particle system, taking account of the fact that the total potential energy includes contributions from pair interactions

$$\Phi(\mathbf{R}_i) = \phi_{12}^{(2)} + \phi_{13}^{(2)} + \phi_{23}^{(2)} + \phi_{123}^{(3)}$$

Given that the $\phi_{ij}^{(2)}$ have been determined previously, the unknown quantity in this equation is $\phi_{123}^{(3)}$.

2.2 Short Range and Long Range Pair Potentials

In general, for particles in the gas, liquid, or solid phases the two-body potential function $\phi^{(2)}(\mathbf{R}_1, \mathbf{R}_2)$ depends on the relative positions and relative orientations of two molecules. For the case of two atoms or ions the spherical symmetry of the particles ensures that $\phi^{(2)}$ is a function of the distance between the centers of the atoms alone, $\phi^{(2)}(r)$. Three regions of interest in the potential energy function of a nonmetallic atom-atom interaction are shown in Fig. 2.1. At short distances where strong repulsive forces arise from the overlap of electron clouds, and at long distances where the electron clouds are widely separated, accurate quantum mechanical calculations of the potential are possible. The third and most important region is near the minimum in the potential where it is extremely difficult to obtain accurate information. Two parameters that are particularly important when discussing atom-atom interactions are indicated in the figure. The effective diameter of an atom, σ, is usually taken to be the position of the first zero in the potential, and the maximum well depth, ε, is the usual measure of the strength of the interaction. Typical values of the parameters ε and σ for the noble gases are given in Table 2.1. A detailed account of the inert gas interactions is given in Chapter 7, results for intermolecular interactions are discussed in

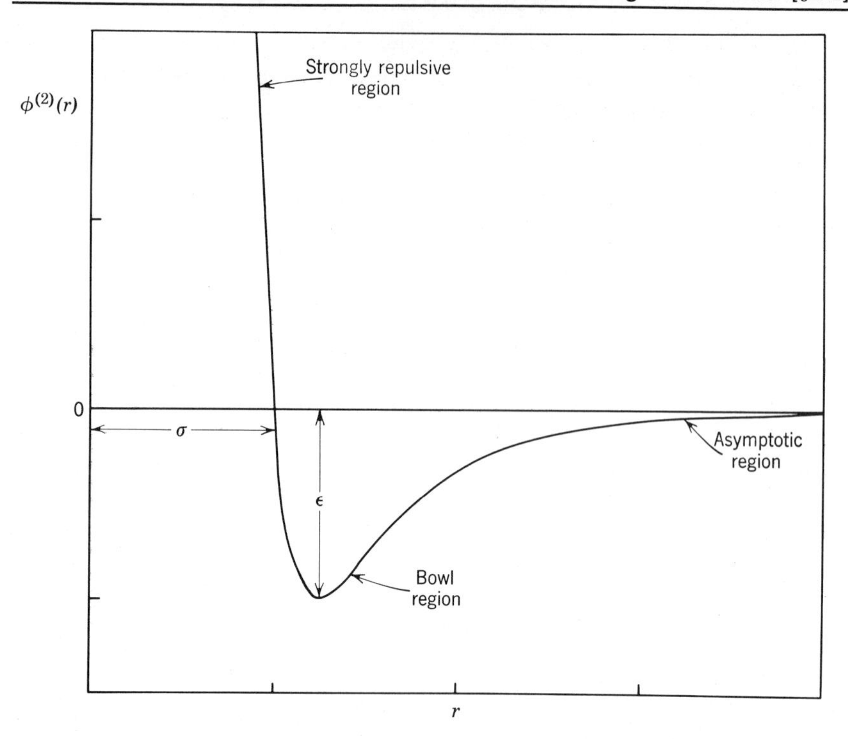

Figure 2.1 Representation of the pair potential for two spherical particles.

Table 2.1 Values of the position of the zero, σ, and depth of the minimum, ε/k, for the inert gas interactions. The values are typical of those used with the Lennard-Jones 12-6 potential

Atoms	He-He	Ne-Ne	Ar-Ar	Kr-Kr	Xe-Xe
σ (Å)	2.556	2.749	3.405	3.60	4.10
ε/k (K)	10.22	35.60	119.8	171.0	221.0

Chapters 8 and 9, and effective pair potentials for liquid metals and molten salts are considered in Chapter 10.

Although the full interaction potential is known with accuracy only for the noble gases, the details of the strongly repulsive and the long range regions are reasonably well documented for a number of systems. High energy molecular beam experiments have been used to measure the interaction potential at close distances [4], and many of the simpler quantum mechanical calculations are accurate in this region also [5]. From a combination of experimental and theoretical results it is now generally accepted that at close distances the repulsion is best expressed by an exponential decay of the form

$$\phi_{\text{rep}}(r) = A \exp(-\alpha r - \beta r^2 - \gamma r^3) \tag{2.9}$$

where A, α, β, and γ are coefficients that are usually obtained by fitting to the available data (either experimental or theoretical). In many cases only A and α are used but β and γ are sometimes included to increase the flexibility.

Considerably more is known about the long range or van der Waals interactions, but no more than a brief summary can be given here. For our purposes these interactions can be divided into three terms [6]: the electrostatic contribution, the induction contribution, and the dispersion contribution. The electrostatic contributions to the total potential energy arise from the interaction of permanent electrostatic moments on the atoms or molecules being considered. Perhaps the simplest example is the Coulomb interaction arising from permanent charges on ions

$$\phi(C, C) = \frac{z_1 z_2 e^2}{r} \tag{2.10}$$

where z_1 and z_2 are the number of charges on each particle and e is the electronic charge; z_i is a positive number for a positively charged species and negative for a negatively charged species.

Suppose now that we have two interacting systems of charges representing two atoms or molecules. The permanent electrostatic interaction potential between the two systems can be written as a sum of Coulombic terms

$$\phi = \sum_{i=1}^{N} \sum_{j=1}^{M} \frac{z_i z_j e^2}{r_{ij}} \tag{2.11}$$

where the sum over i runs over all particles in the first system, the sum over j runs over all particles in the second system, and where r_{ij} is the distance between the two charges. Notice that this equation does not include the interaction between the charges within either system. Now, let the center of one atom or molecule be chosen as the origin of some coordinate system so that the positions of the charges in that atom or molecule are given by the vectors $\mathbf{r}_i$. Let the center of the second system of charges be at $\mathbf{R}$ and its component charges be at $\mathbf{R}+\mathbf{r}_j$. Then we have

$$r_{ij} = |\mathbf{R} + \mathbf{r}_j - \mathbf{r}_i| \tag{2.12}$$

If the distance between the two charge systems $R = |\mathbf{R}|$ is sufficiently large the total potential energy can be expanded in a power series in inverse powers of R [2, 6]. The terms in this expansion represent interactions between the various permanent electrostatic moments on the molecule. The 2^nth electrostatic moment is an n dyadic, given by the sum

$$\boldsymbol{\phi}^{(n)} = \sum_i z_i e \mathbf{r}_i \cdots \mathbf{r}_i \tag{2.13}$$

where there are n outer products forming the tensor and where the sum is taken over all the charges in one of the molecules. As examples, the first three moments are the total charge

$$C = \sum_i z_i e \tag{2.14}$$

the dipole moment

$$\boldsymbol{\mu} = \sum_i z_i e \mathbf{r}_i \tag{2.15}$$

and the quadrupole moment

$$\boldsymbol{\Theta} = \sum_i z_i e \mathbf{r}_i \mathbf{r}_i \tag{2.16}$$

For many purposes it is convenient to redefine $\boldsymbol{\Theta}$ as a traceless tensor

$$\mathbf{Q} = \tfrac{1}{2} \sum_i z_i e (3\mathbf{r}_i \mathbf{r}_i - \mathbf{r}_i^2 \boldsymbol{I}) \tag{2.17}$$

where $\mathbf{I}$ is the unit tensor. Note that this definition of $\mathbf{Q}$ is half that given by Hirschfelder et al. [2].

Returning to the expansion, it can be shown that the leading terms in the multipole expansion of the total energy are as follows

charge-charge

$$\phi(C, C) = \frac{C_1 C_2}{R} \tag{2.18}$$

charge-dipole

$$\phi(C, \mu) = -\frac{C_1 \mu_2}{R^2} \cos \theta_2 \tag{2.19}$$

dipole-dipole

$$\phi(\mu, \mu) = -\frac{\mu_1 \mu_2}{R^3} [2 \cos \theta_1 \cos \theta_2 - \sin \theta_1 \sin \theta_2 \cos(\phi_1 - \phi_2)] \tag{2.20}$$

where the angles θ_1, θ_2, ϕ_1, ϕ_2 are given in Fig. 2.2. The quantities C_i and μ_i represent the total charge and the size of the dipole moment of molecule i, respectively. Formulae for the interaction between other electrostatic moments can be found in the literature [2, 6].

For some purposes it is useful to remove the angle dependence from the dipole-dipole interaction (eq. 2.20). If this term is averaged over all orientations it vanishes. However, such a procedure does not allow for the fact that certain configurations are more stable, and hence more likely to be found than are others. An appropriate method of introducing some weight function is to use the Boltzmann distribution. This distribution function measures the probability of finding two particles in a particular configuration at a given temperature and is a special case of the many-body distribution function derived in the next chapter. We have [7]

$$\bar{\phi}(\mu, \mu) = \frac{\int_0^{2\pi} d\phi_1 \int_0^{2\pi} d\phi_2 \int_0^1 d\cos\theta_1 \int_0^1 d\cos\theta_2 \phi(\mu_1, \mu_2) \exp[-\phi(\mu_1, \mu_2)/kT]}{\int_0^{2\pi} d\phi_1 \int_0^{2\pi} d\phi_2 \int_0^1 d\cos\theta_1 \int_0^1 d\cos\theta_2 \exp[-\phi(\mu_1, \mu_2)/kT]}$$

where k is Boltzmann's constant. At sufficiently high temperatures the exponentials can be written $\exp[-\phi/kT] \sim 1 - \phi/kT$ and the integrals evaluated to give

$$\bar{\phi}(\mu, \mu) = -\frac{2\mu_1^2\mu_2^2}{3kTR^6} \tag{2.21}$$

This result was reported by Keesom [7] and is often referred to as the Keesom potential.

All atoms and molecules are polarizable to some extent in that the electrons are not rigidly bound to the various nuclei. Consequently, the electric field arising from a permanent electrostatic moment on one molecule can induce electric moments in a second [8, 9]. Suppose a neutral particle has isotropic polarizability α, and an ion has charge C, then an electrostatic dipole is induced in the neutral particle and this

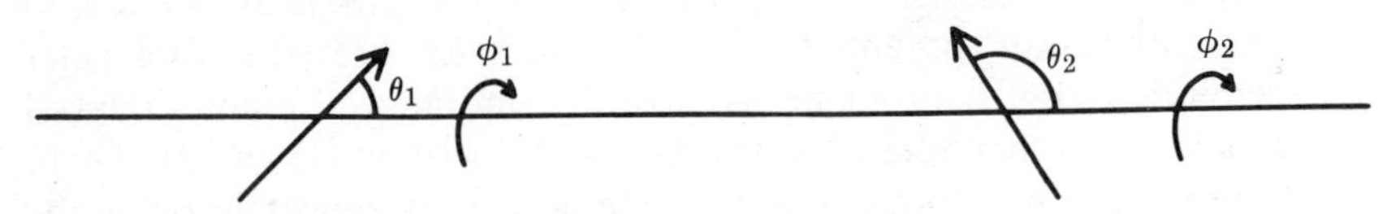

Figure 2.2 Angles used to compute interaction potential of two-point dipoles. The arrows indicate the directions of the dipole vectors.

interacts with the charged particle, a distance R away, to give the potential energy

$$\phi(R) = -\frac{C^2\alpha}{2R^4} \tag{2.22}$$

If the electric moment on one particle is a permanent point dipole then the corresponding expression is

$$\phi(R) = -\frac{\mu^2\alpha(3\cos^2\theta_1 + 1)}{2R^6} \tag{2.23}$$

where θ_1 is the angle between the direction of the dipole and thc linc joining the centers of the particles. If this interaction is averaged over angles, using the Boltzmann weighting as before, the well known expression for the Debye potential is obtained:

$$\phi(R) = -\frac{\mu^2\alpha}{R^6} \tag{2.24}$$

The permanent induced contribution to the intermolecular forces can be extended to groups of three or more molecules and can be quite important in systems with large polarizabilities and strong electrostatic moments.

The third contribution to the long range potential energy is that arising from the London dispersion forces [10]. These terms exist even between neutral atoms and are quantum mechanical in origin. An intuitive model of these interactions can be constructed as follows. At any instant the electrons in an atom will be displaced to one side of the atom giving rise to a temporary dipole moment. This dipole moment will induce a moment on a neighboring particle and then interact with this to give rise to an attractive force. If this idea is explored quantum mechanically it is found that the induced dipole-induced dipole interaction is given approximately for spherical particles by the expression [10]

$$\phi(R) = -\frac{3}{4}\frac{h\nu_1 h\nu_2}{(h\nu_1 + h\nu_2)}\frac{\alpha_1\alpha_2}{R^6} = -\frac{C_6}{R^6} \tag{2.25}$$

where $h\nu_i$ is a characteristic energy of the particle approximately equal to the first ionization potential, and α_1 and α_2 are the molecular polarizabilities. Higher order terms, arising from induced dipole-induced quadrupole and other interactions can also be written down [2]. These terms are in increasing powers of R^{-1}, the next most important being the induced dipole-induced quadrupole term, $-C_8/R^8$, followed by a term $-C_{10}/R^{10}$. At very large separations retardation effects due to the finite velocity of light become apparent [11], but the differences are not important in systems considered in this book.

An important extension of the induced dipole-induced dipole interaction is the triple induced dipole interaction or Axilrod-Teller interaction [12] occurring between three atoms. This term is particularly important for the noble gases at medium and high densities and in the liquid and solid states. It is given by the expression

$$\phi^{(3)}(R_1, R_2, R_3) = \frac{\nu(1 + 3\cos\theta_1\cos\theta_2\cos\theta_3)}{R_1^3 R_2^3 R_3^3} \tag{2.26}$$

where ν, a coefficient related to polarizabilities and ionization potentials, is given approximately by $\nu = \frac{3}{4}\alpha C_6$, and θ_1, θ_2, θ_3, and R_1, R_2, R_3 are the angles and sides of the triangle described by the centers of the atoms. Many body interactions also occur between larger groups of molecules and between higher order induced-induced moments. The long range contribution to the intermolecular potential is obtained by summing contributions from the permanent, induced, and dispersion interactions.

At present there is no information available enabling the intermediate region of the interatomic potential function to be described analytically. Consequently it is usual to assume that in this region the interaction is described by some empirical form with a number of adjustable parameters, the values of the parameters being determined partly from experimental data and partly by the constraint that at short and long distances the model must go smoothly into the known results. In the following section of this chapter and in Chapters 7, 8, 9, and 10 a number of these empirical and semiempirical models are described.

2.3 Analytic Potential Functions

Analytic potential functions are used for two purposes in the theory of fluids and solids. First, simple models containing qualitatively accurate information are used to give some indication of the information that is important in determining the bulk properties of the system and second, flexible multiparametric forms are used to make quantitative statements about real systems. This section gives examples of both types of potential function, beginning with simple spherically symmetric models of the inert gases and continuing with models used for molecular systems such as water and benzene. Properties of systems modeled by these functions are introduced at appropriate places in subsequent chapters.

The simplest spherically symmetric potential useful in liquid state physics is the hard sphere model

$$\phi(r)=\begin{cases}\infty & r<\sigma\\ 0 & r\geq\sigma\end{cases} \tag{2.27}$$

Although there are no attractive terms in the potential, it represents the strongly repulsive region (very idealized!) and has given a great deal of information on both the static structure of fluids [13] and on certain time dependent properties [14]. It is also used as the reference potential in a number of perturbation theories [15] of liquids, as shown in Section 3.9. Properties of the hard sphere system are considered subsequently in several sections.

A related model that does allow for the attractive well of the potential is the square well potential

$$\phi(r)=\begin{cases}\infty & r<\sigma\\ -\varepsilon & \sigma\leq r<a\sigma\\ 0 & a\sigma\leq r\end{cases} \tag{2.28}$$

This is a three parameter model containing the hard sphere diameter σ, the width of the well, $(a-1)\sigma$, and the well depth, ε. It is introduced here for completeness and is not used very widely in this book. Considerably more important as a model of spherically symmetric interactions is the Lennard-Jones 12-6 potential [16]

$$\phi(r)=4\varepsilon\left[\left(\frac{\sigma}{r}\right)^{12}-\left(\frac{\sigma}{r}\right)^{6}\right] \tag{2.29}$$

a two-parameter potential with well depth ε and effective diameter (position of the zero) of σ. Values of ε and σ that are useful for a number of systems are given in Table 2.1. This potential has been very widely used as a model of the noble gas interactions, and results from it are considered in detail in later chapters. The hard sphere, square well (with $a=2.0$) and Lennard-Jones 12-6 models are shown in Fig. 2.3. Several other simple parametric representations of spherically symmetric potentials have been proposed, the best known being the exp-6 potential

$$\phi(r)=A\exp(-\alpha r)-\frac{B}{r^{6}} \tag{2.30}$$

where A, α, and B are adjustable parameters chosen to give good agreement with one or more sets of experimental data. The exp-6 potential is particularly useful in the analysis of structural data for organic crystals, and parameters for the common atom-atom interactions found in such substances have been reported by Williams [17] and Kitaigorodskii [18] among others.

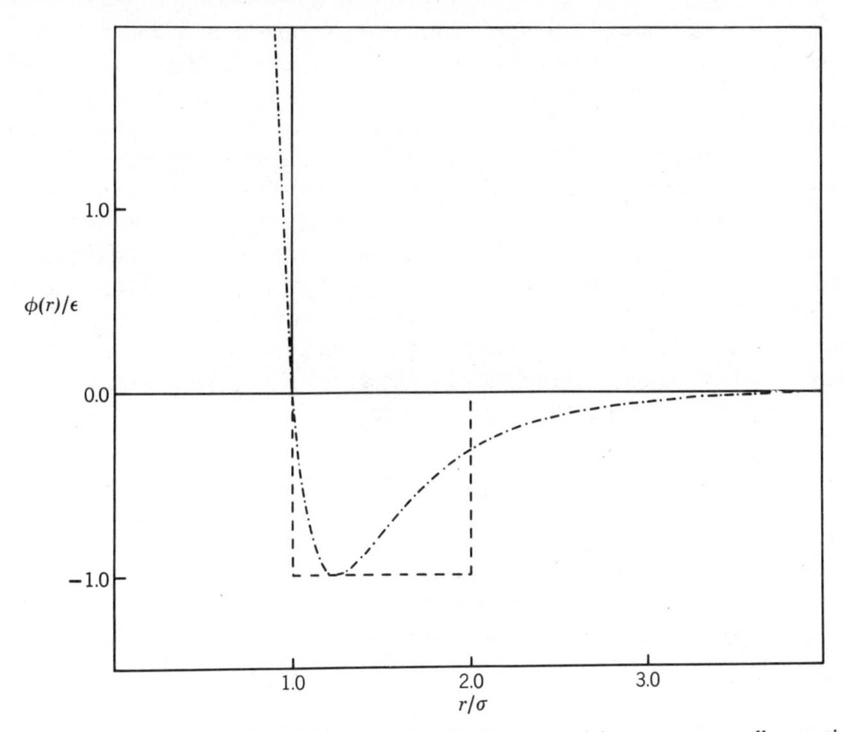

Figure 2.3 Model pair potentials. ——— hard sphere potential; - - - square well potential ($a = 2.0$); -·-·-·-·- Lennard-Jones 12-6 potential.

Although the Lennard-Jones 12-6 potential gives reasonable agreement with a number of properties of the gas, liquid, and solid states, both of the noble gases and of simple molecular systems (e.g., nitrogen, methane), a detailed study shows that it is not possible to obtain one set of potential parameters for a given chemical entity that reproduces all the experimental data for that entity. Furthermore, the coefficient of the term r^{-6} does not agree with quantum mechanical calculations [19]. Consequently, more flexible models have been used in efforts to obtain an accurate pair potential. An extremely successful model for the noble gases is based on an analytic form first suggested for argon by Barker and Pompe [20]

$$\phi(r) = \varepsilon\left[\exp[\alpha(1-R)]\sum_{i=0}^{5} A_i(R-1)^i - \sum_{i=0}^{2}\frac{C_{2i+6}}{(R^{2i+6}+\delta)}\right] \qquad (2.31)$$

where $R = r/R_m$ and R_m is the position of the minimum in the potential. The parameters in this model were chosen initially [20] to give agreement with experimental estimates of the potential energy at short distances (obtained from molecular beam data [4]), with the theoretically known [19] dispersion coefficients C_6, C_8, and C_{10}, and with a number of bulk properties including the viscosity, thermal conductivity, and second and third virial coefficients of the gas. Since that time the potential has been altered [21, 22] to give excellent agreement with solid and liquid state data, and the final form [22] is probably the most accurate representation of the argon pair potential available. Since the initial results for argon, the potential has also been used to fit krypton and xenon data [23], and some attempts have been made to use the model for simple molecular systems [24]. These results are discussed in more detail in Chapter 7.

A great deal of work on the helium pair potential has been reported both using semiempirical models [25] and quantum mechanical calculations [26]. Probably the most accurate semiempirical model was developed by Bruch and McGee [25]. They used piecewise continuous models to fit quantum calculations of the interaction potentials at short distances, the dispersion interactions at long distances, and bulk gas properties around the well of the potential. Calculations of bulk properties of liquid helium [26] have been reported using these models but it is not possible to make an unequivocal choice between them. The situation with neon-neon interactions is much simpler. Accurate experimental measurements of these interactions have been reported using low energy molecular beams [27].

Attempts to introduce nonspherical terms into the potential function have taken two routes. The first is to consider molecules whose repulsive interactions are essentially spherically symmetric and introduce the nonspherical behavior using permanent electrostatic interactions. One of the earliest steps in this direction resulted in the Stockmayer potential [2], a Lennard-Jones 12-6 potential with a point dipole embedded at the center. This gives an interaction potential

$$\phi(r, \theta_1, \theta_2, \phi) = 4\varepsilon\left[\left(\frac{\sigma}{r}\right)^{12} - \left(\frac{\sigma}{r}\right)^{6}\right] - \frac{\mu^2}{r^3}(2\cos\theta_1\cos\theta_2 - \sin\theta_1\sin\theta_2\cos\phi) \tag{2.32}$$

where μ is the value of the dipole moment on the molecule and θ_1, θ_2, and $\phi = \phi_1 - \phi_2$ are the angles used in eq. 2.20. This model has been used to represent the interactions in dipolar substances such as steam, carbon monoxide, and hydrogen chloride. Modifications of this potential replace the dipole-dipole interaction with quadrupole-quadrupole interactions (e.g., in benzene) and octopole-octopole interactions (e.g., in carbon tetrachloride and methane). It has been used to interpret virial coefficient data, in particular, and some results are available for liquid state calculations [28]. A variant of this model is obtained by replacing the Lennard-Jones term with the hard sphere potential; this model has been used to discuss certain dielectric effects [29].

A second method of introducing nonspherical effects is to construct models that contain the repulsive shape of the molecule explicitly. Perhaps the simplest method of doing this was introduced by de Rocco et al. [30] in the form of the spherical shell model. This potential is obtained by considering the interaction of two molecules each of which is made up of a spherical shell of Lennard-Jones 12-6 interactions. The model contains three parameters, ε, σ, and the shell radius d and is spherically symmetric. It has been used for molecules such as methane that are reasonably symmetric.

More substantial attempts to introduce the shape of the molecule have been reported. An example is a model for the interactions between water molecules. It is known from a large number of experiments (on ice and liquid water in particular) that the water-water interaction has deep minima for certain relative orientations when the oxygen atoms are about 2.8 Å apart. These configurations, known as hydrogen bonding configurations, are such that the oxygen and hydrogen atoms on one molecule are collinear with the oxygen on the second molecule, with a given molecule capable of forming bonds to four tetrahedrally distributed neighbors. In addition to these minima, repulsive interactions between atoms are important at short distances, and any successful model must account for these two effects. Both empirical and quantum mechanically based models have been proposed for the water-water interaction. One of the earliest successful models was reported by Rowlinson [31], who constructed an analytic potential function consisting of a spherically symmetric Lennard-Jones 12-6 term acting between the centers of mass of the molecules and four-point charges distributed around each oxygen atom. This gave a total interaction of the form

$$\phi(r, \mathbf{\Omega}_1, \mathbf{\Omega}_2) = 4\varepsilon\left[\left(\frac{\sigma}{r}\right)^{12} - \left(\frac{\sigma}{r}\right)^{6}\right] + \sum_{i=1}^{4}\sum_{j=1}^{4}\frac{q_i q_j}{|\mathbf{S}_{ij}|} \tag{2.33}$$

where the sets of angles $\mathbf{\Omega}_1$ and $\mathbf{\Omega}_2$ describe the relative orientation of the molecules, the sums over i and j refer to the four charges on each molecule and where $\mathbf{S}_{ij}$ is the vector distance between charge q_i on molecule i and charge q_j on molecule j. The parameters in this model were determined by requiring it to have the correct dipole moment of the isolated water molecule (1.84 D), the correct value of the lattice energy of ice when the molecules were placed in the lattice sites of ice *Ic*, and requiring it to give accurate second virial coefficients for steam. A full discussion of this model, together with other representations of water-water interactions, is given in Chapter 9.

A set of analytic interaction potentials has been developed by Kitaigorodskii [18], Williams [17], and others for use in the study of organic crystals. These potentials consist of sums of interactions taken between all the atoms on one molecule and all the atoms on a second molecule. Perhaps the most attractive feature of these interactions is that they are transferable between molecules—that is, the hydrogen-carbon interaction is the same in benzene as it is in phenol. So far these atom-atom interaction models have not been widely used in liquid state calculations, although some results for a similar model will be given in Chapter 8 for studies of liquid benzene. To enable the interested reader to work with atom-atom potentials a comprehensive list of parameters for the interaction

$$\phi(r) = \varepsilon_{ab}\left\{8.28\times 10^5 \exp\left[-0.0368(\sigma_a+\sigma_b)\right] - 3.516\times 10^{-2}\frac{(\sigma_a+\sigma_b)^6}{r^6}\right\} \tag{2.34}$$

is given in Table 2.2. If such interactions are used for fluid state calculations it is probably wise to check their ability to reproduce the second virial coefficient of the substance being considered. A suitable algorithm for performing the multidimensional integration has been given by Evans and Watts [32], and results for benzene vapor are mentioned in Chapter 8.

Table 2.2 Values of σ_a and ε_{ab} to be used with atom-atom models of interactions between organic molecules. The table is adapted from that given by E. L. Eliel, N. L. Allinger, S. J. Angyal, and G. A. Morrison, 1965, *Conformational Analysis*, Wiley, New York

ε_{ab}/k(K)	H	C	N	O	F	S	Cl	Br	I	σ_a (Å)
H	21.1	33.7	31.7	34.7	34.2	57.9	57.9	68.4	81.5	2.4
C		53.8	50.3	55.9	54.3	92.1	92.1	108.0	130.0	3.4
N			47.8	52.8	51.3	86.6	86.6	102.0	122.0	3.0
O				58.4	56.4	95.6	95.6	113.0	135.0	2.8
F					54.9	93.1	93.1	109.0	130.0	2.7
S						158.0	158.0	186.0	222.0	3.7
Cl							158.0	186.0	222.0	3.6
Br								218.0	263.0	3.9
I									314.0	4.3

2.4 Gas Kinetics: An Introduction

Many of the interaction potentials introduced in the previous section have two or more parameters determined from experimental data. The most important bulk data used to determine empirical and semiempirical potential functions are the gas phase viscosities and the equation of state. Expressions for the equation of state of a gas in terms of the virial coefficient expansion are developed in the next chapter. This section gives a short account of the derivation of formulae for the transport properties of the dilute gas. Equations for the transport coefficients are obtained from the Boltzmann equation, a complicated integro-differential equation for the time dependent one-body distribution function of the gas [2]. The equation is derived by assuming that the density of the gas is so low that only binary collisions occur; consequently, we begin by considering the collision of two *spherical* particles whose interactions vanish rapidly at large distances.

Suppose the two particles have masses m_1 and m_2, positions $\mathbf{r}_1$ and $\mathbf{r}_2$, and initial velocities $\mathbf{v}_1$ and $\mathbf{v}_2$. Then the reduced mass is defined by

$$\mu = \frac{m_1 m_2}{m_1 + m_2} \tag{2.35}$$

and the relative velocity before the collision by

$$\mathbf{g} = \mathbf{v}_2 - \mathbf{v}_1 \tag{2.36}$$

We use $\mathbf{g}'$ to represent the relative velocity after the collision. For spherically symmetric particles interacting through a potential $\phi(r)$, where $r = |\mathbf{r}_2 - \mathbf{r}_1|$, the position and velocity vectors can be transformed so that the collision occurs in a plane, the so-called *center-of-mass frame of reference* [2]. In this frame of reference, the positions of the particles are given by two cartesian coordinates, x and y, or alternatively by two polar coordinates, r and θ. It can be shown that the appropriate place to locate the origin of these coordinates is at the center of mass of the two particles [2].

In Fig. 2.4 the trajectories of two particles in the center-of-mass frame are plotted, assuming that the x-axis is defined by the direction of $\mathbf{g}$. The figure introduces two parameters describing the collision, the *impact parameter* b and the *angle of deflection*, χ. The impact parameter is defined as the distance of closest approach in the absence of any interaction potential, and the angle of deflection is the angle between the vectors $\mathbf{g}$ and $\mathbf{g}'$. As we see, χ is an important property of the collision and at this stage we note that it is determined by b and the initial relative speed, g, as well as by $\phi(r)$. One further angle is given in Fig. 2.4, θ_m, the angle

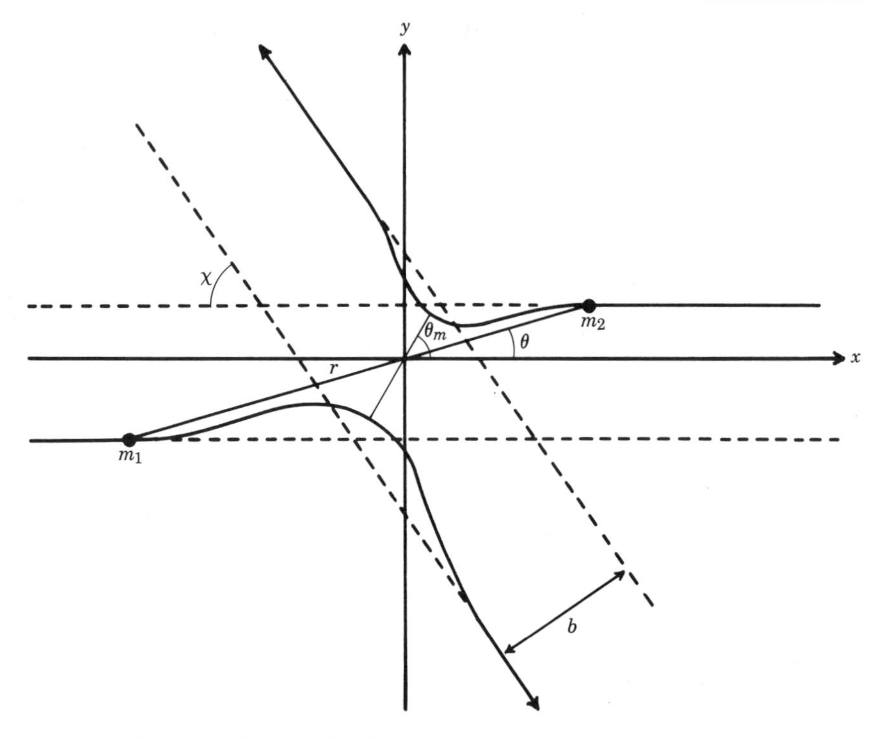

Figure 2.4 Trajectories of two particles in the center-of-mass frame.

between **g** and the vector joining the particles at the distance of closest approach.

If Newton's equations of motion are used to write down the equations for conservation of momentum and energy, in terms of the polar coordinates (r, θ), the two-particle scattering problem can be reduced to that of a single particle, of mass μ, moving in a potential $\phi(r)$ [2]. In particular, conservation of momentum reduces to an equation for conservation of angular momentum of the new particle about the center of mass

$$bg = r^2\dot{\theta} \tag{2.37}$$

where $\dot{\theta} = d\theta/dt$, and conservation of energy gives

$$\tfrac{1}{2}\mu g^2 = \frac{\mu}{2}(\dot{r}^2 + r^2\dot{\theta}^2) + \phi(r) \tag{2.38}$$

If we eliminate the angle derivative between eqs. 2.37 and 2.38

$$\tfrac{1}{2}\mu g^2 = \tfrac{1}{2}\mu\dot{r}^2 + \frac{\tfrac{1}{2}\mu g^2 b^2}{r^2} + \phi(r) \tag{2.39}$$

we obtain the equation of motion of a one-dimensional particle of mass μ moving in an effective pair potential

$$\phi_{\text{eff}}(r) = \phi(r) + \frac{\frac{1}{2}\mu g^2 b^2}{r^2} \tag{2.40}$$

where the term $\frac{1}{2}\mu g^2 b^2/r^2$ is the centrifugal potential. This equation is very important and is the basis of all classical collision calculations for spherically symmetric potentials. Typical effective pair potentials are given in Fig. 2.5, where features of particular interest are the rotational barrier and the minimum that is found for small and intermediate values of the impact parameter.

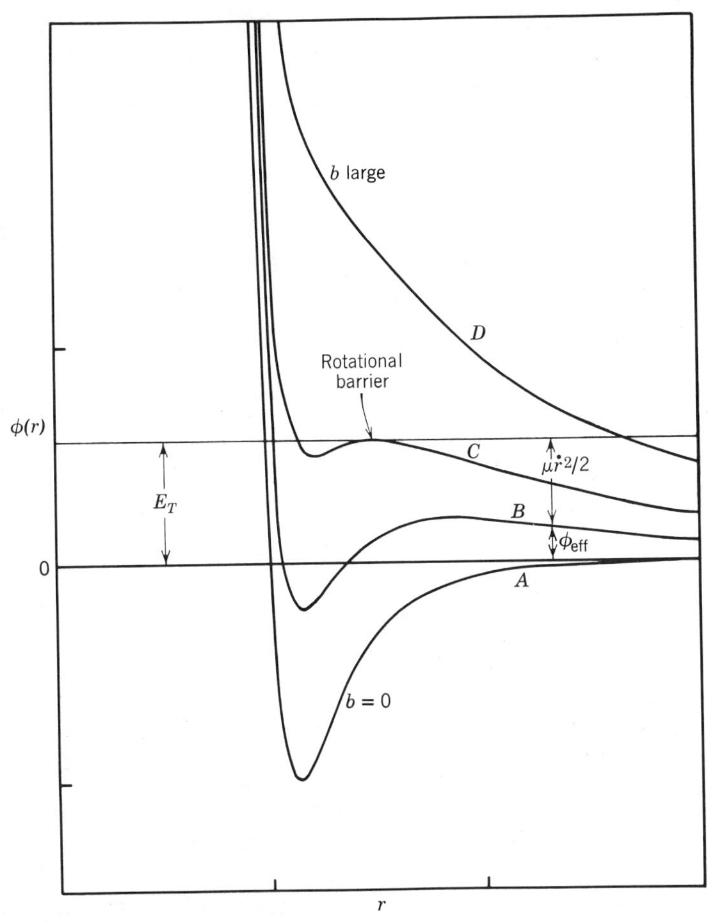

Figure 2.5 Effective pair potentials for several values of the impact parameter.

On the trajectory diagram (Fig. 2.4) the angle of deflection, χ, is shown as the angle between the initial and final relative velocity vectors. This angle is of paramount importance and is related to the angle θ_m by the equation

$$\chi = \pi - 2\theta_m \tag{2.41}$$

where θ_m is the value of θ at the distance of closest approach. Although θ_m could be obtained by eliminating time dependence between eqs. 2.37 and 2.38, it is readily obtained as follows

$$\frac{dr}{d\theta} = \frac{dr/dt}{d\theta/dt} = -\frac{r^2}{b}\left[1 - \frac{\phi(r)}{\frac{1}{2}\mu g^2} - \frac{b^2}{r^2}\right]^{1/2}$$

$$\theta_m = \int_0^{\theta_m} d\theta = \int_{r_m}^{\infty} \frac{(b/r^2)\,dr}{[1 - \phi(r)/\frac{1}{2}\mu g^2 - b^2/r^2]^{1/2}} \tag{2.42}$$

where r_m, the distance of closest approach, is found by solving the equation $dr/d\theta = 0$ and is the outermost zero of the denominator of the integrand in eq. 2.42. If this result is substituted into eq. 2.41 the angle of deflection can be calculated.

The behavior of the angle of deflection as a function of the impact parameter for fixed total energy is of great interest. At very small values of b the particles are meeting head-on whatever the value of the total energy, and their trajectories will be completely reversed with the value of χ being π. As the impact parameter is increased χ will decrease until at very large values of b there will be no scattering and $\chi = 0$. The effective pair potentials in Fig. 2.5 show four possible collisions for a pair of particles interacting through a spherical pair potential with attractive and repulsive regions. All four collisions have the same total energy E_T but each one has a different value of the impact parameter. The division of the total energy into terms arising from the effective pair potential and from the kinetic energy of relative motion is indicated for one possible collision and a similar division exists for the others. All the collisions are discussed using the equation of conservation of energy and angular momentum

$$E_T = \phi_{\text{eff}}(r) + \tfrac{1}{2}\mu \dot{r}^2 \tag{2.43}$$

$$M_\theta = \mu r^2 \dot{\theta} \tag{2.44}$$

Collision A in Fig. 2.5 is such that the impact parameter $b = 0$. As the particles approach, $\phi_{\text{eff}}(r)(= \phi(r)$ for $b = 0)$ becomes more and more negative and the kinetic energy steadily increases to keep E_T constant. Once the particles are separated by a distance less than the minimum in the potential energy, the potential energy becomes steadily more positive and the relative velocity decreases until it vanishes (at $E_T = \phi_{\text{eff}}(r)$, $\dot{r} = 0$ from eq. 2.43). As θ, and hence $\dot{\theta}$, is always zero for $b = 0$ the deflection is π and the particles then travel along the reverse path eventually separating with reversed relative velocity. Collisions B and D behave similarly and they can be discussed together. As the particles approach the potential energy increases and the kinetic energy decreases (and for collision B increases again as the maximum is passed) until as the line of total energy intersects the curve $\phi_{\text{eff}}(r)$ the relative velocity is zero. The particles then reverse their relative velocity, eventually parting. However, during these collisions neither θ nor $\dot{\theta}$ is zero and as r decreases eq. 2.44, for angular momentum conservation, tells us that $\dot{\theta}$ must increase. Consequently, for collisions of type B and D, χ is in general not equal to π and the particles have been deflected from their original courses. Collision C is an extreme example of types B and D and gives rise to the phenomenon of spiraling collisions. As the value of r decreases from infinity the relative velocity $\dot{r}$ decreases; when r is close to the maximum in the rotational barrier $\dot{r}$ is very close to zero. The particle enters the region within the rotational barrier with steadily increasing angular velocity $\dot{\theta}$ and very small relative velocity. Consequently, the particles spend an extremely long time in the vicinity of each other and can make many revolutions about their center of mass. The result is that χ is very large and negative for collisions whose impact parameter and total energy are such that spiraling collisions occur. Figure 2.6 shows qualitatively the behavior expected for the angle of deflection as a function of impact parameter for several energies. We return to the dependence of χ on b and g shortly, when discussing elastic scattering cross-sections.

The Boltzmann equation, which determines the time dependent distribution function of the gas, is very complicated, and it is not possible to give a full treatment here. Generally the equation is solved in the hydrodynamical limit by expanding the distribution function in a power series and using certain conservation equations to determine the first few terms in the series [2]. The result is that transport coefficients such as the viscosity, thermal diffusion coefficient, and self diffusion coefficients are obtained in terms of certain *collision integrals*

$$\Omega^{(l,s)}(T) = \left(\frac{kT}{2\pi\mu}\right)^{1/2} \int_0^\infty \exp(-\gamma^2)\gamma^{2s+3} Q^{(l)}(g)\, d\gamma \tag{2.45}$$

where

$$Q^{(l)}(g) = 2\pi \int_0^\infty (1 - \cos^l \chi) b\, db \tag{2.46}$$

and $\gamma^2 = \mu g^2/2kT$, g being the relative velocity before (or after) a

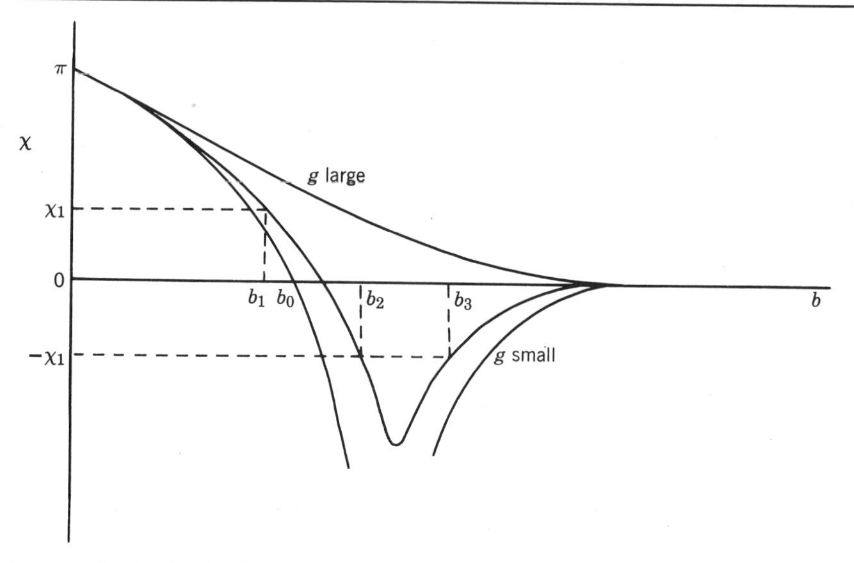

Figure 2.6 Angle of deflection, χ, as a function of the impact parameter, b, for several values of the initial relative speed, g.

collision. We see that the collision integrals can be obtained for a given interaction potential by calculating χ using eqs. 2.41 and 2.42 and substituting the results in these equations.

Approximate expressions for the transport coefficients can be derived in terms of the first few collision integrals and the first order expressions obtained for a one-component gas are as follows

$$\text{Viscosity: } \eta = \frac{5}{8} \frac{kT}{\Omega^{(2,2)}(T)} \tag{2.47}$$

$$\text{Thermal conductivity:} \lambda = \frac{25}{32} \frac{kT}{\Omega^{(2,2)}(T)} \left(\frac{C_v}{\mu}\right) \tag{2.48}$$

$$\text{Self diffusion: } D = \frac{3}{16} \frac{k^2 T^2}{\mu p \Omega^{(1,1)}(T)} \tag{2.49}$$

In these equations C_v is the specific heat at constant volume and p is the pressure of the gas.

We can now see in principle how to use transport coefficient data to determine the interaction potential. Some form for $\phi(r)$ with adjustable parameters is assumed and the parameters are then determined by obtaining good agreement between calculated transport coefficients and the corresponding experimental values over a wide range of temperatures. Methods such as this have been successfully used for the inert gases

[20, 33] and can also be used for electron-atom and ion-atom interactions [34]. Until recently this approach was the only way of determining $\phi(r)$ in the region of the minimum with any accuracy, but during the past few years molecular beam experiments and ultraviolet spectra have been used; we examine the background to these methods in the following sections.

2.5 Differential and Total Scattering Cross-Sections

Suppose N noninteracting particles each of energy $\frac{1}{2}\mu g^2$ are incident per unit area per unit time on a center of force. Let the direction of incidence be the polar axis of a coordinate system and count the number of particles scattered per unit time into an element of solid angle $d\Omega$. In the center-of-mass frame we can use the angle of deflection as the polar angle and hence $d\Omega = \sin\chi\, d\chi\, d\phi$ where ϕ is the azimuthal angle. The number of particles scattered into $d\Omega$, $N_s(\chi, \phi)$, will be proportional to both N and $d\Omega$

$$N_s(\chi, \phi) = I(\chi, \phi, g) N\, d\Omega \qquad (2.50)$$

and the constant of proportionality, which will depend on the energy of the collision as well as on the polar angles, is called the *differential scattering cross-section.* As scattering is isotropic about the polar axis, the collision frequency is proportional to an element of area in (r, ϕ) space, $b\, db\, d\phi$ so that we can write $N_s(\chi, \phi)$ as

$$N_s(\chi, \phi) = Nb\, db\, d\phi \qquad (2.51)$$

Equating $N_s(\chi, \phi)$ from eqs. 2.50 and 2.51 we see that the angle ϕ can be dropped, and the differential scattering cross-section is given by

$$I(\chi, g) = \left| \frac{b}{\sin\chi} \frac{db}{d\chi} \right| \qquad (2.52)$$

the modulus sign arising because $I(\chi, g)$, being an intensity, is always positive.

The differential scattering cross-section is of fundamental importance in all collision phenomena and occurs in various forms in subjects such as nuclear and molecular physics and in experiments such as neutron, electron, and X-ray diffraction and in light scattering. In our problem, notice that if we are given the intermolecular potential function the differential scattering cross-section can be calculated; the inversion of this process gives a direct experimental route to $\phi(r)$.

If we consider the variation of χ with b at various energies, discussed in some detail in the previous section and shown qualitatively in Fig. 2.6, it is apparent that for low and medium energy scattering the angle of deflection is a two-valued function of the impact parameter. An example where two values of the impact parameter, b_2 and b_3, have the same value of χ is given in Fig. 2.6. It is not possible to distinguish experimentally between particles scattered through an angle χ and those scattered through an angle $-\chi$. A good example of this is a head-on collision with $\chi = \pi$ and a half orbit with $\chi = -\pi$. It follows that the values of b, labeled b_1, b_2, and b_3 in Fig. 2.6, cannot be distinguished, and, similarly, scattering when $b = b_0$ cannot be distinguished from scattering when b is infinite. Consequently, eq. 2.52 is modified to read

$$I(\theta, g) = \sum_b \left| \frac{b}{\sin \chi} \frac{db}{d\chi} \right| \tag{2.53}$$

where the measured scattering angle θ lies in the range $(0, \pi)$ and where the sum is taken over all values of b such that $\theta(b) = \chi_{\text{mod } \pi}$ is the same for all b. At energies where orbiting occurs χ tends to $-\infty$ and there will be many contributions to $I(\theta, g)$.

At this stage we can consider the qualitative behavior of the differential scattering cross-section predicted from classical mechanics. There are three main regions of interest, very small angles, angles close to π, and at the intermediate angles where $\chi(b, g)$ has a minimum for some energies. In the region $\theta = 0$, corresponding to $\chi = 0, -2\pi, -4\pi, \ldots$, the term $\sin \chi$ is zero and from eq. 2.53 $I(\theta, g)$ is infinite. Thus the scattering is very much enhanced in the beam direction giving rise to *forward glory scattering*. Experimentally, forward glory scattering cannot be detected as the scattered particles are indistinguishable from those in the unscattered beam. For values of $\chi = \pi, -\pi, -3\pi, \ldots$ $\sin \chi$ again vanishes, $I(\theta, g)$ is infinite, and we have the phenomenon of *backward glory scattering*. At $b = 0$, where $\chi = \pi$, this singularity is suppressed, however. If the energy is reasonably low the χ-b plot has a minimum (under circumstances where orbiting collisions occur this minimum is at $\chi = -\infty$) and we have $d\chi/db = 0$, giving rise to a further infinity. Near the minimum in the χ-b plot there is little change in χ with b and so the intensity of scattered particles is large. This is analogous to the internal reflection of light inside water droplets and is consequently called *rainbow scattering* [35]. If the energy is sufficiently low that orbiting occurs, there is no rainbow scattering but glory scattering still occurs. Experimentally, high energy results are in good agreement with the classical predictions, but the low and medium energy results show considerable differences that can only be resolved using quantum mechanics.

The total scattering cross-section measures the relative intensity of particles scattered with a particular energy and is given by

$$S(g)=2\pi\int_0^{\pi} I(\theta, g)\sin\theta\, d\theta \tag{2.54}$$

or

$$S(g)=2\pi\int_0^{\infty} b\, db=\infty$$

for the classical problem. This result is interpreted in terms of intermolecular forces influencing the relative motion of two particles even for very large values of the impact parameter. However, for the experimental analysis of total scattering cross-sections [4], measurements of the intensity are made beyond some minimum angle θ_m. Under these conditions the zero angle scattering corresponding to $b=\infty$ is not included, and the classical formula is replaced by

$$S^*(g)=\int_{\theta_m}^{\pi} I(\theta, g)\sin\theta\, d\theta \tag{2.55}$$

It is of interest to note that experimental differential scattering cross-sections and total cross-sections do not show the infinities predicted classically. As we see later, quantum calculations do not predict these infinities either.

2.6 Quantum Scattering

The Schrödinger equation for two particles moving under the influence of a central potential function $\phi(r)$ is [35]

$$\left[-\frac{\hbar^2}{2m_1}\nabla_1^2-\frac{\hbar^2}{2m_2}\nabla_2^2+\phi(|\mathbf{r}_2-\mathbf{r}_1|)\right]\Psi_T=E_T\Psi_T \tag{2.56}$$

where $\hbar=h/2\pi$ and $\nabla_i^2=\partial^2/\partial x_i^2+\partial^2/\partial y_i^2+\partial^2/\partial z_i^2$. The total energy of the system is the eigenvalue E_T and Ψ_T is the total wave function. If this equation is written in terms of coordinates $\mathbf{R}$ and $\mathbf{r}$ giving the center-of-mass and relative separations, respectively, it factorizes into an equation that determines the center-of-mass wave function

$$-\frac{\hbar^2}{2(m_1+m_2)}\nabla_R^2\psi_c(\mathbf{R})=E_c\psi_c(\mathbf{R}) \tag{2.57}$$

and an equation that depends on relative coordinates

$$\left[-\frac{\hbar^2}{2\mu}\nabla^2+\phi(r)\right]\psi(\mathbf{r})=E\psi(\mathbf{r}) \tag{2.58}$$

where $\Psi_T = \psi_c\psi$ and $E_T = E_c + E$. Equation 2.58 is the Schrödinger equation for a particle of mass μ moving in the central potential $\phi(r)$ and is the analog of the classical equations of motion, eqs. 2.37 and 2.38. We now make a change of coordinate system to spherical polars (r, θ, ϕ) so that $x = r \sin\theta \cos\phi$, $y = r \sin\theta \sin\phi$, $z = r\cos\theta$, and assume that the wave function can be written as a product of three terms, each depending on only one of the three coordinates

$$\psi(\mathbf{r}) = R(r)\Theta(\theta)\Phi(\phi)$$

It can be shown that the angle dependent terms are given by the spherical harmonics [35]

$$\Theta(\theta)\Phi(\phi) = Y_l^m(\theta, \phi) \tag{2.59}$$

where m is the magnetic quantum number and l is the angular momentum quantum number, with l a nonnegative integer and $m = -l, -l+1, \ldots 0, \ldots l$. The radial part of the wave function is obtained as the solution to the equation

$$-\frac{\hbar^2}{2\mu r^2}\frac{d}{dr}\left(r^2\frac{dR_l}{dr}\right) + \left\{\phi(r) + \frac{l(l+1)\hbar^2}{2\mu r^2} - E\right\}R_l = 0 \tag{2.60}$$

This equation can be simplified by writing $U_l(r) = rR_l(r)$, giving

$$-\frac{\hbar^2}{2\mu}\frac{d^2U_l}{dr^2} + \{\phi_{\text{eff}}(r) - E\}U_l = 0 \tag{2.61}$$

where the effective pair potential is given by

$$\phi_{\text{eff}}(r) = \phi(r) + \frac{l(l+1)\hbar^2}{2\mu r^2} \tag{2.62}$$

This equation should be compared with the corresponding expression for the classical effective pair potential, eq. 2.40. In particular we note that the centrifugal potential is given by $l(l+1)\hbar^2/2\mu r^2$. Figure 2.7 gives the effective pair potential for the system K^+-Ar in several angular momentum quantum states [36].

To proceed, consider the radial wave equation for the case where $\phi(r) = 0$ everywhere

$$-\frac{\hbar^2}{2\mu}\frac{d^2U_l}{dr^2} + \left(\frac{l(l+1)\hbar^2}{2\mu r^2} - E\right)U_l = 0 \tag{2.63}$$

Provided the total energy E is positive, two independent solutions of this equation are the half-integer Bessel functions [35, 37]. We can write $U_l(r)$ as the linear combination

$$U_l(r) = Arj_l(qr) + Brn_l(qr) \tag{2.64}$$

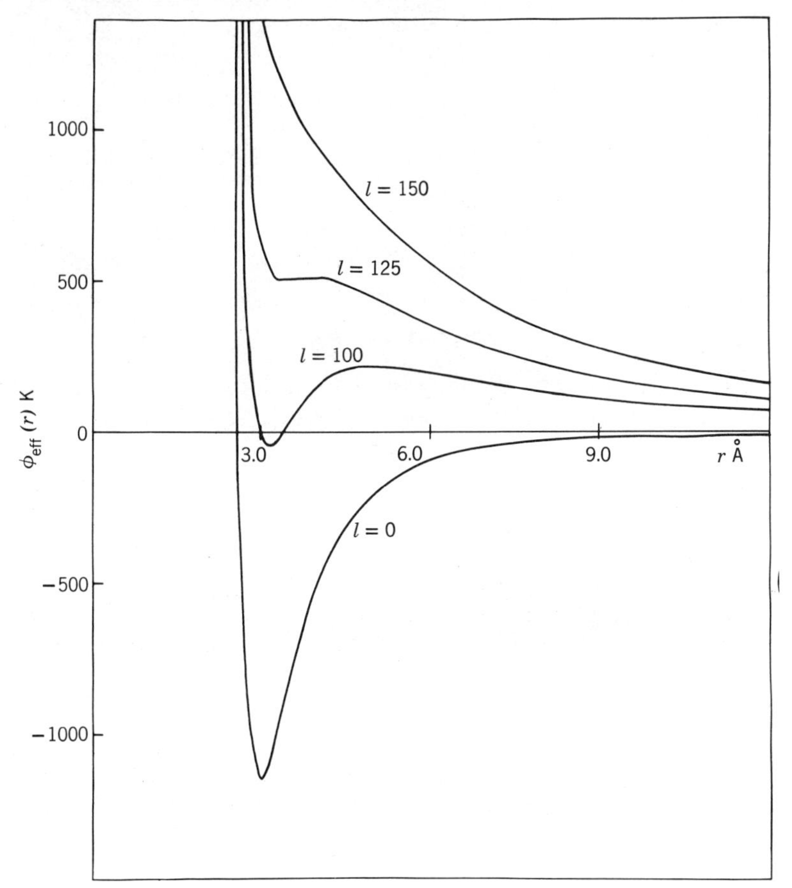

Figure 2.7 Effective pair potential for K^+-Ar as a function of distance for several values of the angular momentum quantum number. (Adapted with permission from *Aust J. Phys.*, **27,** 227, 1974.)

where A and B are parameters to be determined by the boundary conditions and $q=(2\mu E/\hbar^2)^{1/2}$ is the wave number. At small values of their arguments the half-integer Bessel functions have the following properties

$$\left.\begin{aligned} j_l(qr) &\sim (qr)^l \\ n_l(qr) &\sim \frac{1}{(qr)^{l+1}} \end{aligned}\right\} qr \sim 0. \tag{2.65}$$

For physical reasons we wish the wave function to be finite everywhere and it follows that $B=0$; the parameter A can be determined by requiring the wave function to be normalized

$$\int_0^\infty U_l^2(r)\,dr = 1 \tag{2.66}$$

At very large distances the term in $1/r^2$ in eq. 2.63 vanishes leaving the equation

$$\frac{d^2U_l}{dr^2} = -\frac{2\mu E U_l}{\hbar^2}$$

with the solution

$$U_l(r) = A \sin(qr + p)$$

where p is determined by the long range behavior of the Bessel function. In fact, the long range solution to eq. 2.63 is given by the asymptotic form of the Bessel function $j_l(qr)$ so that [37]

$$U_l(r) = \sin(qr - \tfrac{1}{2}\pi l) \tag{2.67}$$

where for convenience the normalizing constant A has been put equal to 1. If we now introduce the required pair potential, provided it vanishes more rapidly than $1/r$ (Coulomb potential), it can be shown that at large distances [35]

$$U_l(r) = \sin(qr - \tfrac{1}{2}l\pi + \eta_l) \tag{2.68}$$

where η_l is the *phase shift.* All the properties of the scattering process that are of interest to us can be calculated from the phase shift.

Consider a wave function $\psi(\mathbf{q}, \mathbf{r})$ of the form

$$\psi(\mathbf{q}, \mathbf{r}) = \exp(i\mathbf{q}\cdot\mathbf{r}) + f(q, \theta, \phi)\frac{\exp(iqr)}{r} \tag{2.69}$$

where $\mathbf{r}$ is the position vector given by spherical polar coordinates (r, θ, ϕ). At large values of r, $\psi(\mathbf{q}, \mathbf{r})$ is a solution to the Schrödinger equation in the center-of-mass frame, eq. 2.58. The function $f(q, \theta, \phi)$ is known as the *scattering amplitude,* for reasons that become obvious shortly. Now, the first term on the right hand side of eq. 2.69 represents a plane wave propagating in the direction $\mathbf{q}$ and can be taken to represent the beam of particles before the collision. It is not difficult to show also that the magnitude of $\mathbf{q}$ is identical to the wave number used in eq. 2.64 [38] so that

$$E = \frac{\hbar^2 q^2}{2\mu} \tag{2.70}$$

We can interpret the second part of the assumed wave function as representing the scattered wave of particles. If we follow standard practice [35] and interpret $\psi\psi^*$ as a probability density, then the probability of scattering into solid angle $d\Omega = \sin\theta\, d\theta\, d\phi$ is given by $|f(q, \theta, \phi)|^2$. Previously, the differential scattering cross-section was interpreted as measuring this property, and it follows that quantum mechanically the differential scattering cross-section is given by

$$I(q, \theta) = |f(q, \theta)|^2 \tag{2.71}$$

where, from eq. 2.70, k measures the total energy of the scattered particles. The ϕ dependence in the differential scattering cross-section has been dropped as it is known that for a central potential scattering is isotropic.

We are now able to obtain an expression for the differential and total scattering cross-section. Equation 2.59 gives an expression for the angle dependence of the wave function in terms of spherical harmonics. As we can ignore the rotation about the incident beam it is sufficient to use the Legendre polynomials, with

$$\Theta(\theta) \sim P_l(\cos\theta) \tag{2.72}$$

Our solution then becomes at large distances

$$\psi(q, \mathbf{r}) = \sum_{l=0}^{\infty} C_l(q) \frac{\sin(qr - l\pi/2 + \eta_l)}{r} P_l(\cos\theta) \tag{2.73}$$

where we have constructed the wave function by a linear sum over the r dependent (eq. 2.68) and θ dependent (eq. 2.72) solutions to the differential equations. This solution has to be identical to that given in eq. 2.69, and the coefficients $C_l(q)$ are used to make this possible. After expanding the function $\exp(i\mathbf{q}\cdot\mathbf{r})$ in a power series in Legendre polynomials and comparing coefficients of $P_l(\cos\theta)$ it is found that [38]

$$f(q, \theta) = \frac{1}{2iq} \sum_{l=0}^{\infty} (2l+1)\{\exp[2i\eta_l(q)] - 1\} P_l(\cos\theta) \tag{2.74}$$

Substituting this expression in eq. 2.71 gives

$$I(q, \theta) = \frac{1}{4q^2}\left\{\left[\sum_l (2l+1)(\cos 2\eta_l - 1) P_l(\cos\theta)\right]^2 + \left[\sum_l (2l+1)\sin 2\eta_l\, P_l(\cos\theta)\right]^2\right\} \tag{2.75}$$

for the differential scattering cross-section. The total scattering cross-section, obtained from the equation

$$S(q) = 2\pi \int_0^{\pi} I(q, \theta) \sin\theta \, d\theta$$

is found to be

$$S(q) = \frac{4\pi}{q^2} \sum_{l=0}^{\infty} (2l+1) \sin^2 \eta_l \tag{2.76}$$

This result requires the use of the orthogonality properties of the Legendre polynomials

$$\int_{-1}^{1} P_l(\cos\theta) P_m(\cos\theta) \, d\cos\theta = \frac{2\delta_{lm}}{2l+1}$$

where $\delta_{lm} = 0,\ l \neq m; = 1,\ l = m$.

We can now consider the qualitative behavior of the quantum results in some detail. A typical theoretical differential scattering cross-section such

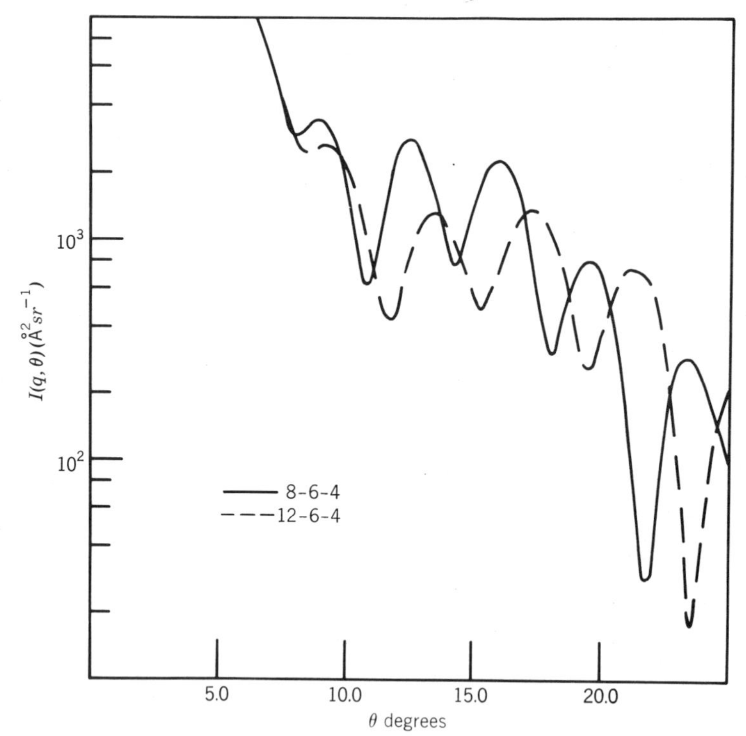

Figure 2.8 Differential scattering cross-section for Cs^+-Ne computed using the 12–6–4 and 8–6–4 potentials. Results are given for $q = 18.31$ Å^{-1}. (Adapted with permission from *Aust. J. Phys.*, **27,** 787, 1974.)

as that shown in Fig. 2.8 is finite at small angles, decays as the angle increases, and shows a large number of finite oscillations due to interference effects between the scattered waves (N.B. throughout this section we have considered particles in their "wave" representation). These oscillations are very marked at angles less that the classical rainbow angle and decrease at larger angles. In addition to the very large amplitude oscillations it is found that the envelope of the cross-section shows an oscillatory character, and by analogy with the classical case these are termed rainbow oscillations. Experiments are unable to follow the fine structure in the cross-section but they do show the envelope oscillations. The important qualitative differences between classical calculations, quantum mechanical calculations, and experiment are that (1) the quantum mechanical and experimental differential scattering cross-sections have no regions in

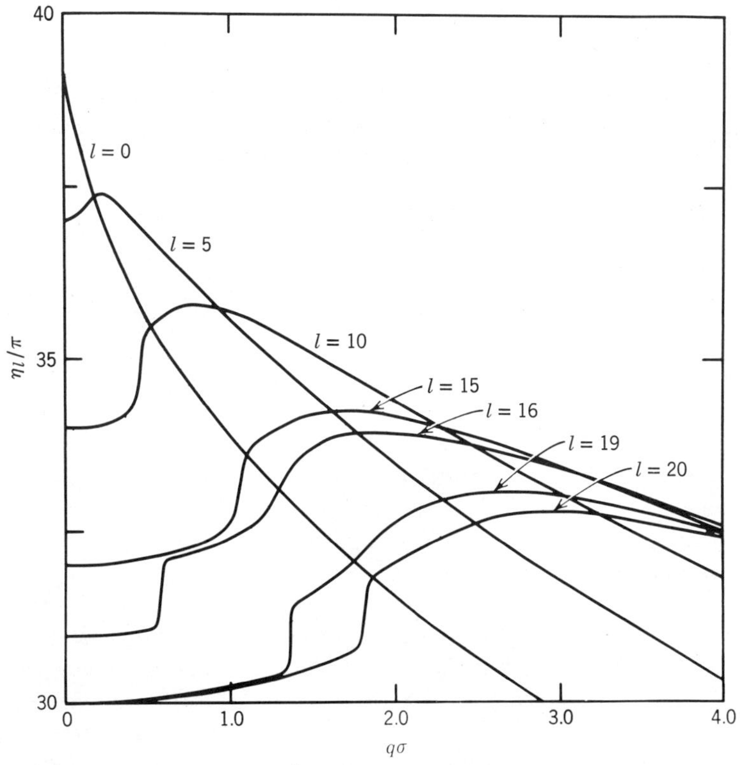

Figure 2.9 Phase shifts η_l for K^+-Ar as a function of wave number q and angular momentum quantum number l. (Adapted with permission from *Aust. J. Phys.*, **27,** 227, 1974.)

which they are infinite and (2) unlike the classical results, low energy quantum calculations are in very good agreement with experiment, provided the potential is accurate.

To understand the total scattering cross-section it is necessary to consider the phase shifts as a function of the angular momentum quantum number l for various values of the wave number q. A description of η as a function of l for various q is given in Fig. 2.9, where, for our purposes, the important features are the maxima that occur before the monotonic fall. Detailed discussion of the figure is left until the next section. Unlike the classical total scattering cross-section the quantum result is finite. It is found that the behavior of $S(q)$ is dominated by the maxima in the $\eta_l - q$ plot. Outside this region a given phase shift is changing fairly rapidly and it makes a relatively small contribution to $S(q)$ as $\sin^2 \eta_l$ is generally small. If the maximum is close to $\eta \sim \pi/2$, or an odd multiple of it, $\sin^2 \eta_l$ will be close to one, and will stay near that value for a range of energies, giving rise to a maximum in $S(q)$. Consequently, $S(q)$ shows oscillations, the glory extrema, as a function of q; the q^{-2} term in eq. 2.76 causes the function to decay. Experimental results for the differential and total scattering cross-sections of the inert gases are discussed in Chapter 7 where it is shown that such measurements give a very accurate description of the pair interaction [39].

2.7 Vibrational Energy Levels

After eq. 2.63 the restriction that E must be positive was introduced, and it was seen that the radial wave equation has solutions for all such values of E. If attempts are made to solve the radial wave equation for negative E it is found that solutions for which $\int \Psi\Psi^* \, d\mathbf{r}$ is finite can be found only for certain discrete values of E. These discrete values correspond to the vibrational energy levels of the molecular dimer formed by two atoms; the negative total energy tells us that the molecule is stable. The total number of energy levels, $E < 0$, for a given value of the angular momentum quantum number l gives the number of *bound states* for that value of l. To determine the values of the energy levels one method [40] is to solve eq. 2.61 for trial values of E and adjust E until a solution that vanishes at large distances is obtained. Experimentally, the energy levels can be measured spectroscopically, and as an example the vibrational levels of the argon dimer Ar_2 have been determined very accurately from the side bands on the ultraviolet spectrum of argon gas [41]. These experiments have two useful purposes—first they confirm the existence of the *molecule* Ar_2, and second, they give sufficient information to enable the bowl of the interatomic potential function to be determined. It has been found experimentally [41] that in the ground rotational state ($l = 0$) there are at least seven bound states for Ar_2 and that other noble gases also form stable dimers. Calculations using accurate interaction potentials predict that all the noble gases have one or more bound states, although if it exists that for the He_2 molecule is very weakly bound.

All effective pair potentials whose angular momentum quantum numbers are low enough for a negative region to exist may have bound states, as is shown schematically in Fig. 2.10. For a given effective pair potential, both the energies of the vibrational levels and their number depend on the value of the angular momentum quantum number. If the pair potential $\phi(r)$ has a deep well, giving very stable dimers, the vibrational levels are rather insensitive to changes in l if l is small and it is possible to assume that there is no coupling between vibrational and rotational states. Examining Fig. 2.10 raises the question of the detection and

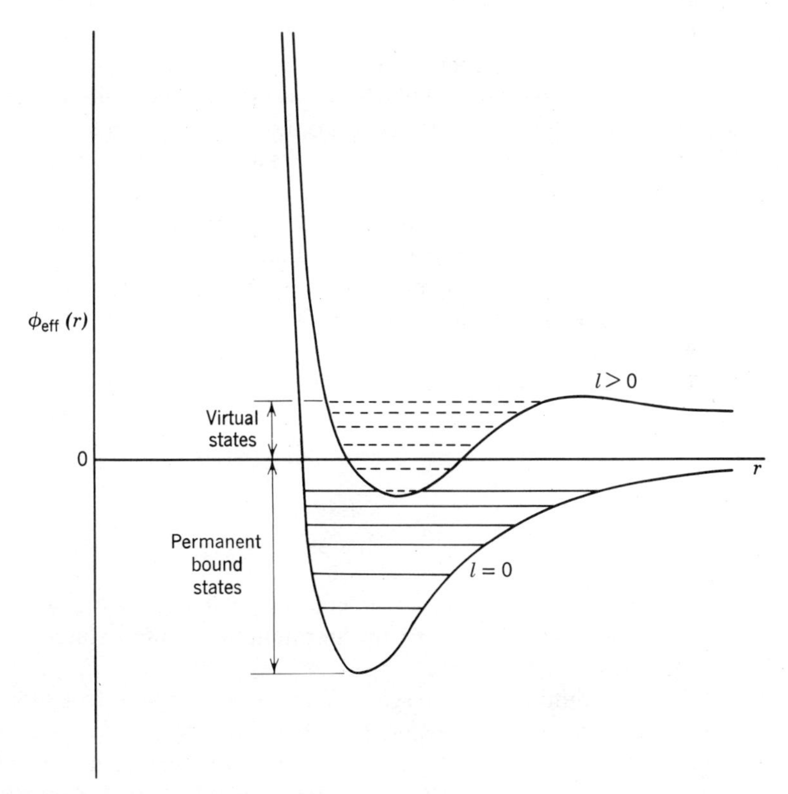

Figure 2.10 Representation of effective pair potentials showing permanent and virtual states.

meaning of the states with positive energy indicated as lying between the repulsive core and the rotational barrier. Classically, conservation of energy ensures that two particles colliding elastically cannot enter this region. However, quantum mechanics allows "tunneling" through such a barrier and the region must be considered. This is best done by examining the behavior of the phase shifts as a function of wave number q for several values of l, as shown in Fig. 2.9.

We see that at $q=0$ the phase shifts all take on integral values of π (e.g., with $l=0$, $\eta_0(0)=39\pi$ in the diagram). This means that the wave function $R_l(r)$ has shifted its phase by 39π rad with respect to the solution for scattering in the absence of an interaction potential. It can be shown using a theorem attributed to Levinson [42] that each shift of π represents one permanent bound state. For example in the figure the $l=0$ state has 39 vibrational levels and the $l=19$ state has 30 vibrational levels. As the value of l increases two effects are apparent. First, the phase shift begins to show a positive maximum and, later, pronounced jumps before it decreases monotonically; second, the value of $\eta_l(0)$ falls towards zero. The second effect reflects the observation made earlier—that the number of bound states falls as l increases until eventually there is no negative region in $\phi_{\text{eff}}(r)$ and so no bound state is possible. It is the first effect, the sharp rises in η_l, that is of interest now. As q increases for given l we see that the phase shift shows one or more sharp increases of π. Given that Levinson's theorem tells us that the number of values of π at $q=0$ corresponds to the number of bound states we can expect a similar interpretation of the increases for $q>0$. In fact, the sharp increases are due to the formation of *virtual states*, where a temporary dimer is formed between the rotational barrier and the repulsive well [43]. These virtual states are inherently unstable and their energies can be calculated from the value of q at the midpoint of the jump, using the equation [44]

$$E_v = \frac{\hbar^2 q^2}{2\mu} \tag{2.77}$$

with v standing for virtual. For some values of l it is possible to have several virtual states in addition to the permanently bound states, as is shown schematically in Fig. 2.10.

As well as determining the energies of the virtual states from phase shift calculations it is possible to calculate their lifetimes. Such states are formed by quantum tunneling through the rotational barrier and have a finite lifetime, for after a period the state may revert to the free particle condition by back tunneling. It can be shown that the lifetime of a virtual state is given by [44]

$$\tau = \frac{\hbar}{2}\left(\frac{\partial \eta}{\partial E}\right)_{\max} \tag{2.78}$$

where the derivative is calculated at the maximum slope in the phase shift jump. It is found that the higher the energy of the virtual state the shorter the lifetime [36]. Virtual states that lie deep in the well have very long lifetimes, but as the barrier is wide they are difficult to form.

References

1. H. Margenau and N. R. Kestner, 1971, *Theory of Intermolecular Forces*, 2nd ed., Pergamon, London.
2. J. O. Hirschfelder, C. F. Curtiss, and R. B. Bird, 1954, *Molecular Theory of Gases and Liquids*, Wiley, New York.
3. M. Born and J. R. Oppenheimer, 1927, *Ann. Phys.* (*Leipzig*), **84,** 457.
4. J. E. Jordan and I. Amdur, 1967, *J. Chem. Phys.*, **46,** 165; I. Amdur and E. A. Mason, 1954, *ibid.*, **22,** 670; I. Amdur and A. L. Harkness, 1954, *ibid.*, **22,** 664.
5. T. L. Gilbert and A. C. Wahl, 1967, *J. Chem. Phys.*, **47,** 3425; P. E. Phillipson, 1962, *Phys. Rev.*, **125,** 1981; N. R. Kestner and O. Sinanoğlu, 1966, *J. Chem. Phys.*, **45,** 194.
6. H. Margenau, 1939, *Rev. Mod. Phys.*, **11,** 1.
7. W. H. Keesom, 1921, *Physik. Z.*, **22,** 129.
8. P. Debye, 1920, *Physik. Z.*, **21,** 178.
9. H. Falkenhagen, 1922, *Physik. Z.*, **23,** 87.
10. F. London, 1930, *Z. Phys.*, **63,** 245; 1930, *Z. Phys. Chem.* (*Leipzig*), *B* **11,** 222.
11. H. B. G. Casimir and D. Polder, 1948, *Phys. Rev.*, **73,** 360.
12. B. M. Axilrod and E. Teller, 1943, *J. Chem. Phys.*, **11,** 299; B. M. Axilrod, 1949, *ibid.*, **17,** 1349; 1951, *ibid.*, **19,** 71.
13. F. H. Ree, 1971, Chapter 3 in *Physical Chemistry—An Advanced Treatise*, H. Eyring, D. Henderson, W. Jost, Eds., Vol. 8, Academic, New York.
14. B. J. Alder and T. E. Wainwright, 1957, *J. Chem. Phys.*, **27,** 1208.
15. W. R. Smith, 1973, "Perturbation Theory in the Classical Statistical Mechanics of Fluids," Chapter 2 in *Specialist Periodical Reports of the Chemical Society: Statistical Mechanics*, K. Singer, Ed., Vol. 1, Chemical Society, London.
16. J. E. Lennard-Jones, 1932, *Trans. Faraday Soc.*, **28,** 334.
17. D. E. Williams, 1966, *J. Chem. Phys.*, **45,** 3770.
18. A. I. Kitaigorodskii, 1960, *Tetrahedron*, **9,** 183.
19. P. J. Leonard, 1968, Masters Thesis, University of Melbourne.
20. J. A. Barker and A. Pompe, 1968, *Aust. J. Chem.*, **21,** 1683.
21. M. V. Bobetic and J. A. Barker, 1970, *Phys. Rev.*, B**2,** 4176.
22. J. A. Barker, R. A. Fisher, and R. O. Watts, 1971, *Mol. Phys.*, **21,** 657.
23. J. A. Barker, R. O. Watts, J. K. Lee, T. P. Schafer, and Y. T. Lee, 1974, *J. Chem. Phys.*, **61,** 3081.
24. J. Ogilvie, 1974, private communication.
25. L. W. Bruch and I. J. McGee, 1967, *J. Chem. Phys.*, **46,** 2959; 1970, *ibid.*, **52,** 5884; D. E. Beck, 1968, *Mol. Phys.*, **14,** 311.

26. I. K. Snook and R. O. Watts, 1972, *Aust. J. Phys.*, **25,** 735; R. D. Murphy and R. O. Watts, 1970, *J. Low Temp. Phys.*, **2,** 507.
27. P. E. Siska, J. M. Parson, T. P. Schafer, and Y. T. Lee, 1971, *J. Chem. Phys.*, **55,** 5762.
28. B. J. Berne and G. D. Harp, 1970, *Adv. Chem. Phys.*, **17,** 63; G. D. Harp and B. J. Berne, 1970, *Phys. Rev.*, **A2,** 975.
29. M. S. Wertheim, 1971, *J. Chem. Phys.*, **55,** 4291.
30. A. G. De Rocco and W. G. Hoover, 1962, *J. Chem. Phys.*, **36,** 916; A. G. De Rocco, T. H. Spurling, and T. S. Storvick, 1967, *ibid.*, **46,** 599.
31. J. S. Rowlinson, 1951, *Trans. Faraday Soc.*, **41,** 120.
32. D. J. Evans and R. O. Watts, 1975, *Mol. Phys.*, **29,** 777; 1974, *ibid.*, **28,** 1233.
33. H. J. M. Hanley and M. Klein, 1972, *J. Phys. Chem.*, **76,** 1743.
34. L. G. H. Huxley and R. W. Crompton, 1974, *The Diffusion and Drift of Electrons in Gases*, Wiley-Interscience, New York.
35. E. Merzbacher, 1970, *Quantum Mechanics*, 2nd ed., Wiley, New York; M. A. D. Fluendy and K. P. Lawley, 1973, *Chemical Applications of Molecular Beam Scattering*, Chapman and Hall, London.
36. R. O. Watts, 1974, *Aust. J. Phys.*, **27,** 227.
37. M. Abramowitz and I. A. Stegun, 1964, *Handbook of Mathematical Functions*, Dover, New York.
38. R. A. Newing and J. Cunningham, 1967, *Quantum Mechanics*, Oliver and Boyd, London.
39. J. M. Parson, P. E. Siska, and Y. T. Lee, 1972, *J. Chem. Phys.*, **56,** 1511; C. H. Chen, P. E. Siska, and Y. T. Lee, 1973, *ibid.*, **59,** 601.
40. S. Imam-Rahajoe, C. F. Curtiss, and R. B. Bernstein, 1965, *J. Chem. Phys.*, **42,** 530; J. K. Cashion, 1968, *ibid.*, **48,** 94.
41. Y. Tanaka and K. Yoshino, 1970, *J. Chem. Phys.*, **53,** 2012.
42. C. A. Levinson, 1949, *Kgl. Dan. Vidensk. Selsk., Mat.-Fys. Medd.*, **25,** 9.
43. R. B. Bernstein, C. F. Curtiss, S. Imam-Rahajoe, and H. T. Wood, 1966, *J. Chem. Phys.*, **44,** 4072.
44. N. F. Mott and H. S. W. Massey, 1965, *Theory of Atomic Collisions*, 3rd. ed., Clarendon, Oxford.

Three

Equilibrium Statistical Mechanics of Fluids

The relationships between various macroscopic properties of a system at equilibrium are dealt with in the subject of thermodynamics. For example, the pressure and volume of a gas are related to its temperature. Similarly, the microscopic properties of a system, such as the intermolecular forces, are treated using quantum mechanics. The subject of statistical mechanics is designed to relate the macroscopic properties of a system to the properties of the particles contained in the system. In this chapter a brief introduction to statistical mechanics is developed with the intention of giving an understanding of the theory of dense fluids. As we saw in the preceding chapter, it is reasonable to assume that the potential energy of two particles is a function of internuclear distances only, and that in the absence of external fields we can write the total potential energy of a system of N particles as a sum of two-body, three-body, ... terms [1]

$$\Phi(\mathbf{r}_1, \ldots \mathbf{r}_N) = \sum_{i<j} \phi^{(2)}(\mathbf{r}_{ij}) + \sum_{i<j<k} \phi^{(3)}(\mathbf{r}_{ij}, \mathbf{r}_{jk}, \mathbf{r}_{ik}) + \cdots \tag{3.1}$$

In this chapter it is usually assumed that the three-body and higher order interactions are not important. It must be stressed, however, that the last assumption is not valid even for the inert gases [2], and it is used here solely for algebraic simplicity.

3.1 Statistical Mechanical Averages

When an experimental measurement is made of some macroscopic property of a system the result is a time average. Consider, for example, the pressure of a gas. If the response time of a barometer was short enough, then the trace of the pressure reading would reflect every collision between the gas molecules and the surface of the barometer. However, few instruments are that sensitive and the result is an average reading over some period of time τ. The magnitude of τ is of considerable interest to mathematical physicists, but for our purposes we assume no more than that it is much longer than the mean lifetime of microscopic fluctuations in the system but that it is short enough to be useful. We return to the idea of an experimental measurement being a time average when we consider computer simulation methods toward the end of this chapter.

If we know the way in which the particles in a system are interacting, we can, in principle, compute the development of that system in time from some initial state and so construct the time average corresponding to any given experimental property. Equilibrium properties of a system are time independent by definition, and if we are interested only in such properties it is not necessary to follow the time behavior of the system. Instead we can use the *ensemble approach* propounded by Gibbs [3]. Consider the following example: we wish to evaluate the possible hands in a game of cards. We proceed (experimentally) by taking a deck of cards, shuffling, dealing, noting the resulting hands, and then repeating the process a large number of times. We can then estimate the probability of finding a particular hand. To model this process mathematically we could describe in detail the mechanics of the shuffle. Alternatively, we could think of a very large number of identical packs, in which the cards are ordered at random, deal the cards out, and count. In the second approach we do not worry about the method by which the various orderings are obtained and, in fact, *construct an ensemble of decks.*

Gibbs introduced a similar approach into the problem of calculating the thermodynamic properties of a system of molecules from the properties of the constituent particles. The ensemble in this case is a collection of replica systems, and the particles of each system are allowed to take up any of the configurations open to the system being modeled. It is assumed that the number of systems in the ensemble is so large that the number of systems having any given configuration is a measure of the probability of finding the real system in that configuration. A basic postulate of statistical mechanics is that the behavior of some property averaged over the ensemble reproduces the average behavior of that property in the real system. It is not possible to prove the equivalence and one can only assume it. To date experience has justified this assumption.

An ensemble can be constructed for all possible thermodynamic systems, three of which are considered here and one of which, *the canonical ensemble*, is examined in detail shortly. An isolated thermodynamic system is one in which neither energy nor matter is exchanged with the surroundings. In the laboratory such a system is approximated by the contents of a stoppered vacuum flask. The appropriate mathematical model is the *microcanonical ensemble*, and as in the corresponding real system, although it has fixed composition and energy, its temperature may fluctuate. A thermodynamic closed system is one for which the composition and temperature are fixed but which allows energy to be exchanged with the surroundings. A typical example would be a stoppered flask in a constant temperature bath. In this case we model the experimental system using the *canonical ensemble*. The third system that is of interest to us is the thermodynamic open system, for which both energy and matter may be exchanged with the surroundings. A typical example experimentally would be an open flask in a room at constant temperature. Such a system is modeled using the *grand canonical ensemble*. Other systems that are sometimes used, including those where the pressure rather than the volume is kept constant, are considered by Hill [4].

3.2 Introduction of Thermodynamics

We have introduced one postulate stating the equivalence of an experimental measurement and an ensemble average. In this section we introduce a second postulate relating an ensemble property to a thermodynamic property. This then gives us a complete formulation of statistical mechanics, as we show by considering in detail the canonical ensemble. It is easier to develop statistical mechanics using concepts taken from quantum mechanics rather than from classical mechanics. We suppose that if the system is in a quantum state ν it has energy E_ν, remembering that ν is in practice a set of quantum numbers. A simple example, that of a system of N harmonic oscillators, makes this clear. A single oscillator of angular velocity ω_1 and in quantum state ν_1 has energy

$$E_{\nu_1} = \hbar\omega_1(\nu_1 + \tfrac{1}{2}) \tag{3.2}$$

where ν_1 is an integer and $\hbar = h/2\pi$, h being Planck's constant. It follows that the system of N independent oscillators has energy

$$E_\nu = \sum_{i=1}^{N} \hbar\omega_i(\nu_i + \tfrac{1}{2}) \tag{3.3}$$

from which we see that ν represents the vector $(\nu_1, \nu_2, \ldots \nu_N)$.

We have previously mentioned that a system has a certain probability of being found in a particular configuration. For a classical system this probability is a function of the positions and momenta of the particles in the system; for a quantum system it depends on the quantum number vector ν. Using the symbol f_ν as the probability of the system being in a particular quantum state ν we introduce the second postulate: the entropy of the system, S, is given by the equation

$$S = -k \sum_\nu f_\nu \log f_\nu \tag{3.4}$$

where k is Boltzmann's constant, the sum is over all states of the system, and where

$$\sum_\nu f_\nu = 1 \tag{3.5}$$

The second equation says that the probability density function is normalized.

This postulate satisfies the notion that the entropy is a measure of the disorder in the system. Suppose we are absolutely certain that the system is in its lowest quantum state. Then as $f_0 = 1$ and $f_\nu = 0$, $\nu \neq 0$, it follows from eq. 3.4 that $S = 0$. Alternatively, let the probability of the system being in the ith state be $1/N$ for $i = 1, \ldots N$, and be zero otherwise. Then

$$S = -k \sum_{i=1}^{N} \frac{1}{N} \log \frac{1}{N} = k \log N \tag{3.6}$$

and as the number of quantum states available to the system increases so does the entropy. It must be emphasized that eq. 3.4 represents a postulate and that it cannot be proven. As with the first postulate, which equates experimental measurements and ensemble averages, so far experience has supported this postulate also.

3.3 The Canonical Ensemble

As was discussed previously, the canonical ensemble is used to model a thermodynamic closed system. In this type of system the temperature, T, the volume, V, and the number of particles, N, are fixed. As the system is in equilibrium with a heat bath the average energy, U, is fixed but the total energy of each system in the ensemble varies. Thus we must determine the probability density function f_ν subject to the constraints that it is normalized to unity and that the average energy is constant. As in a real system we suppose that subject to these constraints the entropy of the system at equilibrium is a maximum. Mathematically, we have to maximize

$$S = -k \sum_\nu f_\nu \log f_\nu$$

subject to the constraints

$$\sum_{\nu} f_{\nu} = 1 \tag{3.7}$$

$$\sum_{\nu} E_{\nu} f_{\nu} = U \tag{3.8}$$

This type of problem is readily solved by introducing Lagrange multipliers α and β, which are later determined by constraints given in eqs. 3.7 and 3.8. The function

$$F = -k\sum_{\nu} f_{\nu} \log f_{\nu} + \alpha \sum_{\nu} f_{\nu} + \beta \sum_{\nu} E_{\nu} f_{\nu}$$

is a maximum when $\partial F/\partial f_{\nu} = 0$ for all ν, that is when

$$-k \log f_{\nu} - k + \alpha + \beta E_{\nu} = 0$$

or

$$f_{\nu} = \exp \frac{(\alpha - k + \beta E_{\nu})}{k} \tag{3.9}$$

Using the constraint given in eq. 3.7 we have

$$\sum_{\nu} \exp \frac{(\alpha - k + \beta E_{\nu})}{k} = 1$$

or

$$\exp \frac{k - \alpha}{k} = Z \tag{3.10}$$

where Z, the *canonical partition function*, is given by

$$Z = \sum_{\nu} \exp \frac{\beta E_{\nu}}{k} \tag{3.11}$$

To determine β we invoke the first postulate and equate the average energy U with the thermodynamic internal energy so that it can be obtained from

$$S = \frac{-A + U}{T} \tag{3.12}$$

where A is the Helmholtz free energy of the system. Inserting eq. 3.9 into eq. 3.4 and using eq. 3.8 we obtain

$$S = (k - \alpha) - \beta U \tag{3.13}$$

and comparing this with eq. 3.12 we see that

$$\alpha - k = \frac{A}{T} \tag{3.14}$$

$$\beta = -\frac{1}{T} \tag{3.15}$$

Comparing eq. 3.14 with eq. 3.10 we see that the Helmholtz free energy of the system is obtained from the canonical partition function

$$A = -kT \log Z \tag{3.16}$$

This is the basic relation connecting the microscopic properties of the particles in a canonical ensemble to the thermodynamic properties of the system being modeled. If we use eq. 3.10 in eq. 3.9 we obtain the probability of finding the system in any given quantum state ν

$$f_\nu = \frac{\exp - E_\nu/kT}{Z} \tag{3.17}$$

Having obtained eq. 3.16 as the basic result for the canonical ensemble we can calculate other thermodynamic properties using standard thermodynamic methods. For example, we can obtain expressions for the internal energy, entropy, and pressure using the equations

$$U = \frac{\partial(A/T)}{\partial(1/T)}; \qquad S = -\left(\frac{\partial A}{\partial T}\right)_V; \qquad p = -\left(\frac{\partial A}{\partial V}\right)_T$$

Similar methods can be used to obtain expressions for the distribution functions and thermodynamic properties in other ensembles [5]. In particular, the equations for the grand canonical ensemble are obtained by minimizing the entropy subject to the constraints that the average energy and the average number of particles are constant. Using $f_{N\nu}$ as the probability that the system contains N particles in quantum state ν it can be shown that

$$f_{N\nu} = \frac{\exp(N\mu - E_{N\nu})/kT}{Z^*} \tag{3.18}$$

where μ is the chemical potential per particle and Z^*, the *grand partition function*, is given by

$$Z^* = \sum_{N,\nu} \exp \frac{(N\mu - E_{N\nu})}{kT} \tag{3.19}$$

Again the connection between the energy states of the system and thermodynamic properties is through the normalizing constant, with [5]

$$pV = kT \log Z^* \tag{3.20}$$

Other thermodynamic quantities may be obtained from eq. 3.20 by using the appropriate algebraic relationships.

We have shown that the relationship between the partition functions and thermodynamic properties are different for different ensembles, and so there may be some confusion over which formulation to use. Fortunately, it can be shown that for very large systems, where N, $V \rightarrow \infty$ in such a way that the ratio N/V is constant, the various ensembles give identical thermodynamic properties [4]. Consequently, we use whichever ensemble is most convenient without worrying about details associated with finite systems.

3.4 Transition to Classical Statistics

The formal transition from quantum statistical mechanics to classical statistical mechanics was discussed in detail by Wigner and Kirkwood [6]. They wrote the equations for the partition function in terms of the Hamiltonian operator and after some rearrangement expanded in powers of the quantity $\lambda^2 = h^2/2\pi mkT$, generating the classical partition function as the first term in the power series. We discuss the importance of higher terms in this series in the chapter on inert gases (Chapter 7), but do not consider the derivation of the expansion [6]. Instead we examine a less rigorous derivation of the classical partition function.

We have derived the partition function as a sum over quantum states. In classical statistics the energy E_ν of a given quantum state becomes the total energy of the system for a particular set of particle coordinates $\mathbf{r}_i$ and momenta $\mathbf{p}_i$. The sum over states is then replaced by an integration over all values of $\mathbf{r}_i$ and $\mathbf{p}_i$ for all the particles in the system. We know from the Heisenberg uncertainty principle that, although adjacent energy states become very close together in the classical region, nevertheless, they remain distinct. In classical mechanics we integrate over the region between quantum states and so overestimate the number of contributions to the partition function. This problem is made more obvious by considering a simple example, that of a one dimensional harmonic oscillator [5].

The total energy of a classical harmonic oscillator is given by

$$H = \frac{p^2}{2m} + \frac{m\omega^2 x^2}{2} \tag{3.21}$$

where p is its momentum, x its position, m its mass and ω its angular frequency. In its νth quantum state the total energy is given quantum mechanically by

$$E_\nu = \hbar\omega(\nu + \tfrac{1}{2}) \tag{3.22}$$

and, although in the classical region ν is very large, formally the energy is still given by eq. 3.22. For a given value of the total energy H, eq. 3.21 represents an ellipse in (x, p) space. The maximum and minimum values of the momenta occur when the displacement is zero and are given by the roots of the equation

$$\frac{p^2}{2m} = \hbar\omega(\nu + \tfrac{1}{2}) \tag{3.23}$$

the right hand side following from the fact that E_ν is the total energy. Equation 3.23 has roots

$$p_M = \pm[2\hbar m\omega(\nu + \tfrac{1}{2})]^{1/2} \tag{3.24}$$

Similarly, the maximum displacements occur when the momentum is zero and are given by

$$x_M = \pm\left[\frac{2\hbar(\nu + \frac{1}{2})}{m\omega}\right]^{1/2} \tag{3.25}$$

By increasing ν we obtain a set of confocal ellipses, each representing a particular quantum state. When we integrate over x and p in the classical limit we are in effect calculating the area of the (x, p) space. The area of the ellipse representing a particular value of ν is

$$\begin{aligned} A(\nu) &= \pi p_M x_M \\ &= h(\nu + \tfrac{1}{2}) \end{aligned} \tag{3.26}$$

and so the difference in area between adjacent ellipses is

$$A(\nu + 1) - A(\nu) = h \tag{3.27}$$

That is, every time we integrate between two ellipses we cover an area h, or to put it another way, there is only one energy state in an area h. Thus the number of quantum states in the whole (x, p) plane is

$$N_s = \frac{\text{area of plane}}{h} = h^{-1}\int_{-\infty}^{\infty} dp \int_{-\infty}^{\infty} dx \tag{3.28}$$

and consequently a sum over the states of a single one-dimensional particle is replaced by an integration over its position and momentum coordinates weighted by the quantity h^{-1}.

We can now return to the problem of converting the sum over states for a system of particles into an integration over the position and momentum coordinates of the particles in that system. Following the simple example of the one-dimensional oscillator we note that we replace E_ν by the total energy of the system

$$H=\sum_{i=1}^{N}\frac{\mathbf{p}_i^2}{2m}+\Phi(\mathbf{r}_1,\ldots\mathbf{r}_N) \tag{3.29}$$

where $\Phi(\mathbf{r}_1,\ldots\mathbf{r}_N)$ is the total potential energy of the system, and replace the sum over states by integrations over all the coordinates ($\mathbf{r}_i$ and $\mathbf{p}_i$) weighting the result by h^{-1} for each position (or momentum) coordinate integrated over. We can apply this formula to the expression for the canonical partition function

$$Z=\sum_{\nu}\exp\frac{-E_\nu}{kT}=h^{-3N}\int\cdots\int\exp\frac{-H}{kT}\,d\mathbf{r}_1\cdots d\mathbf{p}_N \tag{3.30}$$

Although eq. 3.30 has enabled us to replace the sum over states by integrations over position and momentum coordinates, it is still incomplete. Quantum mechanics does not recognize that identical particles may be distinguished, but classical mechanics, by labeling $\mathbf{r}$ and $\mathbf{p}$ with a particle number i, implicitly labels the particles. We must compensate for this deficiency. Suppose our system has its N particles distributed between M different particle quantum states. Let there be n_j particles with energy ε_j so that

$$\sum_j n_j=N$$

$$\sum_j n_j\varepsilon_j=E_\nu$$

the sums going over all M states. We can obtain the system state E_ν in $N!/(n_1!n_2!n_3!\cdots n_M!)$ different ways and so should divide the right hand side of eq. 3.30 by this quantity to allow for the indistinguishability of the particles. When the particle states get closer together, as they do in the classical region, transitions between neighboring states become easier. At the same time, as we are raising the temperature to get to the classical region, the number of available states becomes greater. It follows that we are unlikely to find more than one particle in any given state, and so we make the reasonable approximation that in going to the classical limit all the n_j are either 0 or 1, and so all $n_j!=1$. It follows that to this approximation there are just $N!$ ways of distributing the particles among the available quantum states and as they are indistinguishable we must divide the classical partition function by this quantity. This gives us the required result

$$Z = \frac{h^{-3N}}{N!} \int \cdots \int \exp \frac{-H}{kT} \, d\mathbf{r}_1 \cdots d\mathbf{p}_N \tag{3.31}$$

When writing down the expression for the total energy of the system, eq. 3.29, we noted that the potential energy depends on the positions of the particles but not on their momenta, and we also assumed that the only contribution to the kinetic energy was from the center of mass motion of the particles. With this form of the total energy we can do the momentum integrations in eq. 3.31. Consider the kinetic energy contribution to the partition function (arising from the momenta of the particles)

$$Z_{\text{mom}} = \int_{-\infty}^{\infty} dp_{x1} \int_{-\infty}^{\infty} dp_{y1} \int_{-\infty}^{\infty} dp_{z1} \cdots \int_{-\infty}^{\infty} dp_{zN}$$
$$\times \exp\left(- \sum_{i=1}^{N} \frac{p_{xi}^2 + p_{yi}^2 + p_{zi}^2}{2mkT} \right)$$

or as each integral is of the same functional form

$$Z_{\text{mom}} = \left[\int_{-\infty}^{\infty} \exp \frac{-p^2}{2mkT} \, dp \right]^{3N}$$

This integration is easily performed to give

$$Z_{\text{mom}} = (2\pi mkT)^{3N/2}$$

Substituting this result into eq. 3.31 we obtain the classical partition function in the form

$$Z = \frac{\lambda^{-3N}}{N!} \int \cdots \int \exp \frac{-\Phi(\mathbf{r}_1, \ldots \mathbf{r}_N)}{kT} \, d\mathbf{r}_1 \cdots d\mathbf{r}_N \tag{3.32}$$

where $\lambda^2 = (h^2/2\pi mkT)$ as before. Note that although we have now transformed Z from quantum mechanical to classical mechanical terms we have not altered in any way the relationship between statistical mechanics and thermodynamics; thus all the expressions noted after eq. 3.17 remain valid. Finally, we can transform the expression for the grand canonical partition function (eq. 3.19) in exactly the same way to obtain

$$Z^* = \sum_{N \geq 0} \frac{z^N}{N!} \int \cdots \int \exp \frac{-\Phi(\mathbf{r}_1, \ldots \mathbf{r}_N)}{kT} \, d\mathbf{r}_1 \cdots d\mathbf{r}_N \tag{3.33}$$

where $z = \lambda^{-3} \exp(\mu/kT)$ is called the activity. Once again, the appropriate equations relating Z^* to thermodynamic properties remain valid.

During the remainder of this book we are concerned primarily with classical statistical mechanics. However, in Chapter 7, when we discuss models of the rare gas interactions, we show that the leading terms in the Wigner-Kirkwood expansion mentioned at the beginning of this section are needed to obtain quantitative agreement with experiment.

3.5 Distribution Functions

Although in Section 3.3 we derived expressions for the probability of finding the system in a particular quantum state (eqs. 3.17 and 3.18), we have not considered these functions in any detail. In this section a number of useful probability density functions for use with classical systems are defined and their use in the calculation of thermodynamic quantities considered. We have derived the probability of finding the system in a quantum state E_ν for both the canonical and grand canonical ensembles. At equilibrium, the probability of finding a particle with a given momentum in a classical system is proportional to $\exp(-p^2/2mkT)$, and as we have just seen integrations over the momenta are readily carried out. Consequently, we are interested only in the distribution of the positions of the particles.

Within the canonical ensemble the probability of finding particles at positions $(\mathbf{r}_1, \ldots \mathbf{r}_N)$ is given by

$$f^{(N)}(\mathbf{r}_1, \ldots \mathbf{r}_N)\, d\mathbf{r}_1 \cdots d\mathbf{r}_N = \frac{1}{Z}\frac{\lambda^{-3N}}{N!} \exp \frac{-\Phi(\mathbf{r}_1, \ldots \mathbf{r}_N)}{kT} \, d\mathbf{r}_1 \cdots d\mathbf{r}_N \quad (3.34)$$

This probability density function is of little practical value as it cannot be readily used. Consequently, we consider the distribution of smaller numbers of particles, in particular the one- and two-body density functions. These are defined, respectively, as the probabilities of finding a single particle at position $\mathbf{r}_1$ and of finding two particles simultaneously at $\mathbf{r}_1$ and $\mathbf{r}_2$.

To calculate the single particle density function we integrate the N particle density function $f^{(N)}$ over the coordinates $\mathbf{r}_2, \ldots \mathbf{r}_N$ and multiply the resulting expression by the number of ways of choosing particle 1, namely N. This gives

$$f^{(1)}(\mathbf{r}_1) = \frac{1}{Z}\frac{\lambda^{-3N}}{(N-1)!}\int \cdots \int \exp \frac{-\Phi(\mathbf{r}_1, \ldots \mathbf{r}_N)}{kT}\, d\mathbf{r}_2 \cdots d\mathbf{r}_N \quad (3.35)$$

If we now write out the expression for Z, eq. 3.32 we find

$$f^{(1)}(\mathbf{r}_1)=\frac{N\int\cdots\int \exp\left[-\Phi(\mathbf{r}_1,\ldots \mathbf{r}_N)/kT\right]d\mathbf{r}_2\cdots d\mathbf{r}_N}{\int\cdots\int \exp\left[-\Phi(\mathbf{r}_1,\ldots \mathbf{r}_N)/kT\right]d\mathbf{r}_1\cdots d\mathbf{r}_N}$$

In a uniform system the potential energy is a function of relative coordinates only, and the denominator of $f^{(1)}(\mathbf{r}_1)$ is just the volume, V, multiplied by the integral in the numerator. It follows that

$$f^{(1)}(\mathbf{r}_1)=\frac{N}{V}=\rho \tag{3.36}$$

where ρ is the number of particles per unit volume or number density. That is, the probability of finding a particle in a given volume element of a uniform fluid is independent of the position of that volume element and is given by the number density.

The other distribution function of particular interest is the two-body density function $f^{(2)}(\mathbf{r}_1, \mathbf{r}_2)$. This is obtained by integrating $f^{(N)}$ over the particle coordinates $\mathbf{r}_3, \ldots, \mathbf{r}_N$ and multiplying by the number of ways of selecting two particles from N, given by $N(N-1)$. We find that

$$f^{(2)}(\mathbf{r}_1,\mathbf{r}_2)=\frac{1}{Z}\frac{\lambda^{-3N}}{(N-2)!}\int\cdots\int \exp\frac{-\Phi(\mathbf{r}_1,\ldots \mathbf{r}_N)}{kT}\,d\mathbf{r}_3\cdots d\mathbf{r}_N \tag{3.37}$$

This function is usually considered in the form of the *radial distribution function*, defined by

$$g^{(2)}(\mathbf{r}_1,\mathbf{r}_2)=\frac{f^{(2)}(\mathbf{r}_1,\mathbf{r}_2)}{f^{(1)}(\mathbf{r}_1)f^{(1)}(\mathbf{r}_2)} \tag{3.38}$$

For an isotropic system of spherical particles, $f^{(2)}(\mathbf{r}_1, \mathbf{r}_2)$ depends only on the distance apart of two particles

$$g^{(2)}(r)=\frac{f^{(2)}(r)}{\rho^2} \tag{3.39}$$

The radial distribution function measures the probability of finding a particle at a given distance from a fixed particle in the fluid. At large distances the probability of finding two particles at positions $\mathbf{r}_1$ and $\mathbf{r}_2$ is just the product of the single particle density functions, and we have

$$\lim_{r\to\infty} f^{(2)}(\mathbf{r}_1,\mathbf{r}_2)=f^{(1)}(\mathbf{r}_1)f^{(1)}(\mathbf{r}_2)=\rho^2$$

or

$$\lim_{r\to\infty} g(r)=1 \tag{3.40}$$

The radial distribution function is of paramount importance in liquid state physics. It has two uses—one when discussing the structure of the fluid and the other when obtaining the thermodynamic properties of the system. Experimentally, $g(r)$ may be measured using radiation scattering studies [7], and we devote part of Chapter 5 to a description of such measurements and to a study of the structural information contained in them. During subsequent sections of this chapter we are particularly concerned with the use of $g(r)$ in thermodynamic calculations. In our discussion so far we have assumed that we are dealing with spherical particles, and as a consequence $g(r)$ has been a function of distance alone. For more complicated systems this function may depend on both the distance apart and the relative orientation of two particles, but if the angle averaged function is considered the qualitative features are essentially unaltered. The behavior of the radial distribution function for nonspherical particles is discussed in more detail in Chapters 8 and 9.

The radial distribution function is a two-body function, and it is sometimes necessary to consider higher order generalizations. We are concerned on occasion with the three-body radial distribution function, defined by

$$g^{(3)}(\mathbf{r}_1, \mathbf{r}_2, \mathbf{r}_3) = \frac{f^{(3)}(\mathbf{r}_1, \mathbf{r}_2, \mathbf{r}_3)}{\rho^3} \tag{3.41}$$

In a uniform fluid of spherical particles $g^{(3)}$ depends only on the distance apart of the centers of three particles, $g^{(3)}(r, s, t)$. We make use of this function for two purposes, first when discussing certain approximate equations for $g(r)$, and second, when calculating contributions to the thermodynamic properties from three-body potential functions.

3.6 Thermodynamic Equations

As mentioned in the preceding section, under certain circumstances the thermodynamic properties of a fluid may be calculated using the radial distribution function [4]. We assume in the following derivations that the potential energy of the system is composed of two-body interactions alone, and do no more than quote the corresponding three-body results. The first part of this section deals with equations for the internal energy and the pressure, both of which are most conveniently obtained from the canonical partition function. At the end of the section we use the grand canonical partition function to obtain an expression for the isothermal compressibility.

The internal energy is related to the microscopic properties of the system using the equation

$$U=\frac{\partial(A/T)}{\partial(1/T)}=\frac{\partial(A/kT)}{\partial(1/kT)}=-\frac{\partial \log Z}{\partial(1/kT)}$$

From eq. 3.32 we see that

$$-\log Z=\log N!-\tfrac{3}{2}N\log\frac{2\pi mkT}{h^2}-\log Q$$

where

$$Q=\int\cdots\int \exp\left[-\sum_{i<j}\frac{\phi(r_{ij})}{kT}\right]d\mathbf{r}_1\cdots d\mathbf{r}_N \tag{3.42}$$

and where we have used the assumed pairwise additivity of the potential energy. Differentiating with respect to $1/kT$ we find

$$U=\frac{3}{2}NkT+\frac{1}{Q}\int\cdots\int\sum_{i<j}\phi(r_{ij})\exp\left[-\sum_{i<j}\frac{\phi(r_{ij})}{kT}\right]d\mathbf{r}_1\cdots d\mathbf{r}_N \tag{3.43}$$

The first part of this equation corresponds to the internal energy of an ideal gas and the second part arises from interactions between the particles. As the integrations in eq. 3.43 cover all positions for all the particles the sum contributes $N(N-1)/2$ identical terms, giving

$$U=\frac{3}{2}NkT+\frac{N(N-1)}{2Q}\int\cdots\int\phi(r_{12})\exp\left[-\sum_{i<j}\frac{\phi(r_{ij})}{kT}\right]d\mathbf{r}_1\cdots d\mathbf{r}_N \tag{3.44}$$

Substituting eq. 3.37 for the two-body density function enables us to write eq. 3.44 in the form

$$U=\tfrac{3}{2}NkT+\tfrac{1}{2}\int\int\phi(r_{12})f^{(2)}(\mathbf{r}_1,\mathbf{r}_2)\,d\mathbf{r}_1\,d\mathbf{r}_2$$

Multiplying and dividing the integration by ρ^2 introduces the radial distribution function

$$U=\tfrac{3}{2}NkT+\frac{\rho^2}{2}\int\int\phi(r_{12})g(r_{12})\,d\mathbf{r}_1\,d\mathbf{r}_2$$

We now change the variables of integration to $\mathbf{r}_{12}=(\mathbf{r}_1-\mathbf{r}_2)$ and $\mathbf{R}_{12}=(\mathbf{r}_1+\mathbf{r}_2)/2$, integrate over $\mathbf{R}_{12}$, which gives the volume, V, and write using polar coordinates

$$\int_V d\mathbf{r}_{12}=4\pi\int_0^\infty r^2\,dr$$

to give the final form of the equation

$$U = \tfrac{3}{2}NkT + 2\pi N\rho \int_0^\infty \phi(r) g(r) r^2 \, dr \tag{3.45}$$

When the total potential energy of the system includes contributions from three-body interactions in addition to pairwise additive terms, the thermodynamic internal energy is given by eq. 3.45 along with a third term

$$U^{(3)} = \frac{N\rho^2}{6} \iint \phi^{(3)}(r, s, t) g^{(3)}(r, s, t) \, d\mathbf{r} \, d\mathbf{s} \tag{3.46}$$

where r, s, and t, the sides of the triangle joining the centers of the three molecules, are related by the formula

$$\mathbf{t} = \mathbf{r} - \mathbf{s} \tag{3.47}$$

Similar formulae written in terms of higher order distributions give the contributions from other interactions.

Given the interpretation of the radial distribution function as the distribution of interparticle distances, eq. 3.45 can be written down by inspection. The first term gives the kinetic energy of the system, and it is known from the elementary kinetic theory of gases that every degree of freedom of the system contributes $kT/2$ to this quantity. Assuming that only two-body interactions are present, the potential energy contribution to the internal energy may be calculated as follows. The probability of finding two particles separated by a distance r is equal to the product of the probability of finding a particle at the origin, ρ, and the probability of finding a second one at a distance r from the first, $4\pi r^2 g(r)\, dr$. Multiplying this probability together with the potential energy of the pair at this separation, $\phi(r)$, and integrating over all distances gives the average pair interaction contribution to the total energy of a given particle. There are N particles in the system, but multiplying the integral by N is not sufficient because then every interaction will have been counted twice, once when particle 1 was at the origin and once when particle 2 was at the origin. Consequently, the integral is multiplied by $N/2$ to give the second term in eq. 3.45.

To obtain an expression for the equation of state of a fluid we start from the canonical partition function [4], using the thermodynamic relation between the pressure, p, and the Helmholtz free energy

$$p = -\left(\frac{\partial A}{\partial V}\right)_T = kT\left(\frac{\partial \log Z}{\partial V}\right)_T \tag{3.48}$$

Using eq. 3.32, and assuming pairwise additivity of the potential energy,

$$p = \frac{kT}{Z} \frac{\partial}{\partial V} \frac{\lambda^{-3N}}{N!} \int \cdots \int \exp\left[-\sum_{i<j} \frac{\phi(r_{ij})}{kT}\right] d\mathbf{r}_1 \cdots d\mathbf{r}_N \quad (3.49)$$

It is an experimental observation that the pressure of a system is generally independent of the shape of its container, and so we will assume that our system is contained in a cube of side $V^{1/3}$. The integrations in eq. 3.49 can then be written out in full

$$p = \frac{kT}{Z} \frac{\partial}{\partial V} \frac{\lambda^{-3N}}{N!} \int_0^{V^{1/3}} \cdots \int_0^{V^{1/3}} \exp\left[-\sum_{i<j} \frac{\phi(r_{ij})}{kT}\right] dx_1\, dy_1\, dz_1 \cdots dx_N\, dy_N\, dz_N$$

If we now make the change of variables

$$x_k = V^{1/3} u_k, \qquad y_k = V^{1/3} v_k, \qquad z_k = V^{1/3} w_k \quad (3.50)$$

the pressure is given by an equation in which the volume occurs explicitly

$$p = \frac{kT}{Z} \frac{\partial}{\partial V} \frac{\lambda^{-3N}}{N!} \int_0^1 \cdots \int_0^1 \exp\left[-\sum_{i<j} \frac{\phi(r_{ij})}{kT}\right] \times V^N du_1\, dv_1\, dw_1 \cdots du_N\, dv_N\, dw_N \quad (3.51)$$

where

$$\begin{aligned} r_{ij} &= [(x_j - x_i)^2 + (y_j - y_i)^2 + (z_j - z_i)^2]^{1/2} \\ &= V^{1/3}[(u_j - u_i)^2 + (v_j - v_i)^2 + (w_j - w_i)^2]^{1/2} \end{aligned} \quad (3.52)$$

Using these results we find

$$\begin{aligned} p = \frac{kT}{Z}\Bigg[& N V^{N-1} \frac{\lambda^{-3N}}{N!} \int_0^1 \cdots \int_0^1 \exp\left[-\sum_{i<j} \frac{\phi(r_{ij})}{kT}\right] du_1 \cdots dw_N \\ & - \frac{kT}{Z} V^N \frac{\lambda^{-3N}}{N!} \int_0^1 \cdots \int_0^1 \left(\frac{\partial}{\partial V} \sum_{i<j} \frac{\phi(r_{ij})}{kT}\right) \exp\left[-\sum_{i<j} \frac{\phi(r_{ij})}{kT}\right] \\ & \times du_1\, dv_1\, dw_1 \cdots du_N\, dv_N\, dw_n \Bigg] \end{aligned} \quad (3.53)$$

We now consider

$$\begin{aligned} \frac{\partial}{\partial V} \phi(r_{ij}) &= \frac{dr_{ij}}{dV} \frac{d\phi\,(r_{ij})}{dr_{ij}} \\ &= \frac{V^{-2/3}}{3} [(u_j - u_i)^2 + (v_j - v_i)^2 + (w_j - w_i)^2]^{1/2} \frac{d\phi(r_{ij})}{dr_{ij}} = \frac{r_{ij}}{3V} \frac{d\phi(r_{ij})}{dr_{ij}} \end{aligned}$$

and using the fact that the sum in eq. 3.53 contains $N(N-1)/2$ identical terms obtain

$$p = NkT/V - \frac{1}{6V} \frac{V^N}{Z} \frac{\lambda^{-3N}}{(N-2)!} \int_0^1 \cdots \int_0^1 r_{12} \frac{d\phi(r_{12})}{dr_{12}} \times \exp\left[-\sum_{i<j} \frac{\phi(r_{ij})}{kT}\right] du_1\, dv_1\, dw_1 \cdots dw_N$$

On returning to the original coordinate system and using the definition of the two-body density function, eq. 3.37, we obtain

$$p = \frac{NkT}{V} - \frac{1}{6V}\iint r_{12}\frac{d\phi(r_{12})}{dr_{12}} f^{(2)}(\mathbf{r}_1, \mathbf{r}_2)\, d\mathbf{r}_1\, d\mathbf{r}_2$$

By using the result $f^{(2)}(\mathbf{r}_1, \mathbf{r}_2) = \rho^2 g(\mathbf{r}_1, \mathbf{r}_2)$ this can be written in terms of the radial distribution function, to give for a uniform fluid

$$pV = NkT - \frac{2\pi N\rho}{3}\int_0^\infty r^3 \frac{d\phi(r)}{dr} g(r)\, dr \tag{3.54}$$

As with the internal energy, the equation of state includes the ideal gas term and a contribution from the two-body interactions in the system. When three-body and higher order interactions are present, further terms occur on the right hand side of eq. 3.54 and are written as integrals including many body distribution functions.

The previous two derivations were obtained by working with the canonical partition function. It is instructive to work with the grand canonical ensemble, and we use the corresponding partition function to derive an equation for the isothermal compressibility of the system. Suppose K_T is the isothermal compressibility

$$K_T = -\frac{1}{V}\left(\frac{\partial V}{\partial p}\right)_T = \frac{1}{\rho}\left(\frac{\partial \rho}{\partial p}\right)_T \tag{3.55}$$

where $\rho = N/V$ is the number density. Using the equation relating the grand partition function to the thermodynamic properties of the system being modeled, eq. 3.20, $pV = kT \log Z^*$, we have

$$\begin{aligned} \frac{V}{kT}\left(\frac{\partial p}{\partial \rho}\right)_{T,V} &= \left(\frac{\partial \log Z^*}{\partial \rho}\right)_{T,V} = \frac{1}{Z^*}\frac{\partial Z^*}{\partial \rho} \\ &= \frac{1}{Z^*}\frac{\partial}{\partial \rho}\sum_{N\geq 0}\frac{z^N}{N!}\int\cdots\int \exp\frac{-\Phi(\mathbf{r}_1, \ldots \mathbf{r}_N)}{kT}\, d\mathbf{r}_1 \cdots d\mathbf{r}_N \end{aligned} \tag{3.56}$$

where z is the activity of the system. To continue we note that

$$\frac{\partial Z^*}{\partial \rho} = \frac{\partial Z^*/\partial z}{\partial \rho/\partial z} \tag{3.57}$$

where ρ, the number density, is in general the one-particle density function. We consider first the numerator of eq. 3.57 and obtain

$$\frac{1}{Z^*}\frac{\partial Z^*}{\partial z} = \frac{1}{Z^*}\frac{\partial}{\partial z}\sum_{N\geq 0}\frac{z^N}{N!}\int\cdots\int \exp\frac{-\Phi(\mathbf{r}_1, \ldots \mathbf{r}_N)}{kT}\, d\mathbf{r}_1 \cdots d\mathbf{r}_N$$

or, if I_N is the term involving the integrals,

$$\frac{1}{Z^*}\frac{\partial Z^*}{\partial z}=\frac{1}{Z^*}\frac{\partial}{\partial z}\sum_{N\geq 0}\frac{z^N}{N!}I_N=\frac{1}{Z^*}\sum_{N\geq 1}\frac{z^{N-1}}{(N-1)!}I_N$$

$$=\frac{1}{z}\int\rho\,d\mathbf{r}_1=\frac{V\rho}{z} \tag{3.58}$$

where we have used the definition of the s particle density function in the grand canonical ensemble

$$f^{(s)}(\mathbf{r}_1,\ldots\mathbf{r}_s)=\frac{1}{Z^*}\sum_{N\geq s}\frac{z^N}{(N-s)!}\times\int\cdots\int\exp\frac{-\Phi(\mathbf{r}_1,\ldots\mathbf{r}_N)}{kT}\,d\mathbf{r}_{s+1}\cdots d\mathbf{r}_N \tag{3.59}$$

The denominator in eq. 3.57 is given by

$$\frac{\partial\rho}{\partial z}=\sum_{N\geq 1}\frac{\partial}{\partial z}\left(\frac{z^{N-1}}{(N-1)!}z\cdot\frac{1}{Z^*}\right)\int\cdots\int\exp\frac{-\Phi(\mathbf{r}_1,\ldots\mathbf{r}_N)}{kT}\,d\mathbf{r}_2\cdots d\mathbf{r}_N$$

$$=\left(\sum_{N\geq 2}\frac{z^{N-2}}{(N-2)!}\frac{z}{Z^*}+\sum_{N\geq 1}\frac{z^{N-1}}{(N-1)!}\frac{1}{Z^*}-\frac{1}{Z^{*2}}\frac{\partial Z^*}{\partial z}\sum_{N\geq 1}\frac{z^N}{(N-1)!}\right)\int\cdots\int\exp\frac{-\Phi(\mathbf{r}_1,\ldots\mathbf{r}_N)}{kT}\,d\mathbf{r}_2\cdots d\mathbf{r}_N \tag{3.60}$$

so that

$$\frac{\partial\rho}{\partial z}=\frac{\rho^2}{z}\int g(r)\,d\mathbf{r}+\frac{\rho}{z}-\frac{\rho^2}{z}\int d\mathbf{r}=\frac{\rho}{z}\left\{1+\rho\int[g(r)-1]\,d\mathbf{r}\right\} \tag{3.61}$$

the last result following from eq. 3.59 and the definition of $g(r)$, eq. 3.38. Combining eq. 3.58 and eq. 3.61 we obtain from eq. 3.56 and eq. 3.57

$$\frac{V}{kT}\left(\frac{\partial p}{\partial\rho}\right)_T=\frac{V}{1+\rho\int[g(r)-1]\,d\mathbf{r}} \tag{3.62}$$

The isothermal compressibility is then given by

$$\frac{1}{\rho}\left(\frac{\partial\rho}{\partial p}\right)_T=\frac{1}{\rho kT}\left(kT\frac{\partial\rho}{\partial p}\right)_T$$

or

$$K_T=\frac{1}{\rho kT}\left\{1+4\pi\rho\int_0^\infty r^2[g(r)-1]\,dr\right\} \tag{3.63}$$

We have derived this expression without assuming the pairwise additivity of the total potential energy of the system, and it is valid even for fluids in which many body interactions are significant.

3.7 Virial Expansion of the Equation of State

Kammerlingh Onnes [8] introduced the idea of fitting the equation of state of a gas to a power series in its density, giving an equation of the form

$$\frac{pV}{NkT} = 1 + B_2\rho + B_3\rho^2 + \cdots \tag{3.64}$$

The coefficients B_2, B_3, and so on are known as the second virial coefficient, third virial coefficient, and so on. We can calculate the coefficients B_i from a knowledge of the intermolecular potential function using statistical mechanics. The expression relating the pair potential to the second virial coefficient is commonly used to determine parameters in an empirical potential function from dilute gas data.

Using statistical mechanics, the virial coefficients are usually given in terms of *cluster diagrams* [9, 10], a convenient shorthand for writing down very complicated integrals. Consider the derivation of expressions for B_2 and B_3. Using the grand canonical ensemble we can calculate the equation of state from

$$\frac{pV}{kT} = \log Z^* = \log\left[\sum_{N\geq 0}\frac{z^N}{N!}\int\cdots\int \exp\frac{-\Phi(\mathbf{r}_1,\ldots\mathbf{r}_N)}{kT}\,d\mathbf{r}_1\cdots d\mathbf{r}_N\right]$$

Writing the total potential energy of the system as sums over two-body, three-body, and higher order interactions, this equation can be written explicitly

$$\frac{pV}{kT} = \log\left(1 + z\int d\mathbf{r}_1 + \frac{z^2}{2}\iint \exp\frac{-\phi(r_{12})}{kT}\,d\mathbf{r}_1\,d\mathbf{r}_2 + \frac{z^3}{6}\right.$$
$$\left.\times\iiint \exp\left\{\frac{-[\phi(r_{12})+\phi(r_{13})+\phi(r_{23})+\phi^{(3)}(r_{12}, r_{13}, r_{23})]}{kT}\right\} d\mathbf{r}_1\,d\mathbf{r}_2\,d\mathbf{r}_3 + \cdots\right) \tag{3.65}$$

We now make the substitutions [10]

$$f_{12} = \exp\frac{-\phi(r_{12})}{kT} - 1$$
$$f_{123} = \exp\frac{-\phi^{(3)}(r_{12}, r_{13}, r_{23})}{kT} - 1 \tag{3.66}$$

and assume that the fluid is uniform, whereupon eq. 3.65 becomes

$$\frac{pV}{kT}=\log\left[1+zV+\frac{z^2}{2}(V^2+VI_2)\right.$$
$$\left.+\frac{z^3}{6}(V^3+3V^2I_2+3VI_2^2+VI_3)+\cdots\right] \tag{3.67}$$

where

$$I_2=\int f_{12}\,d\mathbf{r}_{12} \tag{3.68}$$

and

$$I_3=\iint[f_{12}f_{23}f_{13}+f_{123}(1+f_{12})(1+f_{23})(1+f_{13})]\,d\mathbf{r}_{12}\,d\mathbf{r}_{13} \tag{3.69}$$

The second integral covers all values of the vector $\mathbf{r}_{23}$ as $\mathbf{r}_{23}=\mathbf{r}_{13}-\mathbf{r}_{12}$. We now assume that the activity, z, is small and expand the logarithm in a power series in z, retaining terms up to z^3

$$\frac{pV}{kT}=zV+\frac{z^2}{2}(V^2+VI_2)+\frac{z^3}{6}(V^3+3V^2I_2+3VI_2^2+VI_3)$$
$$-\frac{z^2V^2}{2}-\frac{z^3V}{2}(V^2+VI_2)+\frac{z^3V^3}{3}+\cdots \tag{3.70}$$

and collecting terms we find

$$\frac{pV}{kT}=V\left[z+\frac{z^2}{2}I_2+\frac{z^3}{6}(3I_2^2+I_3)+\cdots\right] \tag{3.71}$$

This equation gives the equation of state as a power series in the activity.

We now obtain the density as a power series in the activity, invert the series, and substitute for z in eq. 3.71—a procedure that is not as difficult as it sounds. The density is given by the one-particle function

$$\rho=\frac{1}{Z^*}\sum_{N\geqslant1}\frac{z^N}{(N-1)!}\int\cdots\int\exp\frac{-\Phi(\mathbf{r}_1,\ldots\mathbf{r}_N)}{kT}\,d\mathbf{r}_2\cdots d\mathbf{r}_N \tag{3.72}$$

Rather than expand this sum, including Z^*, in powers of z we note the relation

$$\rho=\frac{z}{V}\frac{d\log Z^*}{dz} \tag{3.73}$$

We prove this by differentiating

$$\rho=\frac{z}{V}\frac{1}{Z^*}\sum_{N\geqslant1}\frac{z^{N-1}}{(N-1)!}\int\cdots\int\exp\frac{-\Phi(\mathbf{r}_1,\ldots\mathbf{r}_N)}{kT}\,d\mathbf{r}_1\cdots d\mathbf{r}_N$$

an equation that can be readily identified with eq. 3.72 after integrating over $\mathbf{r}_1$. In eq. 3.71 we already have the expansion of log Z^* in powers of z and so we may use it together with eq. 3.73 to find

$$\rho = \frac{z}{V} \cdot V\left[1 + zI_2 + \frac{z^2}{2}(3I_2^2 + I_3) + \cdots\right]$$

$$= z + z^2 I_2 + \frac{z^3}{2}(3I_2^2 + I_3) + \cdots \tag{3.74}$$

To obtain the activity as a power series in the density we assume

$$z = a_0 + a_1\rho + a_2\rho^2 + a_3\rho^3 + \cdots$$

substitute this in eq. 3.74, and equate powers of the density on both sides. After some algebra this procedure gives the following equations for the coefficients

$$a_0 = 0$$
$$a_1 = 1$$
$$a_2 = -I_2$$
$$a_3 = \frac{(I_2^2 - I_3)}{2}$$

or

$$z = \rho - I_2\rho^2 + \frac{(I_2^2 - I_3)\rho^3}{2} + \cdots \tag{3.75}$$

We substitute this into eq. 3.71 to give the equation of state as a power series in the density

$$\frac{p}{kT} = \rho - \frac{I_2\rho^2}{2} - \frac{I_3\rho^3}{3} \cdots \tag{3.76}$$

from which we see that the second and third virial coefficients are given by

$$B_2 = -\tfrac{1}{2}\int f_{12}\, d\mathbf{r}_{12} \tag{3.77}$$

$$B_3 = -\tfrac{1}{3}\iint [f_{12}f_{23}f_{13} + f_{123}(1 + f_{12})(1 + f_{13})(1 + f_{23})]\, d\mathbf{r}_{12}\, d\mathbf{r}_{23} \tag{3.78}$$

The cluster diagrams referred to at the beginning of this section are constructed as follows. A line joining two open circles represents the function f_{ij}

$$f_{12} = \exp\frac{-\phi(r_{12})}{kT} - 1 \equiv \underset{1}{\circ}\!\!-\!\!-\!\!-\!\!\underset{2}{\circ}$$

and a line without the open circles represents an integral over f_{ij}

$$\int f_{12}\, d\mathbf{r}_{12} \equiv$$

Similarly, the three-body function is represented as a shaded triangle, with open circles

$$f_{123} \equiv$$

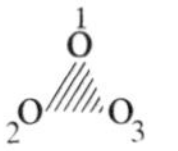

Using this notation the second and third virial coefficients can be written

$$B_2 = -\tfrac{1}{2}$$

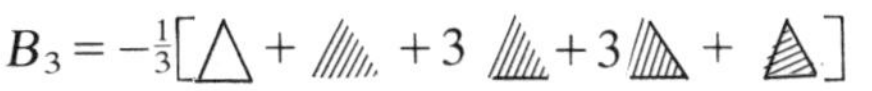

$$B_3 = -\tfrac{1}{3}[\ \ + \ \ + 3 \ \ + 3 \ \ + \ \ ]$$

Some idea of the power of this notation can be obtained by considering a system interacting through a two-body potential alone. Using graph theoretical methods [11] a general expression can be obtained for the Mth virial coefficient

$$B_M = \frac{1}{M(M-2)!} \sum \begin{bmatrix} \text{all labeled irreducible} \\ \text{diagrams of } M \text{ points} \end{bmatrix} \tag{3.79}$$

The term "irreducible" means that the diagram cannot be split into two or more parts by removing one vertex. A simple example may reduce the confusion

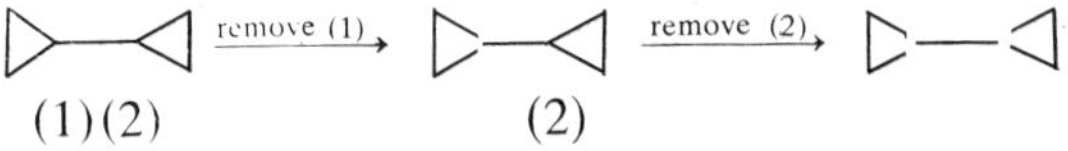

remove (1) → remove (2) →

(1)(2) (2)

This diagram is reducible and so will not contribute to the virial coefficient. We can use eq. 3.79 to write down expressions for the fourth and fifth virial coefficients

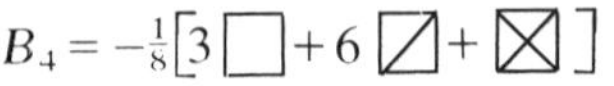

$$B_4 = -\tfrac{1}{8}[3 \ \ + 6 \ \ + \ \ ] \tag{3.80}$$

Here we see the reason for including the word "labeled" in eq. 3.79, for we can label the first diagram in three distinct ways:

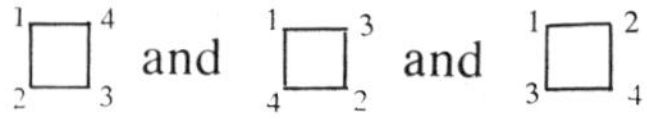

and and

Any other set of labels can be converted into one of the above by a combination of rotations. The fifth virial coefficient is given by

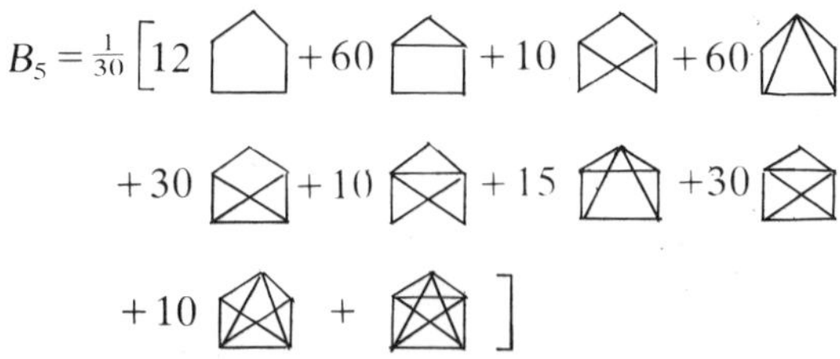

$$B_5 = \tfrac{1}{30}\Big[12 \ \ + 60 \ \ + 10 \ \ + 60 \ \ + 30 \ \ + 10 \ \ + 15 \ \ + 30 \ \ + 10 \ \ + \ \ \Big] \tag{3.81}$$

The problem of calculating these coefficients is formidable, although one or two of the lower order diagrams can be evaluated analytically for some simple model potential functions [12].

3.8 Approximate Theories of the Radial Distribution Function

Although the first few virial coefficients can be calculated for most simple potentials, the corresponding density expansion of the equation of state begins to diverge from experimental results at densities typical of those near the critical point [12]. The equations for thermodynamic properties derived in Section 3.6 are valid at all densities but require values for the radial distribution function. In this section several approximate equations for this function are described [13].

The first type of integral equation for the radial distribution function is obtained by introducing an approximate closure to an infinite hierarchy of equations relating the s body density function to the $(s+1)$ body function [14]. Suppose we assume that the total potential energy of the system is pairwise additive and differentiate the two-body density function with respect to the position vector $\mathbf{r}_1$. From eq. 3.37 we find

$$\frac{\partial f^{(2)}(\mathbf{r}_1,\mathbf{r}_2)}{\partial \mathbf{r}_1}=-\frac{\partial \phi(r_{12})}{\partial \mathbf{r}_1}\frac{f^{(2)}(\mathbf{r}_1,\mathbf{r}_2)}{kT}-\frac{1}{kT}\int \frac{\partial \phi(r_{13})}{\partial \mathbf{r}_1} f^{(3)}(\mathbf{r}_1,\mathbf{r}_2,\mathbf{r}_3)\, d\mathbf{r}_3 \quad (3.82)$$

an exact equation that relates the two-body and three-body density functions. A similar equation can be obtained by differentiating the three-body density function, this time relating $f^{(3)}$ to $f^{(4)}$. Dividing eq. 3.82 by ρ^2 throughout and using the definitions of $g^{(2)}(\mathbf{r}_1, \mathbf{r}_2)$ and $g^{(3)}(\mathbf{r}_1, \mathbf{r}_2, \mathbf{r}_3)$, eqs. 3.38 and 3.41, we obtain the exact relation

$$\frac{\partial g^{(2)}(\mathbf{r}_1,\mathbf{r}_2)}{\partial \mathbf{r}_1}=-\frac{\partial \phi(r_{12})}{\partial \mathbf{r}_1}\frac{g^{(2)}(\mathbf{r}_1,\mathbf{r}_2)}{kT}-\frac{\rho}{kT}\int \frac{\partial \phi(r_{13})}{\partial \mathbf{r}_1} g^{(3)}(\mathbf{r}_1,\mathbf{r}_2,\mathbf{r}_3)\, d\mathbf{r}_3 \quad (3.83)$$

This equation is the first member of the Born-Green-Yvon hierarchy [15] and is exact. We return to it shortly.

A similar hierarchy of equations was obtained by Kirkwood [16] using a different approach. He assumed that the particle labeled "1" was coupled to the rest of the system by a parameter ξ, so that the total potential energy could be written

$$\Phi(\mathbf{r}_1,\ldots \mathbf{r}_N;\xi)=\xi\sum_{i\geq 2}\phi(r_{1i})+\sum_{i>j\geq 2}\phi(r_{ij}) \quad (3.84)$$

When $\xi=0$ particle "1" is effectively removed from the system and when $\xi=1$ the coupling is complete. Consequently, varying this parameter between 0 and 1 enables an additional particle to be added to the system. Introducing this potential energy into the equation for $f^{(2)}(\mathbf{r}_1, \mathbf{r}_2)$ (eq. 3.37) and differentiating with respect to ξ (remember Z is also a function of ξ), we obtain after using the definitions of $g^{(2)}$ and $g^{(3)}$

$$kT\frac{\partial \log g^{(2)}(\mathbf{r}_1, \mathbf{r}_2)}{\partial \xi} = -\phi(r_{12}) - \rho\int \phi(r_{13})\left[\frac{g^{(3)}(\mathbf{r}_1, \mathbf{r}_2, \mathbf{r}_3)}{g^{(2)}(\mathbf{r}_1, \mathbf{r}_2)} - g^{(2)}(\mathbf{r}_1, \mathbf{r}_3)\right] d\mathbf{r}_3 \quad (3.85)$$

Again, higher order density functions can be related using a similar procedure, and the resulting exact series of equations is known as the Kirkwood hierarchy. It can be shown that the Born-Green-Yvon and Kirkwood hierarchies are equivalent [17, 18].

Although the two hierarchies provide exact relations between $g^{(2)}(\mathbf{r}_1, \mathbf{r}_2)$ and $g^{(3)}(\mathbf{r}_1, \mathbf{r}_2, \mathbf{r}_3)$, they are not useful unless one of the distribution functions is known. As none are, some approximate relation between the distribution functions must be assumed. Kirkwood [16] showed that at low densities, when it is unusual to get more than two particles close together, the three-body density function is given by a product of pair density functions so that

$$g^{(3)}(r, s, t) = g(r)g(s)g(t) \quad (3.86)$$

Formally, this is the leading term in a density expansion of $g^{(3)}$, but if all the higher terms in the expansion are ignored we have the approximate relation needed to truncate the hierarchies. The approximation given in eq. 3.86 is called the Kirkwood superposition approximation. Similar approximations can be written down relating any two successive density functions and so either hierarchy can be truncated at any level. The Kirkwood superposition principle asserts that the three-body density function depends on the two-body correlations in the system alone and does not recognize that three-body correlations might be important. Once the superposition approximation is inserted into the Born-Green-Yvon and Kirkwood hierarchies, two approximate integral equations for the radial distribution function are obtained, the Born-Green and Kirkwood equations. Unlike the hierarchical equations these integral equations are not equivalent and so they have different solutions.

It is not necessarily true that any approximate closure of the hierarchies will give an integral equation whose solution is also a solution that satisfies the hierarchies. Raveché and Green [19] have examined the conditions necessary for solutions of the hierarchies to be consistent. Let the potential of mean force for the system be defined by

$$W^{(2)}(r) = -kT \log g(r) \tag{3.87}$$

Then for a fluid at equilibrium it can be shown that the potential of mean force is a conserved property, and this is true for the two hierarchies. The distribution functions obtained from the Born-Green and Kirkwood integral equations also satisfy the condition that the potential of mean force is conserved. Raveché and Green [19] used this conservation law to construct a condition which any given closure of the two hierarchies must satisfy before the resulting approximate integral equations have solutions that satisfy the truncated hierarchies.

The physical justification for the Born-Green and Kirkwood approximations lies in the superposition approximation, that is, in the assumption that three-body effects are unimportant. As we see in later chapters neither approximation is very accurate and a much better radial distribution function can be obtained from other approximations, for example, the hypernetted chain equation (HNC) [20–23]. Unfortunately, it does not seem possible to justify the approximations inherent in the HNC equation on physical grounds although the derivation of this equation introduces the idea of the radial distribution function having a density expansion. In addition the derivation throws some light on the ways by which a number of approximations can be improved. We begin the derivation by introducing a new two-particle function, the *direct correlation function* $c(r)$.

The *total correlation function* is defined by

$$h(r) = g(r) - 1 \tag{3.88}$$

and measures the total correlations between two particles. To see this, we note that according to eq. 3.40 $g(r)$ tends to the value 1 at large distances, from which it follows that $h(r) = 0$—that is, there are no correlations between two particles when they are widely separated. The total correlation between two particles can be separated into two contributions. First, there are correlations transmitted directly between two particles and, without specifying the mechanism by which the correlations are transmitted, we describe these correlations by the function $c(r)$. The second contribution comes from correlations transmitted through a third particle and is made up of direct correlations between particles 1 and 3 multiplied by the total correlation function between particles 3 and 2. Before adding the second term to $c(r)$ it must be multiplied by the probability of finding the third particle in the volume element $d\mathbf{r}_3$, given by the number density ρ, and then we must integrate over all possible positions of particle 3. The result is the Ornstein-Zernike equation [24]

$$h(\mathbf{r}_{12}) = c(\mathbf{r}_{12}) + \rho \int c(\mathbf{r}_{13}) h(\mathbf{r}_{32})\, d\mathbf{r}_3 \tag{3.89}$$

Although we have used a physical argument to justify writing down this equation, it can be treated simply as a definition of the direct correlation function. To see this, we define the Fourier transform of a function $f(\mathbf{r})$ by

$$\tilde{f}(\mathbf{q}) = \int f(\mathbf{r}) \exp(i\mathbf{q} \cdot \mathbf{r})\, d\mathbf{r} \tag{3.90}$$

and its inverse by

$$f(\mathbf{r}) = \int \tilde{f}(\mathbf{q}) \frac{\exp(-i\mathbf{q} \cdot \mathbf{r})\, d\mathbf{q}}{(2\pi)^3} \tag{3.91}$$

We now take the Fourier transform of the Ornstein-Zernike equation, making use of the fact that the right hand side contains a convolution integral [25], to obtain

$$\tilde{h}(\mathbf{q}) = \tilde{c}(\mathbf{q}) + \rho \tilde{c}(\mathbf{q}) \tilde{h}(\mathbf{q})$$

or on solving for $\tilde{c}(\mathbf{q})$

$$\tilde{c}(\mathbf{q}) = \frac{\tilde{h}(\mathbf{q})}{1 + \rho \tilde{h}(\mathbf{q})} \tag{3.92}$$

Consequently, if we know $h(r)$ we can obtain $c(r)$ uniquely through eqs. 3.90 to 3.92, and so the Ornstein-Zernike equation is a definition of the direct correlation function. The Ornstein-Zernike equation was originally derived in an attempt to develop a viable theory of fluctuations in the neighborhood of the critical point and was used in an explanation of the phenomenon of critical opalescence.

Returning to the definition of the two-body density function we see that we can take out a factor $\exp[-\phi(r_{12})/kT]$ from either eq. 3.37 (the canonical ensemble) or eq. 3.59 (the grand canonical ensemble). We can now write the radial distribution function in the form

$$g(r_{12}) = \exp \frac{-\phi(r_{12})}{kT}\, y(r_{12}) \tag{3.93}$$

where $y(r)$ is a function to be determined. Expanding the equation for $f^{(2)}(\mathbf{r}_1, \mathbf{r}_2)$ in powers of the activity, z, as we did with the equation of state, eventually gives a density expansion [11] for $g(r)$. The coefficients in the expansion are cluster diagrams, but unlike those occurring in the equation of state they are a function of position. After completing the calculation it is found that [11]

$$g(r_{12}) = \exp \frac{-\phi(r_{12})}{kT} \exp\left[\sum_{m \geq 3} \rho^{m-2} g_{m-2}\right] \tag{3.94}$$

where the coefficients g_m are irreducible cluster integrals with two root points

$$g_{m-2} = \frac{1}{(m-2)!} \int \cdots \int \Pi^{m-2/1,2} f_{ij} \, d\mathbf{r}_3 \cdots d\mathbf{r}_m \tag{3.95}$$

The sum in eq. 3.95 is over all simply connected diagrams with two root points labeled 1 and 2, no f_{12} bond, and with $m-2$ field points.

The number of technical terms in the last paragraph is frightening! For our purposes we can consider the following definitions:

1. *Simply connected.* This means that there is no more than one factor of $f_{ij} = \exp[-\phi(r_{ij})/kT] - 1$ for every pair of points (i, j). It also implies that all the points in the diagram are connected to at least one other point. In Fig. 3.1 all diagrams other than (a) and (b) are simply connected; (a) is disconnected and (b) has one doubly connected bond.
2. *Root point.* The diagram is a function of the vector positions of the labeled open circles, or roots. As an example all the graphs in Fig. 3.1 are functions of the position vectors $\mathbf{r}_1$ and $\mathbf{r}_2$.
3. *Field point.* Such points represent volume integrals and are the unlabeled vertices in the figure.
4. *Irreducible.* A diagram is irreducible if on removing one point every disconnected part of the diagram is attached to at least one root point and if its parts are only connected at the root points. Graphs (c) and (d) show the meaning of this term. Graph (d) has two distinct contributions. The term reducible implies that the corresponding integral can be factorized into two or more terms.

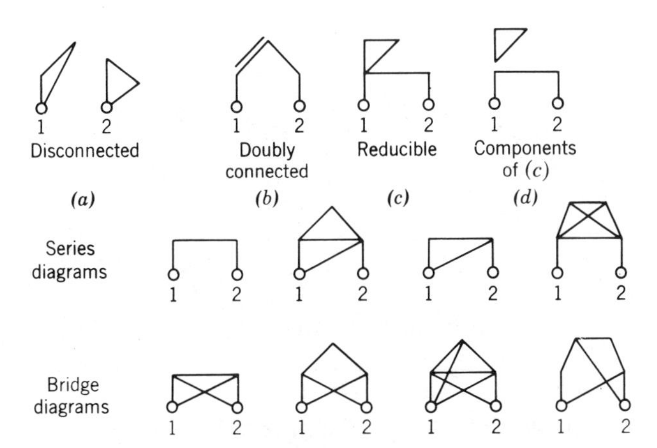

Figure 3.1 Typical doubly rooted graphs some of which occur in the density expansion of the radial distribution function. The various terms are explained in the text.

The irreducible diagrams can be divided into two classes, the *Series diagrams* and the *Bridge diagrams.* Examples of both classes are shown in Fig. 3.1. We see that in the series diagrams all possible paths from one root to the other have at least one point in common, a node point, whereas the Bridge diagrams provide at least two alternative paths between roots.

As all the simply connected diagrams in eq. 3.94 must fall into either the Series or Bridge classes we can rewrite the density expansion for $g(r_{12})$

$$g(r_{12}) = \exp\left(\frac{-\phi(r_{12})}{kT} + S(r_{12}) + B(r_{12})\right) \tag{3.96}$$

where the function $S(r_{12})$ is the sum of all integrals corresponding to the Series diagrams, together with their weights ($\rho^{m-2}/(m-2)!$, for example) and $B(r_{12})$ is the corresponding function representing the sum of Bridge diagrams. From eq. 3.96 it is trivial to write down the expression for $h(r) = g(r) - 1$. We can also obtain the direct correlation function as a density expansion [11] and find that it is given by the equation

$$h(r_{12}) = S(r_{12}) + c(r_{12}) \tag{3.97}$$

Consequently, we see that $c(r_{12})$ consists of all the terms in the density expansion of $h(r_{12})$ except for those contained in the function $S(r_{12})$. We can justify this expression by considering a typical term in $S(r_{12})$. All the diagrams representing this function have at least one node. We can write a nodal diagram as an integral over the product of two other diagrams. Consider the function

$$\underset{1\quad 2}{\text{(diagram)}} \equiv \iiint f_{13}f_{34}f_{32}f_{35}f_{45}f_{25}\, d\mathbf{r}_3\, d\mathbf{r}_4\, d\mathbf{r}_5$$

$$\equiv \int (f_{13})\left[\iint f_{34}f_{32}f_{35}f_{45}f_{25}\, d\mathbf{r}_4\, d\mathbf{r}_5\right] d\mathbf{r}_3$$

$$\equiv \int \left(\underset{1\qquad 3}{\circ\!\!-\!\!\!-\!\!\!-\!\!\circ}\right)\left(\underset{3\ \ 2}{\text{(diagram)}}\right) d\mathbf{r}_3 \tag{3.98}$$

It follows that the function $S(r_{12})$ can be written as an integral over the product of two other functions

$$S(r_{12}) = \rho \int D_1(r_{13}) D_2(r_{23})\, d\mathbf{r}_3 \tag{3.99}$$

where the diagrams contributing to $D_1(r_{13})$ have no nodes between the roots 1 and 3 and where all possible diagrams are included in $D_2(r_{23})$. The density term arises because a diagram in $S(r_{12})$ contains a multiple of ρ for every field point. Now, according to eq. 3.97, the total correlation function is composed of all possible diagrams and so $D_2(r_{23}) = h(r_{23})$. Also from eq. 3.97 we see that the direct correlation function is the sum of all possible nonnodal diagrams, and as $D_1(r_{13})$ was deliberately chosen to be the sum of nonnodal diagrams we must have $D_1(r_{13}) = c(r_{13})$. Bringing these identities together in eqs. 3.99 and 3.97 we have

$$h(r_{12}) = c(r_{12}) + \rho \int c(r_{13}) h(r_{32}) \, d\mathbf{r}_3 \tag{3.100}$$

which is the Ornstein-Zernike equation.

The hypernetted chain (HNC) approximation to the radial distribution function is obtained by making the approximation [20–23]

$$B(r) = 0 \tag{3.101}$$

Introducing this into eq. 3.96 we see that

$$\log [g(r)] = \frac{-\phi(r)}{kT} + S(r)$$

or using the Ornstein-Zernike equation in the form of eq. 3.97

$$c(r) = h(r) - \log \left(g(r) \exp \frac{+\phi(r)}{kT} \right)$$

which gives

$$c(r) = h(r) - \log y(r) \tag{3.102}$$

using eq. 3.93.

The HNC approximation has proven to be superior to both the Born-Green and Kirkwood approximations [13]. To obtain values for the radial distribution function, the direct correlation function is eliminated between eq. 3.102 and the Ornstein-Zernike equation. This gives an integral equation for $g(r)$ that must be solved numerically.

The reader should note that we have not made any physical arguments when deriving the HNC approximation. It is possible to derive the approximation using a variational principle to minimize the free energy, as has been shown by Morita and Hiroike [21], although the term $B(r)$ is still ignored for reasons of computational difficulty. Consequently, the HNC theory is justified on purely mathematical grounds: it is the most comprehensive approximation that can be easily derived using cluster expansions.

We can make other approximations for the radial distribution function and a particularly important one is attributed to Percus and Yevick [26, 27]. This approximation can be obtained by again putting $B(r) = 0$ in eq. 3.96 and then by expanding the exponential

$$g(r) = \exp \frac{-\phi(r)}{kT} [1 + S(r) + \tfrac{1}{2} S^2(r) + \cdots]$$

and ignoring all terms beyond the first power in $S(r)$. This gives

$$g(r) = \exp \frac{-\phi(r)}{kT} [1 + S(r)] \tag{3.103}$$

which when combined with eq. 3.97 can be rearranged to give the Percus-Yevick (PY) approximation

$$c(r) = \left\{1 - \exp \frac{+\phi(r)}{kT}\right\} g(r) = \left(\exp \frac{-\phi(r)}{kT} - 1\right) y(r) \tag{3.104}$$

At first sight this approximation would appear to have even less justification than the HNC approximation. However, it can be obtained from physical arguments and is superior to the HNC approximation for many problems [13]. Thus we have the somewhat surprising result that an approximation that on mathematical grounds is less comprehensive is nevertheless superior. In Fig. 3.2 the diagrams contributing to the second and third terms in the density expansion of $g(r)$ are given for the PY and HNC approximations and for the exact expression.

The physical justification for the PY approximation was given in its original derivation [26] and resembles the harmonic analysis of a crystal. In a well known theory of solids, the Born-von Karman theory [28], the potential energy of the crystal is expanded in a Taylor series about the static lattice sites, resulting in a power series in terms of the displacements. Apart from the static lattice energy, the first important term in the series is that in the square of the displacement and the coefficients in that term are given by the second derivative of the potential energy function evaluated at the lattice sites. If we ignore higher order terms the potential energy has now been represented as a quadratic function, and the atoms in the crystal are to a very good approximation in simple harmonic motion about the lattice sites. The quadratic form in the displacements can be diagonalized to give $3N$ independent simple harmonic oscillators, called phonons, that represent collective vibrations of the particles. All the thermodynamic properties of the low temperature solid may be calculated in terms of the frequencies of these harmonic oscillators. Thus the statistical thermodynamics of the crystal is dealt with very accurately using a set of coordinates describing the collective vibrations—the collective coordinates.

Percus and Yevick [26] attempted to construct a similar theory for fluids, although there is no regular lattice, and, rather than oscillate about a lattice site, particles move easily through the system. They assumed that the fluid could be approximated by a periodic system constructed from very large unit cells and attempted to find the collective coordinates for

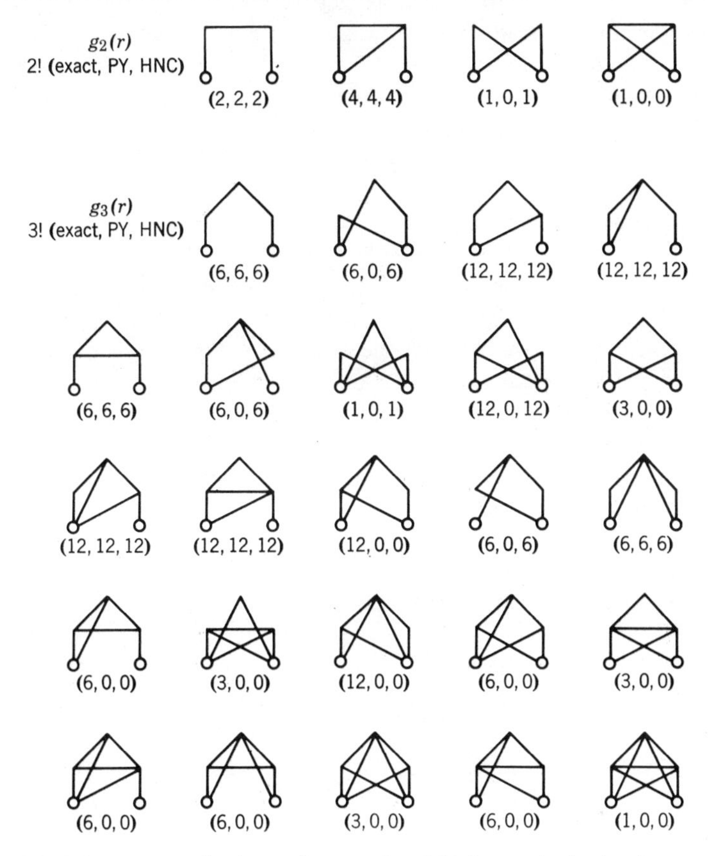

Figure 3.2 Diagrams contributing to the second and third terms in the density expansion of $g(r)$. Numbers give the contributions to the exact expression and to the PY and HNC approximations. (Adapted with permission from *J. Chem. Phys.*, **44,** 3407, 1966.)

this system. In a liquid there are no lattice sites and consequently the Taylor expansion mentioned earlier cannot be performed. Percus and Yevick assumed the existence of some potential function $V(\mathbf{r}_1, \ldots \mathbf{r}_N)$ that enabled a phonon analysis of the fluid to be constructed and determined the $3N$ allowed collective modes by minimizing the function

$$F = |V(\mathbf{r}_1, \ldots \mathbf{r}_N) - \Phi(\mathbf{r}_1, \ldots \mathbf{r}_N)| \tag{3.105}$$

The result was an approximate set of collective coordinates that could be used to determine the structure and thermodynamic properties of the fluid. Unfortunately, the mathematical analysis of this problem is complicated and so we do not reproduce it here.

The integral equation theories discussed here have been the most important during the past few years. A number of other approximations have been proposed based on one or more of the above theories and an extensive account can be found elsewhere [13].

3.9 Perturbation Theories

Integral equation theories are not based on any prior knowledge of the fluid being considered. However, if we know the properties of a simple system, perturbation theory may be used to estimate the properties of other systems. In this section we examine several of these theories. As a simple example we begin by outlining a theory attributed to Zwanzig [29] and use this to obtain the van der Waals theory [30].

Using an assumption that is basic to all perturbation theories Zwanzig [29] wrote the pair potential for two particles in the form

$$\phi(r) = u_0(r) + \eta u_1(r) \tag{3.106}$$

where η is a parameter that when set to zero gives the reference potential $u_0(r)$ and when set to one gives the required potential $\phi(r)$. We introduce this assumption into the expression for the partition function in the canonical ensemble to give

$$Z(\eta) = \frac{\lambda^{-3N}}{N!} \int \cdots \int \exp\left[-\sum_{i<j} \frac{u_0(r_{ij}) + \eta u_1(r_{ij})}{kT} \right] d\mathbf{r}_1 \cdots d\mathbf{r}_N \tag{3.107}$$

The exponential is now expanded as a power series in η to give

$$Z(\eta) = Z_0 \left[1 - \frac{\eta}{Z_0} \frac{\lambda^{-3N}}{N!} \int \cdots \int \sum_{i<j} \frac{u_1(r_{ij})}{kT} \times \exp\left[-\sum_{i<j} \frac{u_0(r_{ij})}{kT} \right] d\mathbf{r}_1 \cdots d\mathbf{r}_N + O(\eta^2) \right] \tag{3.108}$$

where Z_0 is the partition function for the reference system

$$Z_0 = \frac{\lambda^{-3N}}{N!} \int \cdots \int \exp\left[-\sum_{i<j} \frac{u_0(r_{ij})}{kT} \right] d\mathbf{r}_1 \cdots d\mathbf{r}_N \tag{3.109}$$

The sum over pairs in eq. 3.108 gives $N(N-1)/2$ identical contributions to the integral, and we may use this together with the definition of the radial distribution function to write

$$Z(\eta) = Z_0 \left[1 - \eta \frac{2N\pi\rho}{kT} \int_0^\infty u_1(r) g_0(r) r^2 \, dr + O(\eta^2) \right] \tag{3.110}$$

where $g_0(r)$ is the radial distribution function of the reference system. This general procedure was used in Section 3.6 to obtain expressions for the internal energy and the equation of state of the fluid. Taking the logarithm of both sides of eq. 3.110 gives an expression for the free energy of the system

$$\begin{aligned} A(\eta) &= -kT \log Z(\eta) \\ &= A_0 - kT \log\left[1 - \eta \frac{2N\pi\rho}{kT}\int_0^\infty u_1(r) g_0(r) r^2 \, dr + O(\eta^2)\right] \end{aligned}$$

where A_0 is the Helmholtz free energy of the reference system. Expanding the logarithm in powers of η using the expression

$$\log(1-x) = -x - \frac{x^2}{2} - \cdots$$

gives for the original system, where $\eta = 1$,

$$A = A_0 + 2N\pi\rho \int_0^\infty u_1(r) g_0(r) r^2 \, dr + O(\eta^2) \tag{3.111}$$

which is the desired expression. All other thermodynamic properties of the system can be obtained using appropriate thermodynamic relationships [31].

The important point to note about eq. 3.111 is that if the thermodynamic and structural properties of the reference system are known then the thermodynamic properties of the original system can be estimated. Higher order terms in the expansion can be calculated and a similar expansion is possible for the radial distribution function. We examine these quantities later, after obtaining the van der Waals theory as an interesting example of the applications of the perturbation theory.

The van der Waals theory is obtained by using as a reference system a fluid of hard spheres [30]. For such a system the reference potential is given by

$$u_0(r) = \begin{cases} \infty & r < \sigma \\ 0 & r \geq \sigma \end{cases} \tag{3.112}$$

where σ is the diameter of the sphere. This is a very simple interaction potential but it is capable of giving a good qualitative description of the structure of a fluid, as we illustrate in a later chapter. We begin our derivation of the van der Waals theory by considering the first term in eq. 3.111. The partition function for an ideal gas is given by

$$Z_{id} = \frac{\lambda^{-3N}}{N!}\int \cdots \int d\mathbf{r}_1 \cdots d\mathbf{r}_N = \frac{\lambda^{-3N} V^N}{N!}$$

where V is the volume of the system. In a system of close packed hard spheres there is a certain amount of waste space into which other spheres will not fit. The total volume occupied by the spheres is Nb, where $b = 2\pi\sigma^3/3$ is the *covolume* or effective volume of one sphere. We approximate the partition function of a fluid of hard spheres by replacing V in the ideal gas expression by $(V - Nb)$. The resulting approximation for the Helmholtz free energy of the reference system, to be used in eq. 3.111, is

$$\begin{aligned} A_0 &= -kT \log\left[\frac{\lambda^{-3N}}{N!}(V - Nb)^N\right] \\ &= -kT[N \log(V - Nb) - 3N \log \lambda - \log N!] \end{aligned} \tag{3.113}$$

To obtain the second term in the perturbation expansion we use the value of the radial distribution function at zero density. If we refer back to eq. 3.94 we see that when the density is zero we have

$$g(r) = \exp\frac{-\phi(r)}{kT} = \begin{cases} 0 & r < \sigma \\ 1 & r \geqslant \sigma \end{cases} \tag{3.114}$$

the second result being specific for the hard sphere potential. When this radial distribution function is used to calculate the second term in eq. 3.111 we find that the first order approximation to the free energy is given by

$$A = -kT[N \log(V - Nb) - 3N \log \lambda - \log N!] + N\rho a \tag{3.115}$$

where the constant a is given by

$$a = 2\pi\int_\sigma^\infty u_1(r) r^2\, dr \tag{3.116}$$

The equation of state of the system is obtained by differentiating eq. 3.115 with respect to V at constant T:

$$p = -\left(\frac{\partial A}{\partial V}\right)_T = \frac{NkT}{V - Nb} - \frac{N^2}{V^2}a$$

or on rearranging

$$\left(p + \frac{N^2 a}{V^2}\right)(V - Nb) = NkT \tag{3.117}$$

which is the van der Waals equation in its usual form. Thus the simple perturbation theory given above can be used to obtain this well known result.

The main obstacle to using the Zwanzig perturbation theory successfully is that the reference system must be chosen with care. If it is as difficult to perform calculations for $u_0(r)$ as it is for the system of interest nothing is gained by using perturbation theory. Probably the simplest reference system that can be used successfully is the hard sphere fluid. Calculations of the free energy and radial distribution function are readily performed and have been widely published [32–36]. The major difficulty with this reference system is the choice of hard sphere diameter σ. In a realistic potential function the repulsive term at small distances is not infinitely steep and as a consequence such a function cannot be written in the form of eq. 3.106. If we approximate the real potential by eq. 3.106 there are several choices for the reference hard sphere diameter. One choice would be to choose σ as the distance to the first zero in the potential, as indicated in Fig. 3.3. An alternative would be to use σ to minimize the Helmholtz free energy calculated from the truncated

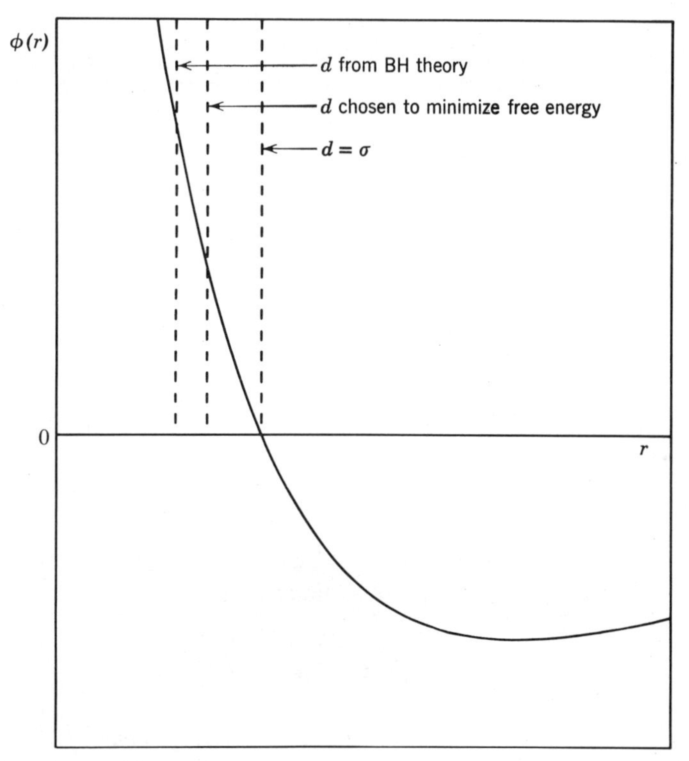

Figure 3.3 Schematic representation of possible choices of hard sphere diameter in perturbation theory.

perturbation series [37]. Such a choice usually gives a value of σ less than the first zero in the potential as is also shown in Fig. 3.3. Several other possible values of σ may be obtained, but there is no clear method for determining which is the most suitable.

Barker and Henderson [38] have described a very successful perturbation theory that has the advantage of determining the appropriate hard sphere diameter satisfactorily. They defined a new potential, $v(r)$, with two perturbation parameters α and γ

$$\exp\frac{-v(r)}{kT} = \left[1 - H(d'-\sigma\right]\exp\frac{-\phi(d')}{kT}$$

$$+H(d'-\sigma)+H(r-\sigma)\left[\exp\frac{-\gamma\phi(r)}{kT}-1\right] \quad (3.118)$$

where $d' = d + (r-d)/\alpha$ and $H(s)$ is the Heaviside step function

$$H(s) = \begin{cases} 0 & s < 0 \\ 1 & s \geqslant 0 \end{cases}$$

The parameter d is the diameter of the hard sphere reference fluid and σ is the position of the zero in the unperturbed potential $\phi(r)$. When α and γ are set equal to one the new potential is identical to the unperturbed potential; when they are both zero $v(r)$ becomes the hard sphere potential. The free energy of the fluid interacting with potential $v(r)$ is now expanded in a double Taylor series in terms of α and γ

$$A = A_0 + \alpha\left(\frac{\partial A}{\partial \alpha}\right)_0 + \gamma\left(\frac{\partial A}{\partial \gamma}\right)_0 + \cdots$$

to give to first order in α and γ [38]

$$A = A_0 - 2\pi\alpha NkT\rho d^2 g_0(d)\left\{d + \int_0^\sigma \left[\exp\frac{-\phi(r)}{kT} - 1\right] dr\right\}$$

$$+2\pi N\rho\gamma\int_\sigma^\infty r^2\phi(r)g_0(r)\,dr \quad (3.119)$$

We can now specify the diameter of the particles in the hard sphere reference system by making the first order term in α vanish

$$d = -\int_0^\sigma \left[\exp\frac{-\phi(r)}{kT} - 1\right] dr \quad (3.120)$$

This gives a temperature dependent hard sphere diameter, leaving a theory that to first order is essentially the same as that of Zwanzig except that the hard sphere diameter is specified. Barker and Henderson have given methods for calculating the second order term in γ^2, the most recent of which [39] is based on very extensive computer simulation results for the hard sphere system [36]. The Barker-Henderson theory can also be used to calculate the radial distribution function and gives very satisfactory results.

A second very successful perturbation theory has been described by Weeks, Chandler, and Andersen [40], who wrote the potentials $u_0(r)$ and $u_1(r)$

$$u_0(r) = \begin{cases} \phi(r) + \varepsilon, & r < r_m \\ 0 & r \geqslant r_m \end{cases} \tag{3.121}$$

and

$$u_1(r) = \begin{cases} -\varepsilon & r < r_m \\ \phi(r) & r \geqslant r_m \end{cases} \tag{3.122}$$

where ε is the value of $\phi(r)$ at its minimum and r_m is the position of the minimum. If this division is used in eq. 3.111 the free energy becomes to first order

$$A = A_0 + 2N\pi\rho \int_0^\infty u_1(r) g_0(r) r^2 \, dr \tag{3.123}$$

where A_0 and $g_0(r)$ are properties of the reference fluid defined by eq. 3.121. In principle this is a very good choice of $u_0(r)$ and $u_1(r)$ because the perturbing term, $u_1(r)$, varies slowly, particularly within the core of the unperturbed potential. This has the property of reducing the size of the second order terms, which depend on fluctuations. The major disadvantage of this method is that the properties of the reference system are not well known. Weeks, Chandler, and Andersen approximated A_0 and $g_0(r)$ using the corresponding properties for hard spheres:

$$A_0 = A_{\mathrm{HS}} \tag{3.124}$$

$$g_0(r) = \exp \frac{-u_0(r)}{kT} \, y_{\mathrm{HS}}(r) \tag{3.125}$$

where $y(r)$ is the function defined in eq. 3.93. The hard sphere diameter was determined by the condition

$$\int_d^{r_m} r^2 y_{\mathrm{HS}}(r) \, dr = \int_0^{r_m} r^2 \exp \frac{-u_0(r)}{kT} \, y_{\mathrm{HS}}(r) \, dr \tag{3.126}$$

Both the Barker-Henderson theory and the Weeks, Chandler, and Andersen theory have proven to be very successful for spherical potentials. Although the Barker-Henderson theory requires terms in γ^2 to be included, this does not cause difficulties now as the appropriate functions are tabulated [39]. Consequently, either method may be used to calculate the properties of simple fluids.

3.10 Computer Simulation Methods

Two computer simulation methods are available to be used with dense fluids, the Monte Carlo method [41, 42] and the method of molecular dynamics [43]. The first method is basically a very powerful multidimensional integration procedure and is used to calculate the equilibrium properties of a system. The method of molecular dynamics is a numerical integration of Newton's equations of motion and is used to calculate both the equilibrium properties of a fluid and various transport properties such as the viscosity. Both methods have several features in common, in particular the basic description of the system to be modeled.

It is not possible to use either method to deal with systems containing more than a few hundred particles. This gives two potentially large sources of error. First, a real system generally contains about 10^{23} particles and so we have the possibility of a significant size effect. Second, if we have a three-dimensional system of a few hundred particles most of the particles will be found on the outside of the cluster. This gives a potentially significant wall effect. Both of these sources of error can be reduced by making use of periodic boundary conditions. Figure 3.4 shows a few particles in a two-dimensional box, together with images in all four walls. The system consists of an infinite number of replicas of the central box. We allow the walls of the box to be completely permeable so that if a particle passes through a wall one of its images enters through the opposite wall. It follows that for fixed box size the number density of the system is also fixed and consequently it is usual to model statistical mechanical ensembles that have this property. The periodic image convention reduces considerably the size and wall effects, and gives an accurate model except in situations where collective fluctuations involving large numbers of particles may be important (e.g., near critical points).

The Monte Carlo method as used in the theory of dense fluids is a multidimensional integration procedure. Some idea of the approach can be obtained by considering the simple example of a one-dimensional integration [44]

$$I = \int_a^b g(x)\, dx \tag{3.127}$$

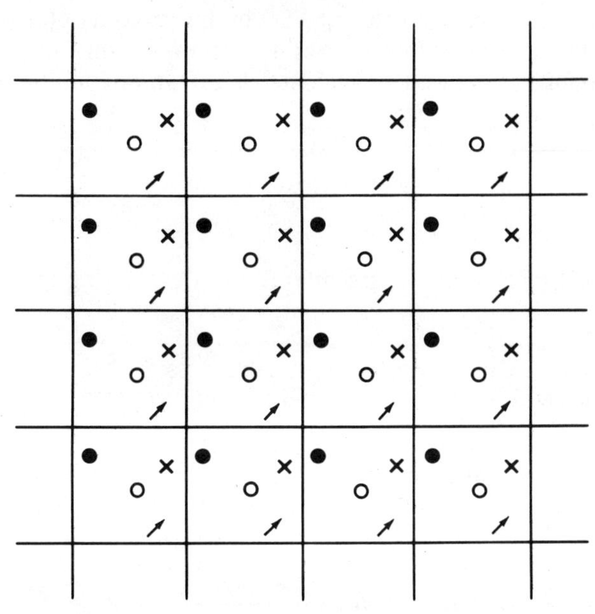

Figure 3.4 Periodic boundary conditions used in computer simulation methods.

A numerical method of calculating I replaces the integration by a weighted sum over values of $g(x)$ for x lying between a and b

$$I \approx \sum_{i=1}^{N} \omega_i g(x_i)$$

and if we are using the trapezoidal rule, for example, with step size δ and $x_i = a + (i-1)\delta$

$$\omega_i = \begin{cases} \delta/2 & i = 1 \text{ or } N \\ \delta & \text{otherwise} \end{cases}$$

Now we write eq. 3.127 in another form

$$I = \int_a^b \frac{g(x)}{f(x)} \cdot f(x)\, dx = \int_a^b h(x) f(x)\, dx \tag{3.128}$$

where $f(x) > 0$ for x in the range $[a, b]$ and where $f(x)$ is integrable, and treat $f(x)$ as a probability density function. Then we can write

$$\langle h(x) \rangle = \frac{I}{\int_a^b f(x)\, dx} \tag{3.129}$$

where $\langle h(x)\rangle$ is the average value of $h(x)$ in the range $[a, b]$ assuming that x is distributed as $f(x)$. It follows that if we choose points in the range $[a, b]$ with probability $f(x)$ we can evaluate $\langle h(x)\rangle$, and by multiplying by the normalizing factor

$$F=\int_a^b f(x)\,dx$$

calculate I.

In statistical mechanics we are interested in evaluating integrals of the form

$$\langle G\rangle=\frac{1}{Z}\cdot\frac{\lambda^{-3N}}{N!}\int\cdots\int G(\mathbf{r}_1,\ldots\mathbf{r}_N)\exp\frac{-\Phi(\mathbf{r}_1,\ldots\mathbf{r}_N)}{kT}\,d\mathbf{r}_1\cdots d\mathbf{r}_N \tag{3.130}$$

where Z, the partition function, is a normalizing factor. Consequently, if we choose position vectors $(\mathbf{r}_1,\ldots\mathbf{r}_N)$ according to the distribution function

$$f(\mathbf{r}_1,\ldots\mathbf{r}_N)=\exp\frac{-\Phi(\mathbf{r}_1,\ldots\mathbf{r}_N)}{kT} \tag{3.131}$$

we can evaluate integrals of the type given in eq. 3.130. Unfortunately, we cannot evaluate the normalizing constant, Z, unless an appropriate many dimensional integral exists [45–47]. However, we can determine a method by which the thermodynamic properties and structure can be obtained.

We use the method described by Metropolis et al. [41] based on the theory of Markov chains [48]. Starting from some initial configuration of particles, a very long chain of configurations typical of the equilibrium state is generated. The system of particles is taken from one configuration to the next in such a way that the transition probabilities tend to a finite limit, namely, $\exp[-\Phi(\mathbf{r}_1,\cdots\mathbf{r}_N)/kT]$. Let $P_{ij}^{(M)}$ be the probability of getting from state i to state j in M steps and $P_{ij}^{(1)}=P_{ij}$. Then we have

$$P_{ij}^{(M)}=\sum_k P_{ik}^{(M-1)}P_{kj} \tag{3.132}$$

and we are interested in $\lim_{M\to\infty}P_{ij}^{(M)}$. For a Markov process it can be shown that provided all possible states are attainable from any given state in a finite number of steps [48] then

$$\lim_{M\to\infty}P_{ij}^{(M)}=\omega_j \tag{3.133}$$

independently of i. That is, after a sufficiently large number of steps the state of the system is independent of its starting configuration. The limits are subject to certain constraints, namely

$$\omega_i > 0 \tag{3.134}$$

$$\sum_i \omega_i = 1 \tag{3.135}$$

$$\omega_i = \sum_j P_{ji}\omega_j \tag{3.136}$$

The first two constraints state that the limiting transition probabilities must be positive and that they are normalized, and the third constraint says that each state is accessible from the others. If we know the P_{ij} we can solve eq. 3.136 for the limiting probabilities ω_i. We choose

$$\frac{P_{ij}}{P_{ji}} = \exp\frac{-(\Phi_j - \Phi_i)}{kT} \tag{3.137}$$

where Φ_i is the potential energy of the system in configuration i, and in addition choose

$$\omega_i = \text{const.}\exp\frac{-\Phi_i}{kT} \tag{3.138}$$

It follows from eqs. 3.137 and 3.138 that

$$\omega_i P_{ij} = \omega_j P_{ji} \tag{3.139}$$

If we substitute this result into eq. 3.136 we find

$$\omega_i = \sum_j \omega_i P_{ij} = \omega_i \sum_j P_{ij}$$

or $\Sigma_j P_{ij} = 1$, which must be true as the system is in one of the states j after any one move. Thus all the requirements for a Markov process are satisfied and by choosing P_{ij} such that

$$P_{ij} = \exp\frac{-(\Phi_j - \Phi_i)}{kT}\, P_{ji}$$

the required limiting transition probabilities are obtained. It remains to choose appropriate values for the P_{ij}. Metropolis et al. [41] used the following criterion

$$P_{ij} = A_{ij},\ i \neq j,\ \Phi_i > \Phi_i; \quad = A_{ij}\exp\frac{-(\Phi_j - \Phi_i)}{kT}\quad i \neq j,\ \Phi_i < \Phi_j$$

$$P_{ii} = 1 - \sum_{j \neq i} P_{ij}$$

Every state is accessible from any other and the constants are chosen so that $A_{ij} = A_{ji}$. To determine the constants A_{ij} the following rules are followed:

1. A molecule is selected at random.
2. A change is selected in the position, $(\delta x, \delta y, \delta z)$.
3. $A_{ij} = 0$, if any one of $|\delta x|, |\delta y|, |\delta z| > \Delta$; otherwise $A_{ij} = 1/(2\Delta)^3$.

The calculation is initiated by giving the N particles contained in the periodic box position vectors $\mathbf{r}_i$, $i = 1, \ldots N$ according to some rule. Examples of such rules could be: choose N position vectors at random; place the particles on the lattice sites of an appropriate crystal structure; and use the final configuration from a previous run, scaling the position vectors appropriately if there has been a change in density.

A large number of configurations are then generated using the rules given above with the variable Δ chosen such that the number of accepted new configurations is about half the total number of configurations generated. The preliminary configurations in the chain, usually around 200,000, are discarded to reduce the influence of the starting point. Properties of interest are calculated using the remaining configurations, acceptable averages usually being obtained from about 500,000 configurations. It is sometimes necessary to use many more configurations to obtain good estimates of properties such as the specific heat that depend on the second or higher moments of quantities (in this case, the potential energy). This method can be generalized to handle nonspherical molecules, in which case the new configuration also includes a possible change in the coordinates specifying the orientations of the molecules.

The Monte Carlo procedure outlined above is appropriate for calculating the properties of a system using the canonical ensemble. From eq. 3.32 and the equation relating Z to the free energy, $A = -kT \log Z$, we can obtain expressions for various thermodynamic properties. As an example, the internal energy is given by

$$U = \frac{\partial(A/T)}{\partial(1/T)}$$

$$= U_K + \frac{1}{Z} \cdot \frac{\lambda^{-3N}}{N!} \int \cdots \int \Phi(\mathbf{r}_1, \ldots \mathbf{r}_N) \exp \frac{-\Phi(\mathbf{r}_1, \ldots \mathbf{r}_N)}{kT} \, d\mathbf{r}_1 \cdots d\mathbf{r}_N$$

or

$$U = U_K + \langle \Phi(\mathbf{r}_1, \ldots \mathbf{r}_N) \rangle \tag{3.140}$$

where U_K is the kinetic energy ($= 3NkT/2$ for a system of spherical particles) and arises from the term in λ^{-3N}. Consequently, we can estimate the internal energy of the system by averaging the potential energy of all the configurations in the Monte Carlo chain. A further differentiation gives us the specific heat at constant volume

$$C_V = \left(\frac{\partial U}{\partial T}\right)_V = C_V(K) + \frac{1}{kT^2}[\langle\Phi^2\rangle - \langle\Phi\rangle^2] \tag{3.141}$$

where $C_V(K)$ is the contribution to the specific heat from the kinetic energy. The pressure of the system is obtained using the virial theorem formula [4]

$$\frac{pV}{NkT} = 1 - \frac{1}{3NkT}\left\langle \sum_i \mathbf{r}_i \cdot \nabla_i \Phi(\mathbf{r}_1, \ldots \mathbf{r}_N)\right\rangle \tag{3.142}$$

where ∇_i is the gradient operator for the ith particle. This equation is equivalent to eq. 3.54.

In addition to calculating thermodynamic properties of the system the Monte Carlo method can also be used to obtain distribution functions. We showed earlier that the radial distribution function measures the probability of finding a particle at a distance r from a given particle. It follows that the average number of particles separated by a distance in the range $(r, r+dr)$ is given by

$$\langle N(r, r+dr)\rangle = 4\pi\rho r^2 g(r)\, dr \tag{3.143}$$

a formula that enables us to calculate $g(r)$.

The method of molecular dynamics is more powerful than the Monte Carlo method as it can be used to calculate both equilibrium and transport properties of a fluid. If we are using classical mechanics the evolution in time of a system of particles is given by Newton's equations of motion. Suppose $\mathbf{p}_i$ is the momentum of the ith particle, then the force acting on the particle at any time t is

$$\mathbf{F}_i = \frac{d\mathbf{p}_i}{dt} = -\nabla_i \Phi(\mathbf{r}_1, \ldots \mathbf{r}_N) \tag{3.144}$$

and the momentum of the particle is related to its position by

$$\mathbf{p}_i = m\frac{d\mathbf{r}_i}{dt} \tag{3.145}$$

When we introduce eq. 3.145 into eq. 3.144 we have a set of coupled second order differential equations which can be solved numerically to give the positions $\mathbf{r}_i$, and momenta $\mathbf{p}_i$, $i = 1, \ldots N$ over some period of time. As with the Monte Carlo procedure the N particles are assumed to be contained in a periodic box thus enabling a relatively small number of particles to be considered. The numerical method used to solve the coupled differential equations usually involves the finite difference approximation [49]

$$-\boldsymbol{\nabla}_i\Phi = m\ddot{\mathbf{r}}_i \simeq \frac{m}{\delta^2}[\mathbf{r}_i(t+\delta)-2\mathbf{r}_i(t)+\mathbf{r}_i(t-\delta)] \qquad (3.146)$$

where δ is the step size to be used in the time integration. Predictor-corrector methods have also been used to solve these equations [43]. To begin the numerical solution it is usual to select a set of initial coordinates $\mathbf{r}_i$ in much the same way as is done for the Monte Carlo method. The particles are then assigned initial momenta according to the Boltzmann distribution law

$$f(\mathbf{p}) = (2\pi mkT)^{-3/2} \exp\frac{-\mathbf{p}\cdot\mathbf{p}}{2mkT}$$

and the second set of position coordinates calculated using the result

$$\mathbf{r}_i(t) = \mathbf{r}_i(t-\delta) + \frac{\delta\mathbf{p}_i(t-\delta)}{m}$$

Subsequent coordinates are then calculated using eq. 3.146. The step size in the time integration, δ, is dependent on the system being considered. When calculations are being performed for spherically symmetric systems such as the inert gases, $\delta \approx 10^{-14}$ s is the usual choice. However, when rotational motions are important, as in the simulation of liquid water [50], a step size of the order of 10^{-16} s is needed. The length of time during which the system must be followed to obtain satisfactory estimates of the thermodynamic and transport properties is usually greater than 10^{-11} s, a period that approximates the time when hydrodynamic equations become valid.

The molecular dynamics method, by solving Newton's equations of motion, conserves the total energy of the system. As the interaction energy changes continuously with time it follows that the kinetic energy must also vary. The connection with thermodynamics is made by using the first postulate given in Section 3.1, namely, that an experimental measurement is a time average of some microscopic quantity. The temperature of the system is then obtained by averaging the kinetic energy

$$T = 2\sum_i \frac{\langle\mathbf{p}_i\cdot\mathbf{p}_i\rangle}{3Nmk} \qquad (3.147)$$

where the average is a time average rather than the ensemble average used in the Monte Carlo method. Other properties of the system can be calculated using eqs. 3.140–3.143 with the angle brackets implying time averages rather than ensemble averages. It is important to remember that in the molecular dynamics method the temperature is a property to be estimated and that the fixed quantities are the total energy and the volume. The statistical mechanical model of such a situation is the microcanonical ensemble. This leads to certain formal difficulties with using eqs. 3.140–3.143 as they are appropriate to the canonical ensemble. Lebowitz et al. [51] have examined this problem and shown that the equations for the pressure and energy are satisfactory. However, it must be noted that as the total energy is fixed the specific heat can be calculated using either the variance of the interaction energy or that of the kinetic energy [52]. We consider the use of the molecular dynamics method in the theory of transport processes in later chapters.

Both the Monte Carlo and molecular dynamics methods have the property of being, in principle, exact calculations for a given model interaction potential. In practice it is possible to calculate most properties of interest to an accuracy of 1 or 2% by using a fast computer. This gives the methods two main uses, first as a standard by which approximate theories may be judged and second as a method for obtaining accurate semiempirical intermolecular potential functions. The use of computer simulation methods as a standard is important when we consider various approximate theories in later chapters, and the second use is important in the chapter on inert gases (Chapter 7) when a method of using experimental properties to determine interaction potentials becomes apparent.

References

1. D. D. Fitts, 1966, *Ann. Rev. Phys. Chem.*, **17,** 59.
2. J. A. Barker and A. Pompe, 1968, *Aust. J. Chem.*, **21,** 1683.
3. J. W. Gibbs, 1902, *Elementary Principles of Statistical Mechanics*, Yale University Press.
4. T. L. Hill, 1956, *Statistical Mechanics*, McGraw-Hill, New York.
5. P. M. Morse, 1969, *Thermal Physics*, Benjamin, New York.
6. J. G. Kirkwood, 1931, *Phys. Rev.*, **44,** 31; E. P. Wigner, 1932, *ibid.*, **40,** 749; H. S. Green, 1951, *J. Chem. Phys.*, **19,** 955; L. D. Landau and E. M. Lifshitz, 1958, *Statistical Physics*, Addison-Wesley, Reading, Mass.
7. C. J. Pings, 1968, Chapter 10 in *Physics of Simple Liquids*, H. N. V. Temperley, J. S. Rowlinson, and G. S. Rushbrooke, Eds., North Holland, Amsterdam.
8. H. Kammerlingh-Onnes, 1901, *Comm. Phys. Lab. Leiden*, No. 71.
9. E. E. Salpeter, 1958, *Ann. Phys. (N.Y.)*, **5,** 183.
10. J. E. Mayer, 1942, *J. Chem. Phys.*, **10,** 629.
11. G. Stell, 1964, "Cluster Expansions for Classical Systems in Equilibrium," in *The Equilibrium Theory of Classical Fluids*, H. L. Frisch and J. L. Lebowitz, Eds., Benjamin, New York.
12. G. E. Uhlenbeck and G. W. Ford, 1962, in *Studies in Statistical Mechanics*, J. de Boer and G. E. Uhlenbeck, Eds., Vol. 1, North Holland, Amsterdam; F. H. Ree and W. G. Hoover, 1964, *J. Chem. Phys.*, **40,** 939; J. A. Barker and D. Henderson, 1967, *Can. J. Phys.*, **45,** 3959; D. Henderson and M. Chen, 1970, *ibid.*, **48,** 634.

13. R. O. Watts, 1973, "Integral Equation Approximations in the Theory of Fluids," Chapter 1 in *Specialist Periodical Reports of the Chemical Society: Statistical Mechanics*, K. Singer, Ed., Vol. 1, Chemical Society, London.
14. H. S. Green, 1956, "The Structure of Liquids," *Handbuch der Physik*, Vol. 10, p. 1, Springer-Verlag, Berlin.
15. H. S. Green, 1952, *The Molecular Theory of Fluids*, Dover, New York; M. Born and H. S. Green, 1946, *Proc. Roy. Soc. (London), Ser. A.*, **188,** 10; J. Yvon, 1935, *Actualities Scientifiques et Industriel,* Herman et Cie, Paris.
16. J. G. Kirkwood, 1935, *J. Chem. Phys.*, **3,** 300.
17. J. M. H. Levelt and E. D. G. Cohen, 1964, in *Studies in Statistical Mechanics*, J. de Boer and G. E. Uhlenbeck, Eds., Vol. 2, p. 111, North Holland, Amsterdam.
18. A. Bellemans, 1969, *J. Chem. Phys.*, **50,** 2784.
19. H. J. Raveché and M. S. Green, 1969, *J. Chem. Phys.*, **50,** 5334.
20. J. M. J. van Leeuwen, J. Groeneveld, and J. de Boer, 1959, *Physica*, **25,** 792.
21. T. Morita and K. Hiroike, 1960, *Progr. Theor. Phys. (Kyoto)*, **23,** 1003.
22. E. Meeron, 1960, *J. Math. Phys.*, **1,** 192.
23. M. S. Green, 1959, Hughes Aircraft report.
24. L. S. Ornstein and F. Zernike, 1914, *Proc. Acad. Sci. (Amsterdam)*, **17,** 793.
25. I. N. Sneddon, 1951, *Fourier Transforms*, McGraw-Hill, New York.
26. J. K. Percus and G. J. Yevick, 1958, *Phys. Rev.*, **110,** 1.
27. J. K. Percus, 1962, *Phys. Rev. Lett.*, **8,** 462.
28. M. Born and K. Huang, 1954, *Dynamical Theory of Crystal Lattices*, Oxford University Press.
29. R. Zwanzig, 1954, *J. Chem. Phys.*, **22,** 1420; see also R. Peierls, 1933, *Z. Phys.*, **80,** 763 and L. D. Landau and E. Lifshitz, 1969, *Statistical Physics*, p. 93, Pergamon, Oxford.
30. R. O. Watts, 1971, *Rev. Pure and Appl. Chem.*, **21,** 167.
31. A. B. Pippard, 1957, *Elements of Classical Thermodynamics*, Cambridge University Press.
32. G. J. Throop and R. J. Bearman, 1965, *J. Chem. Phys.*, **42,** 2408.
33. F. Mandel, R. J. Bearman, and M. Y. Bearman, 1970, *J. Chem. Phys.*, **52,** 3315.
34. W. R. Smith and D. Henderson, 1970, *Mol. Phys.*, **19,** 411.
35. L. Verlet and J. J. Weis, 1972, *Phys. Rev.*, **A5,** 939.
36. J. A. Barker and D. Henderson, 1971, *Mol. Phys.*, **21,** 187.
37. J. Rasaiah and G. Stell, 1970, *Mol. Phys.*, **18,** 249.
38. J. A. Barker and D. Henderson, 1967, *J. Chem. Phys.*, **47,** 2856, 4714.
39. J. A. Barker and D. Henderson. 1972, *Ann. Rev. Phys. Chem.*, **23,** 439.
40. H. C. Andersen. D. Chandler, and J. D. Weeks, 1972, *J. Chem. Phys.*, **56,** 3812.
41. N. Metropolis, A. W. Rosenbluth, M. N. Rosenbluth, E. Teller and A. H. Teller, 1953, *J. Chem. Phys.*, **21,** 1087.
42. I. R. McDonald and K. Singer, 1970, *Quart. Revs.*, **24,** 238.
43. A. Rahman, 1964, *Phys. Rev.*, **136,** A405; B. J. Alder and T. E. Wainwright, 1957, *J. Chem. Phys.*, **27,** 1208; 1959, *ibid.*, **31,** 459; 1960, *ibid.*, **33,** 1439.
44. Yu. A. Schreider, 1967, *Monte Carlo Methods*, Pergamon, London.
45. It is possible to evaluate Z if an efficient importance sampling method [46] is available. One possible approach has been described by Torrie and Valleau [47] and has proven useful for simple spherical interaction potentials.
46. J. M. Hammersley and D. C. Handscomb, 1964, *Monte Carlo Methods*, Methuen and Co., London.
47. G. M. Torrie and J. P. Valleau, 1974, *Chem. Phys. Lett.*, **28,** 578.

48. W. Feller, 1950, *An Introduction to Probability Theory and its Applications*, Vol. 2, Wiley, New York.
49. L. Verlet, 1967, *Phys. Rev.*, **159**, 98; 1968, *ibid.*, **165,** 201.
50. A. Rahman and F. H. Stillinger, 1971, *J. Chem. Phys.*, **55,** 3336.
51. J. L. Lebowitz, J. K. Percus, and L. Verlet, 1967, *Phys. Rev.*, **153,** 250.
52. H. J. M. Hanley and R. O. Watts, 1975, *Physica*, **79A,** 351.

Four

Theory of Time-Dependent Processes

Both in thermodynamics and in the Gibbs statistical mechanics we are concerned with systems that are at equilibrium. Many properties of a system can be studied using equilibrium theory, including thermodynamic properties such as pressure, specific heat, and entropy and structural properties such as the radial distribution function. A number of properties of a system are associated with processes that occur after the system has been disturbed in some way. Typical of these properties are the diffusion coefficient, the shear viscosity, and the thermal conductivity. The first of these describes the rate at which a concentration gradient is dissipated, the second measures the rate at which momentum gradients are dissipated, and the third measures the rate at which energy is transferred through the system. Many attempts have been made to develop theories of such processes, the most successful being the kinetic theory of dilute monatomic gases [1, 2]. Using this theory properties, such as the shear viscosity, can be calculated from a knowledge of the pair interaction, as has been mentioned in Section 2.4. Kinetic theory in its most successful form relies on the assumption that no more than two molecules are involved in any collision. Several attempts have been made to include higher order collisions, but the success of such theories has been limited [3]. In atomic gases, such as argon, and in the simpler molecular gases, such as methane, the number of ternary and higher order collisions occurring is primarily a function of density. However, in more complicated molecular gases, where there can be significant transfer of energy between translational states and intramolecular degrees of freedom, particularly rotational states, it is possible for very long lived collisions to occur. Under these conditions the kinetic theory must be modified to allow for both multiple [3] collisions and multiple encounters within one collision [4–7]. Although several attempts have been made to develop such modifications to the theory [4–7], relatively little computational work has been done toward testing the adequacy of these theories [8]. Most studies have either examined the nature of the collision process itself [9, 10] or have calculated transport properties using approximations such as fixed relative orientation of two molecules during the collision [11] and that the molecules are rigid and convex [12].

In dense systems the kinetic theory of transport phenomena is less well developed. Numerous attempts have been made to obtain expressions for time dependent distribution functions. A hierarchy of differential equations for such functions, similar to the Born-Green-Yvon hierarchy [13] discussed in Section 3.8, has been derived by several workers [14–17]. However, attempts to obtain transport coefficients for dense fluids have usually been based either on a modification of Brownian motion theory developed by Kirkwood [16, 18] or on the Enskog theory of hard sphere gas kinetics [2, 19]. Although the Kirkwood approach appeared to be very successful for many years, culminating in the Rice-Allnatt theory of transport [20–22], recent work has shown that there are serious flaws in the theory [23–25]. Modifications of the Enskog theory have been more successful [26–28], but are generally based on phenomenological arguments rather than on mathematical theory.

Fortunately there is an alternative approach that is useful in a dense system, the linear response theory of Green [29, 30], Kubo [31], and others [32, 33]. This theory is based on the assumption that the disturbance causing the transport process is such that the system is close to its equilibrium state. Under these circumstances the transport process may be treated using first order perturbation theory, with the equilibrium state as reference system. As Gibbs statistical mechanics gives the information required about the equilibrium state this transport problem is, at least in principle, solved. Linear response theory is closely related to the nonequilibrium thermodynamics of Onsager [34, 35], and is a microscopic theory developed along similar lines to the Onsager theory. Consequently, the following two sections of this chapter develop first the macroscopic theory of transport and second the microscopic equivalent.

4.1 Nonequilibrium Thermodynamics

The thermodynamics of systems at equilibrium is well established and a host of relationships between various bulk properties have been derived [36]. Once a system is removed from equilibrium the situation is less satisfactory. Irreversible thermodynamics [34, 35] is constructed on the postulate of local equilibrium [37]. This postulate claims that for a system in which irreversible phenomena are taking place all thermodynamic functions of state exist for each volume element of the system. Furthermore, these thermodynamic quantities are the same function of the local state variables as the corresponding equilibrium properties. It is important to understand the concept of a volume element of the disturbed system. The element has to be large enough for bulk properties to be definable but small enough for inhomogeneities in those properties to be ignored. Such a situation exists if the system is close to equilibrium rather than in systems having large gradients.

Irreversible thermodynamics gives a formal description of the coupling between various fluxes in the system. For example, if there is a temperature gradient across a fluid at least two phenomena can be observed. First heat is transferred through the fluid at a rate determined by the thermal conductivity. Second, particles move through the system at a rate determined by the thermal diffusion coefficient. The first of these coefficients measures the heat flux due to a temperature gradient, and the second measures a mass flux caused by a temperature gradient. Not all fluxes can couple with a given gradient, or driving force, and irreversible thermodynamics provides a description of the couplings that are allowable. A full treatment of irreversible thermodynamics is properly a subject for a complete book and excellent texts exist [37, 38]. In this section we outline steps in the development of irreversible thermodynamics using as an example a two-component system in which temperature and concentration gradients may be found.

Consider a two-component system in which there is established a temperature gradient and a concentration gradient. Suppose that the gradients are sufficiently small that we can assume each local volume element to be in equilibrium. Then in the region of some point $\mathbf{r}$ the Gibbs equation [36] will hold:

$$dU = TdS - pd\left(\frac{1}{\rho}\right) + \mu_1\, dx_1 + \mu_2\, dx_2 \tag{4.1}$$

where U, S, p, ρ are the internal energy, entropy, pressure, and mass density, respectively, and μ_i is the chemical potential and x_i the mass fraction of component i. (The reader should note that *in this section only*, ρ represents the mass density rather than the number density.) The quantities U and S are intensive quantities, and they represent the internal energy and entropy *per unit mass*. In order to obtain equations for the various fluxes in the system we must consider the rate of change of entropy with time:

$$\frac{\rho\, dS}{dt} = \frac{\rho}{T}\frac{dU}{dt} - \frac{p}{\rho T}\frac{d\rho}{dt} - \frac{\rho}{T}\left[\mu_1 \frac{dx_1}{dt} + \mu_2 \frac{dx_2}{dt}\right] \tag{4.2}$$

We assume for simplicity that, although the relative concentrations of particles 1 and 2 differ at various positions in the fluid, the total mass density of the system, ρ, is constant so that $d\rho/dt = 0$. To proceed further we must introduce hydrodynamic equations relating the derivatives dU/dt and dx_i/dt to the heat and mass fluxes in the system.

Each volume element in the system, dV, contains a quantity of thermodynamic internal energy $dU = \rho U\,dV$ where U is the local internal energy density. If the volume element is moving with center-of-mass velocity $\mathbf{u}$ then it also has associated with it a kinetic energy

$$dK = \tfrac{1}{2}\rho\mathbf{u}\cdot\mathbf{u}\,dV$$

so that the total energy content of the volume element is $dU + dK$. The total energy content of an arbitrary volume V, which is fixed with respect to some laboratory frame of reference, is $\int_V (dU + dK)$ and its rate of change with time is given by

$$\frac{\partial}{\partial t}\int_V (dU + dK) = \int_V \frac{\partial}{\partial t}\rho(U + \tfrac{1}{2}\mathbf{u}\cdot\mathbf{u})\,dV \tag{4.3}$$

Energy changes in the arbitrary volume arise from two effects, a convection term due to the bulk flow through the surfaces of the element and a conduction term. If $d\mathbf{A}$ is a vector that is perpendicular to, and pointing out of, an element of the surface of the arbitrary volume and whose magnitude is equal to the surface area of the element, the volume of fluid passing through $d\mathbf{A}$ in unit time is $\mathbf{u}\cdot d\mathbf{A}$. The convective energy change associated with this volume is given by the kinetic energy and internal energy content $\rho(U + \frac{1}{2}\mathbf{u}\cdot\mathbf{u})\mathbf{u}\cdot d\mathbf{A}$. The conduction term consists of two quantities, the energy change due to particles diffusing through the surface of the volume and the energy change due to pure heat conduction, and is represented by the flux $\mathbf{j}_E$; the heat change due to this term is then $\mathbf{j}_E\cdot d\mathbf{A}$. Integrating over the surface of the arbitrary volume gives another expression for the rate of change of total energy content

$$-\int_A [\mathbf{j}_E + \rho(U + \tfrac{1}{2}\mathbf{u}\cdot\mathbf{u})\mathbf{u}]\cdot d\mathbf{A} \tag{4.4}$$

where the negative sign arises because, if the vectors $\mathbf{j}_E$ and $\mathbf{u}$ are in the same direction as $d\mathbf{A}$, energy is lost through the surface. Equating eq. 4.3 to eq. 4.4 gives the equation for energy conservation

$$\int_V \frac{\partial}{\partial t}\rho(U + \tfrac{1}{2}\mathbf{u}\cdot\mathbf{u})\,dV + \int_A [\mathbf{j}_E + \rho(U + \tfrac{1}{2}\mathbf{u}\cdot\mathbf{u})\mathbf{u}]\cdot d\mathbf{A} = 0$$

Using a theorem attributed to Gauss a surface integral can be written in terms of a volume integral

$$\int_A \mathbf{J} \cdot d\mathbf{A} = \int_V \boldsymbol{\nabla} \cdot \mathbf{J}\, dV \tag{4.5}$$

where $\mathbf{J}$ is any vector and $\boldsymbol{\nabla}$ is the vector operator $(\partial/\partial x, \partial/\partial y, \partial/\partial z)$. Consequently, the equation for energy conservation becomes

$$\int_V \left\{\frac{\partial}{\partial t}\rho(U+\tfrac{1}{2}\mathbf{u}\cdot\mathbf{u})+\boldsymbol{\nabla}\cdot[\mathbf{j}_E+\rho(U+\tfrac{1}{2}\mathbf{u}\cdot\mathbf{u})\mathbf{u}]\right\} dV=0 \tag{4.6}$$

As the volume element is arbitrary, this equation is satisfied only if the integrand vanishes everywhere. The integrand can be simplified if the identity

$$\frac{d}{dt}=\frac{\partial}{\partial t}+\frac{dx}{dt}\frac{\partial}{\partial x}+\frac{dy}{dt}\frac{\partial}{\partial y}+\frac{dz}{dt}\frac{\partial}{\partial z} \tag{4.7}$$

or

$$\frac{d}{dt}=\frac{\partial}{\partial t}+\mathbf{u}\cdot\boldsymbol{\nabla}$$

is used to eliminate the partial time derivative

$$\frac{\partial}{\partial t}\rho(U+\tfrac{1}{2}\mathbf{u}\cdot\mathbf{u})+\boldsymbol{\nabla}(U+\tfrac{1}{2}\mathbf{u}\cdot\mathbf{u})\rho\mathbf{u}$$

$$=\frac{d}{dt}\rho(U+\tfrac{1}{2}\mathbf{u}\cdot\mathbf{u})+\rho(U+\tfrac{1}{2}\mathbf{u}\cdot\mathbf{u})\boldsymbol{\nabla}\cdot\mathbf{u} \tag{4.8}$$

Introducing eq. 4.8 into the integrand of eq. 4.6 and using our initial assumption that there are no velocity gradients in the system gives

$$\rho\frac{dU}{dt}=-\boldsymbol{\nabla}\cdot\mathbf{j}_E \tag{4.9}$$

which is the hydrodynamic equation for energy transport.

The conduction flux $\mathbf{j}_E$ contains two terms, one due to energy associated with the diffusive flow of particles through the surface of the volume and the other due to pure heat flow (e.g., radiation). In terms of the second law of thermodynamics [37], particle diffusion is associated with an enthalpy flow, given by $H_i\mathbf{j}_i$ where H_i is the enthalpy per unit mass associated with particle i and $\mathbf{j}_i$ is the diffusive flux of particle i. If $\mathbf{u}_i-\mathbf{u}$ is the velocity of component i relative to the center-of-mass velocity $\mathbf{u}$ and ρ_i is the mass density of component i, the diffusive flux is given by

$$\mathbf{j}_i=\rho_i(\mathbf{u}_i-\mathbf{u}) \tag{4.10}$$

If the heat flux due to pure heat flow is represented by the vector $\mathbf{q}$ we have

$$\mathbf{j}_E = \mathbf{q} + H_1\mathbf{j}_1 + H_2\mathbf{j}_2 \tag{4.11}$$

for the two component system, and substituting this in eq. 4.9 gives

$$\rho\frac{dU}{dt} = -\nabla\cdot\mathbf{q} - \nabla\cdot H_1\mathbf{j}_1 - \nabla\cdot H_2\mathbf{j}_2 \tag{4.12}$$

The rate of change of energy is now given in terms of thermodynamic quantities. Remember that this equation is based on the assumption that equilibrium thermodynamics can be extended to systems experiencing a small perturbation.

An expression for the rate of change of composition, dx_i/dt, is obtained in a similar way to that for the energy. Consider the flow of component i into, and out of, an arbitrary volume. Following the development given between eq. 4.3 and eq. 4.9 we can write [37]

$$\rho\frac{dx_i}{dt} = -\nabla\cdot\mathbf{j}_i \tag{4.13}$$

where $\mathbf{j}_i$ is the diffusive flux defined by eq. 4.10.

Introducing eqs. 4.12 and 4.13 into the entropy equation (eq. 4.2) and using the equation

$$H_i = TS_i + \mu_i$$

we can write

$$\rho\frac{dS}{dt} = \frac{1}{T}[-\nabla\cdot\mathbf{q} - \nabla\cdot\mathbf{j}_1(TS_1+\mu_1) - \nabla\cdot\mathbf{j}_2(TS_2+\mu_2) + \mu_1\nabla\cdot\mathbf{j}_1 + \mu_2\nabla\cdot\mathbf{j}_2] \tag{4.14}$$

If the entropy flux $\mathbf{j}_S$ is defined by

$$\mathbf{j}_S = \frac{\mathbf{q}}{T} + \mathbf{j}_1S_1 + \mathbf{j}_2S_2 \tag{4.15}$$

we can use the results

$$\nabla\cdot\mathbf{q} = T\nabla\cdot\left(\frac{\mathbf{q}}{T}\right) + \mathbf{q}\cdot\nabla\log T \tag{4.16}$$

and

$$\nabla\mu_i = \nabla_T\mu_i + \frac{\partial\mu_i}{\partial T}\nabla T = \nabla_T\mu_i - S_i\,\nabla T \tag{4.17}$$

together with a small amount of algebra to obtain the equation for the rate of change of entropy

$$\rho \frac{dS}{dt} = -\boldsymbol{\nabla} \cdot \mathbf{j}_S - \frac{1}{T} \mathbf{q} \cdot \boldsymbol{\nabla} \log T - \frac{\mathbf{j}_1}{T} \cdot \boldsymbol{\nabla}_T \mu_2 - \frac{\mathbf{j}_2}{T} \cdot \boldsymbol{\nabla}_T \mu_2 \tag{4.18}$$

Notice that the operator $\boldsymbol{\nabla}_T$ implies that the dependence of temperature on position has been subtracted from the chemical potential gradient, as is shown by eq. 4.17.

The several contributions to the rate of change of entropy are of particular interest. The first term describes a change caused by the existence of an entropy flux and represents entropy flows that can be attributed to both diffusion and heat transfer. The second, third, and fourth terms are all products of fluxes with the gradients of scalar quantities. If we observe that a temperature gradient is responsible for a heat flux and a chemical potential gradient for a mass flux, it is reasonable to think of the gradient terms as driving forces. It is usual to refer to the gradient terms as *thermodynamic forces*, so that the equation for the rate of change of entropy takes the form

$$\rho \frac{dS}{dt} = -\boldsymbol{\nabla} \cdot \mathbf{j}_S + \frac{1}{T} \sum_i \mathbf{F}_i \cdot \mathbf{J}_i \tag{4.19}$$

where $\mathbf{F}_i$ is the thermodynamic driving force and $\mathbf{J}_i$ is the *conjugate flux*. Equation 4.18 may be extended to include many types of gradients, including those involving linear and angular momenta, chemical reactions, and gravitational and magnetic fields [38]. The forces and fluxes are not necessarily vector quantities, as in eq. 4.18, but may be higher order tensors. For example, in the case of a linear momentum gradient the thermodynamic force is the dyadic $\boldsymbol{\nabla}\mathbf{u}$

$$\boldsymbol{\nabla}\mathbf{u} = \begin{pmatrix} \dfrac{\partial u_x}{\partial x} & \dfrac{\partial u_y}{\partial x} & \dfrac{\partial u_z}{\partial x} \\ \dfrac{\partial u_x}{\partial y} & \dfrac{\partial u_y}{\partial y} & \dfrac{\partial u_z}{\partial y} \\ \dfrac{\partial u_x}{\partial z} & \dfrac{\partial u_y}{\partial z} & \dfrac{\partial u_z}{\partial z} \end{pmatrix}$$

and the conjugate flux is the quantity $\boldsymbol{\sigma} + p\mathbf{I}$ where p is the hydrostatic pressure, I is the unit dyadic, and σ is the stress tensor. In an isotropic fluid the components of the stress tensor are related to the shear and bulk viscosities. An extensive account of the thermodynamic forces together with their conjugate fluxes can be found in De Groot and Mazur [38].

Having established that the entropy change can be written as a sum of products of conjugate forces and fluxes it is pertinent to investigate these quantities in more detail. Experimental measurements have shown that for systems subject to small disturbances the resulting fluxes are linearly related to the gradients established in the system. For example, Fick's first law states that the flux of matter is proportional to the concentration gradient in a binary system

$$\mathbf{j}_i = -D\boldsymbol{\nabla}\rho_i$$

where the constant of proportionality is the diffusion coefficient. Similar results have been obtained for the flow of heat caused by a temperature gradient, the constant of proportionality being the thermal conductivity. In addition, it is known that the coupling effects between various gradients also satisfy linear relationships. If there is a temperature gradient in a system the flux of matter that occurs is proportional to $\boldsymbol{\nabla}T$, and the proportionality constant is the thermal diffusion coefficient. Similarly (but not so well known), a concentration gradient can give rise to a heat flux (the Dufour effect), and again the flux is directly proportional to the gradient. Using such experimental observations as a guide it is reasonable to introduce a postulate relating fluxes and forces, namely, that the fluxes $\mathbf{J}_i$ are linear homogeneous functions of the forces $\mathbf{F}_i$:

$$\mathbf{J}_i = \sum_j L_{ij}\mathbf{F}_j \tag{4.20}$$

The coefficients L_{ii} relate conjugate fluxes and forces and the coefficients L_{ij} describe cross phenomena.

Equation 4.20 implies that all the thermodynamic forces occurring in the equation for entropy production can give rise to any of the fluxes. This is not so, and for isotropic systems (e.g., fluids) it is possible to prove Curie's theorem [37–39]: fluxes and forces whose tensorial characters differ by an odd integer cannot interact. On the basis of this theorem we can expect no coupling between the mass diffusion flux which is a first order tensor (vector) and velocity gradients, where the appropriate thermodynamic force is a dyadic or second order tensor. However, we can expect coupling between mass diffusion and thermal gradients and between the concentration gradients and mass diffusivities in systems containing several components.

On general grounds it can be expected that there will be some relation between the off-diagonal coefficients L_{ij}. Onsager [34,35] introduced the postulate that for thermodynamic forces and fluxes appearing in the equation for entropy production, eq. 4.19, the matrix of phenomenological coefficients is symmetric. This postulate gives rise to the *Onsager reciprocal relations*, $L_{ij} = L_{ji}$, and can be supported on the basis of microscopic arguments. Under certain circumstances the matrix of phenomenological coefficients may be antisymmetric, $L_{ij} = -L_{ji}$. The most important examples of couplings described by an antisymmetric matrix are associated with systems in which magnetic fields or Coriolis forces are found.

It is important to note that the Onsager relation will only hold if both fluxes and forces are those appearing in the equation for entropy production. Equation 4.20, which embodies the postulate that there is a linear relation between the fluxes and forces, does not by itself guarantee a symmetric matrix of phenomenological coefficients. Any standard textbook on irreversible thermodynamics [37, 38] discusses this point in some detail. Similarly, the Onsager reciprocal relation is not a proven property of the flux-force relation but is a postulate. A number of experiments have been carried out to test the validity of the postulate [40] and it appears that it holds. In addition the validity of the reciprocal relation has been established using statistical mechanical arguments by several workers [31–33, 41].

The equation for entropy production in a two-component system with temperature and chemical potential gradients (eq. 4.18) can be used to obtain empirical transport laws. As both $\nabla \log T$ and $\nabla_T \mu_i$ are vector quantities we can write down the linear relations expressed in eq. 4.20 as

$$\begin{aligned}
-\mathbf{q} &= L_{00}\nabla \log T + L_{01}\nabla_T\mu_1 + L_{02}\nabla_T\mu_2 \\
-\mathbf{j}_1 &= L_{10}\nabla \log T + L_{11}\nabla_T\mu_1 + L_{12}\nabla_T\mu_2 \\
-\mathbf{j}_2 &= L_{20}\nabla \log T + L_{21}\nabla_T\mu_1 + L_{22}\nabla_T\mu_2
\end{aligned} \tag{4.21}$$

To relate the phenomenological coefficients L_{ij} to empirically defined quantities we can consider eq. 4.21 when the various gradients vanish. If there are no chemical potential gradients the equations reduce to Fourier's law

$$\mathbf{q} = -L_{00}\nabla \log T = -\frac{L_{00}}{T}\nabla T$$

or

$$q = -\lambda \nabla T \tag{4.22}$$

where λ is the thermal conductivity. If there is no temperature gradient the equations reduce to

$$\begin{aligned}
-\mathbf{j}_1 &= L_{11}\nabla_T\mu_1 + L_{12}\nabla_T\mu_2 \\
-\mathbf{j}_2 &= L_{21}\nabla_T\mu_1 + L_{22}\nabla_T\mu_2
\end{aligned} \tag{4.23}$$

These equations can be simplified using certain conservation relations. From the definition of the diffusive fluxes $\mathbf{j}_i$, eq. 4.10, we can write

$$\mathbf{j}_1 + \mathbf{j}_2 = \rho_1(\mathbf{u}_1 - \mathbf{u}) + \rho_2(\mathbf{u}_2 - \mathbf{u}) = 0 \tag{4.24}$$

as the center-of-mass velocity is given by $\rho\mathbf{u} = \rho_1\mathbf{u}_1 + \rho_2\mathbf{u}_2$. Using eq. 4.24 in eq. 4.23 it is easy to show that

$$\mathbf{j}_1 = -\left[L_{11} - \frac{L_{12}(L_{11}+L_{21})}{(L_{12}+L_{22})}\right]\nabla_T\mu_1$$

$$\mathbf{j}_2 = -\left[L_{22} - \frac{L_{21}(L_{22}+L_{12})}{(L_{21}+L_{11})}\right]\nabla_T\mu_2$$

If we replace the chemical potential gradient by the concentration gradient, to give

$$\mathbf{j}_1 = -D\nabla\rho_1 \tag{4.25}$$

we obtain Fick's first law of diffusion [37] with the diffusion coefficient given in terms of the phenomenological coefficients by

$$D = \left[L_{11} - \frac{L_{12}(L_{11}+L_{21})}{(L_{12}+L_{22})}\right]\frac{\partial\mu_1}{\partial\rho_1} \tag{4.26a}$$

or

$$D = \left[L_{22} - \frac{L_{21}(L_{22}+L_{12})}{(L_{21}+L_{11})}\right]\frac{\partial\mu_2}{\partial\rho_2} \tag{4.26b}$$

The second equality follows from the result given in eq. 4.24 together with the result $\nabla\rho_1 = \nabla(\rho - \rho_2) = -\nabla\rho_2$ as ρ is constant.

To obtain Fick's second law we use the continuity of mass equation, eq. 4.13, together with the definition of the mass density of component i, $\rho_i = \rho x_i$, to write

$$\frac{d\rho_1}{dt} = -\nabla\cdot\mathbf{j}_1 = -\nabla\cdot(-D\nabla\rho_1)$$

or, for a system in which the spatial variation of D is negligible

$$\frac{d\rho_1}{dt} = D\nabla^2\rho_1 \tag{4.27}$$

This equation expresses Fick's second law [37] and is known as the diffusion equation.

Equation 4.26 shows that there is only one diffusion coefficient in a binary system. In a one component system it is possible to measure the *self-diffusion coefficient*, a quantity that occurs frequently in subsequent chapters. This coefficient can only be measured if some of the atoms in the system are labeled either isotopically or by exciting them in some way. Once the labeling is complete the fluid is formally a two-component system.

4.2 Linear Response Theory

Irreversible thermodynamics is based on the assumption that the system is close to equilibrium. Consequently, a statistical mechanical theory that parallels irreversible thermodynamics must provide a quantitative description of the deviation from equilibrium. In Chapter 3 we saw that at equilibrium the probability of finding a system in a particular configuration is

$$f^{(N)}(\mathbf{r}_1, \ldots \mathbf{r}_N) = \exp\left[\frac{-H(\mathbf{r}_1, \ldots \mathbf{p}_N)}{kT}\right] \bigg/ Z \tag{4.28}$$

where $H(\mathbf{r}_1, \ldots \mathbf{p}_N)$ is the total energy of the system and Z is the canonical partition function. The concept of a distribution of positions and momenta can be extended to systems that are changing with time, but the density function is then also a function of time, $f^{(N)}(\mathbf{r}_1, \ldots \mathbf{p}_N; t)$. At equilibrium the time dependence of $f^{(N)}$ vanishes and for a canonical ensemble it is given by eq. 4.28. It is not possible to write down an exact formula for the time dependent density function of a dense fluid. However, we can obtain a partial differential equation for $f^{(N)}$ and attempt to obtain approximate solutions.

Consider the change in the density function due to infinitesimal changes in $\mathbf{p}_i$, $\mathbf{r}_i$, and t

$$df^{(N)} = \frac{\partial f^{(N)}}{\partial t} dt + \sum_{i=1}^{N} \left[\frac{\partial f^{(N)}}{\partial \mathbf{p}_i} \cdot d\mathbf{p}_i + \frac{\partial f^{(N)}}{\partial \mathbf{r}_i} \cdot d\mathbf{r}_i\right]$$

From this equation it is apparent that

$$\frac{df^{(N)}}{dt} = \frac{\partial f^{(N)}}{\partial t} + \sum_{i=1}^{N} \left[\frac{\partial f^{(N)}}{\partial \mathbf{p}_i} \cdot \frac{d\mathbf{p}_i}{dt} + \frac{\partial f^{(N)}}{\partial \mathbf{r}_i} \cdot \frac{d\mathbf{r}_i}{dt}\right] \tag{4.29}$$

In the previous section we considered the flows of energy and mass through an arbitrary volume of a fluid. If we consider the density function, $f^{(N)}$, to represent the density of points in a $6N$ dimension space (phase space) whose coordinates are the three momenta and three spatial components of the N particles, it can be shown [42] that the density $f^{(N)}$ is a conserved quantity so that

$$\frac{df^{(N)}}{dt} = 0 \tag{4.30}$$

For systems in which the potential energy depends only on the spatial coordinates it is straightforward to show that Hamilton's equations of motion apply [43]

$$\frac{d\mathbf{p}_i}{dt} = -\frac{\partial H}{\partial \mathbf{r}_i}; \qquad \frac{d\mathbf{r}_i}{dt} = \frac{\partial H}{\partial \mathbf{p}_i}$$

These equations also hold in a modified form if the interaction potential depends on orientational coordinates, for example, Euler angles [43], but it is necessary that the angular momentum derivatives, corresponding to $d/d\mathbf{p}_i$, be replaced by derivatives with respect to coordinates conjugate to the Euler angles (see ref. 43 and Chapter 8 for a more detailed discussion).

If eq. 4.30 and Hamilton's equations are substituted into eq. 4.29, the time dependent density function is determined by the Liouville equation [42]

$$\frac{\partial f^{(N)}}{\partial t} = \sum_{i=1}^{N} \left[\frac{\partial f^{(N)}}{\partial \mathbf{p}_i} \cdot \frac{\partial H}{\partial \mathbf{r}_i} - \frac{\partial f^{(N)}}{\partial \mathbf{r}_i} \cdot \frac{\partial H}{\partial \mathbf{p}_i} \right] \tag{4.31}$$

This equation is fundamental to nonequilibrium statistical mechanics. We are interested in a solution to the Liouville equation for a system that is close to equilibrium. At equilibrium the solution is given by eq. 4.28. Consequently, we can obtain a solution to the Liouville equation, for systems that are close to equilibrium, by using first order perturbation theory.

Let the total energy of the system be given by

$$H = H_0 + H_1 \tag{4.32}$$

where H_0 is the total energy of a system at equilibrium and H_1 is the energy of the system due to its interaction with a small disturbing force. Write the time dependent density function in the form

$$f^{(N)} = f_0^{(N)} + f_1^{(N)} \tag{4.33}$$

where $f_0^{(N)}$, the equilibrium density function, is given by eq. 4.28 with $H = H_0$. Then we assume that H_1 and f_1 are small. To proceed we define the Liouville operator, L, by

$$\mathsf{L} = \sum_{i=1}^{N} \left[\frac{\partial H}{\partial \mathbf{p}_i} \cdot \frac{\partial}{\partial \mathbf{r}_i} - \frac{\partial H}{\partial \mathbf{r}_i} \cdot \frac{\partial}{\partial \mathbf{p}_i} \right] \tag{4.34}$$

so that the Liouville equation is written

$$\frac{\partial f^{(N)}}{\partial t} = -\mathsf{L} f^{(N)}$$

If eqs. 4.32–4.34 are introduced into the Liouville equation we can write

$$\frac{\partial f_0^{(N)}}{\partial t}+\frac{\partial f_1^{(N)}}{\partial t}=-\mathsf{L}_0 f_0^{(N)}-\mathsf{L}_1 f_0^{(N)}-\mathsf{L}_0 f_1^{(N)}-\mathsf{L}_1 f_1^{(N)} \tag{4.35}$$

where L_0 is the Liouville operator with H_0 as the total energy, and L_1 is the Liouville operator with the perturbing term H_1 as the total energy term. At equilibrium we have

$$\frac{\partial f_0^{(N)}}{\partial t}=-\mathsf{L}_0 f_0^{(N)}=0$$

so that after dropping the second order term $\mathsf{L}_1 f_1^{(N)}$, eq. 4.35 becomes

$$\frac{\partial f_1^{(N)}}{\partial t}=-\mathsf{L}_1 f_0^{(N)}-\mathsf{L}_0 f_1^{(N)}$$

If the conditions are such that at time $t=-\infty$ the system is at equilibrium $(f^{(N)}(t=-\infty)=f_0^{(N)})$, the solution of this equation is

$$f_1^{(N)}=-\int_{-\infty}^{t}\exp\left[(s-t)\mathsf{L}_0\right]\mathsf{L}_1 f_0^{(N)}\,ds \tag{4.36}$$

where the exponential of the operator is defined by its series expansion. This solution can be checked by differentiating with respect to t

$$\begin{aligned}\frac{\partial f_1^{(N)}}{\partial t}&=-\mathsf{L}_1 f_0^{(N)}-\mathsf{L}_0\int_{-\infty}^{t}\exp\left[(s-t)\mathsf{L}_0\right]\mathsf{L}_1 f_0^{(N)}\,ds\\&=-\mathsf{L}_1 f_0^{(N)}-\mathsf{L}_0 f_1^{(N)}\end{aligned}$$

Equation 4.36 gives an explicit expression for the perturbation from equilibrium in terms of the equilibrium distribution and the perturbing potential.

Consider the term $-\mathsf{L}_1 f_0^{(N)}$ in more detail:

$$-\mathsf{L}_1 f_0^{(N)}=-\sum_{i=1}^{N}\left[\frac{\partial f_0^{(N)}}{\partial \mathbf{r}_i}\cdot\frac{\partial H_1}{\partial \mathbf{p}_i}-\frac{\partial f_0^{(N)}}{\partial \mathbf{p}_i}\cdot\frac{\partial H_1}{\partial \mathbf{r}_i}\right] \tag{4.37}$$

Assume that the perturbing force acting on a given molecule is a function of its position but not its velocity, so that

$$\frac{\partial H_1}{\partial \mathbf{p}_i}=0$$

Further, assume that the disturbance acts on the molecules individually, so that

$$H_1=\sum_{i=1}^{N}\psi(\mathbf{r}_i)$$

Then we can use the fact that the equilibrium density function has the form

$$f_0^{(N)} = \frac{1}{Z} \exp - \frac{\Phi(\mathbf{r}_1, \ldots \mathbf{r}_N) + \sum_{i=1}^{N} \mathbf{p}_i^2/2m}{kT}$$

to write

$$\frac{\partial f_0^{(N)}}{\partial \mathbf{p}_i} = -\frac{1}{mkT} f_0^{(N)} \mathbf{p}_i$$

These results enable us to simplify eq. 4.37

$$-\mathsf{L}_1 f_0^{(N)} = -\frac{1}{mkT} f_0^{(N)} \sum_{i=1}^{N} \mathbf{p}_i \cdot \frac{\partial \psi(\mathbf{r}_i)}{\partial \mathbf{r}_i} \tag{4.38}$$

We can write eq. 4.38 in the form

$$-\mathsf{L}_1 f_0^{(N)} = -\frac{1}{mkT} f_0^{(N)} \sum_{i=1}^{N} \int \mathbf{p}_i \cdot \frac{\partial \psi}{\partial \mathbf{r}} \delta(\mathbf{r} - \mathbf{r}_i)\, d\mathbf{r} \tag{4.39}$$

where $\delta(\mathbf{r})$ is the Dirac delta function. This function is zero everywhere except in the neighborhood of $\mathbf{r} = \mathbf{0}$, and has the properties

$$\int \delta(\mathbf{r})\, d\mathbf{r} = 1; \qquad \int f(\mathbf{r})\, \delta(\mathbf{r})\, d\mathbf{r} = f(\mathbf{0})$$

The effect of introducing the delta function is to focus attention on a particular point in the fluid, $\mathbf{r}$, rather than on the positions of the particles, $\mathbf{r}_i$. Given that $\delta(\mathbf{r} - \mathbf{r}_i)$ is zero except in the neighborhood of $\mathbf{r} = \mathbf{r}_i$, the local fluid velocity at $\mathbf{r}$, which is the sum of velocities of all particles in that neighborhood, is given by

$$\mathbf{j}(\mathbf{r}) = \sum_{i=1}^{N} \frac{\mathbf{p}_i}{m} \delta(\mathbf{r} - \mathbf{r}_i) \tag{4.40}$$

The force creating the disturbance from equilibrium at $\mathbf{r}$ is given by

$$\mathbf{F} = -\frac{\partial \psi}{\partial \mathbf{r}} \tag{4.41}$$

and so eq. 4.39 becomes

$$-\mathsf{L}_1 f_0^{(N)} = \frac{1}{kT} f_0^{(N)} \int \mathbf{j}(\mathbf{r}) \cdot \mathbf{F}\, d\mathbf{r}$$

If this result is used in eq. 4.36 the perturbation from equilibrium is given by

$$f_1^{(N)}(\mathbf{r}_1, \ldots \mathbf{p}_N; t) = \frac{1}{kT} \int_{-\infty}^{t} \exp[(s-t)\mathsf{L}_0] f_0^{(N)} \int \mathbf{j}(\mathbf{r}) \cdot \mathbf{F}\, d\mathbf{r} \tag{4.42}$$

The series expansion of the operator has the form

$$\exp[(s-t)\mathsf{L}_0] = 1 + (s-t)\mathsf{L}_0 + \frac{(s-t)^2}{2!}\mathsf{L}_0\mathsf{L}_0 + \cdots$$

The local velocity, $\mathbf{j}$, and force due to the disturbance, $\mathbf{F}$, at position $\mathbf{r}$ are average quantities over some neighborhood. It follows that they are independent of the molecular positions, $\mathbf{r}_1, \ldots \mathbf{r}_N$, and momenta, $\mathbf{p}_1, \ldots \mathbf{p}_N$, so that

$$\mathsf{L}_0\mathbf{j}(\mathbf{r}) = 0; \qquad \mathsf{L}_0\mathbf{F}(\mathbf{r}) = 0$$

Consequently, only the first term in the series expansion contributes to $f^{(N)}$ and the nonequilibrium density function is

$$f^{(N)} = f_0^{(N)} + f_1^{(N)} = f_0^{(N)}\left[1 + \frac{1}{kT}\int_{-\infty}^{t} ds \int \mathbf{j}(\mathbf{r}, s) \cdot \mathbf{F}(\mathbf{r}, s)\, d\mathbf{r}\right] \tag{4.43}$$

It is emphasized that both $\mathbf{j}$ and $\mathbf{F}$ are in general functions of time as well as position.

In the preceding section (on nonequilibrium thermodynamics) the phenomenological coefficients were introduced as proportionality constants in a linear relation between the various fluxes in a system and the conjugate forces. The fluxes are measured as system properties and so a microscopic theory of transport must examine fluxes defined as averages over an ensemble. Consider the average value of some flux $\mathbf{i}(\mathbf{r}, t)$

$$\langle \mathbf{i}(\mathbf{r}, t)\rangle = \int \mathbf{i}(\mathbf{r}, t) f^{(N)}(t)\, d\mathbf{r}_1 \cdots d\mathbf{p}_N$$

where the average is taken over the full position and momentum space available to the system. If we assume that the system is close to equilibrium we can use eq. 4.43 for the time dependent distribution function. At equilibrium the average flux in the system vanishes

$$\int \mathbf{i}(\mathbf{r}, t) f_0^{(N)}\, d\mathbf{r}_1 \cdots d\mathbf{p}_N = 0$$

so that only the second term in eq. 4.43 contributes

$$\langle \mathbf{i}(\mathbf{r}, t)\rangle = \frac{1}{kT}\int_{-\infty}^{t} ds \int \langle \mathbf{i}(\mathbf{r}, t)\mathbf{j}(\mathbf{r}', s) \cdot \mathbf{F}(\mathbf{r}', s)\rangle\, d\mathbf{r}' \tag{4.44}$$

where the average under the time and space integrals has been taken over the equilibrium distribution. If we remember that the force $\mathbf{F}(\mathbf{r}, t)$ was responsible for the disturbance from equilibrium it is clear that eq. 4.44 can provide a linear relation between the flux and the force.

The disturbing force does not depend on the particle coordinates and so can be removed from the ensemble average. In addition, we can assume that the correlation between the two fluxes **i** and **j** is only important over a distance similar to the range of intermolecular forces. Over such a distance any external disturbance is effectively constant, so that $\mathbf{F}(\mathbf{r}', s)$ can be replaced by $\mathbf{F}(\mathbf{r}, s)$ and removed from the volume integral. However, there is no reason to suppose that the disturbing force is constant with time—it may be a high frequency electromagnetic field, for example. Consequently, it is not legitimate to take $\mathbf{F}(\mathbf{r}, s)$ outside the time integration. Consider the case of a uniform fluid, where the ensemble averaged flux is independent of position. The spatial correlation between fluxes in volume elements surrounding the positions $\mathbf{r}'$ and $\mathbf{r}$ depends on their relative positions $\mathbf{r}'-\mathbf{r}$ only. In addition, once the system has relaxed sufficiently from the disturbing pulse used at $t=-\infty$ to distort the initial equilibrium, the correlation between the fluxes only depends on the time difference between them, that is, the flux-flux correlations are independent of the initial time. Consequently, we can change time variables so that $s=t-\tau$ and write eq. 4.44 as

$$\langle \mathbf{i}(t)\rangle = \frac{1}{kT}\int_0^\infty d\tau \mathbf{F}(t-\tau)\cdot\left\langle \int \mathbf{j}(\mathbf{r}', -\tau)\mathbf{i}(0)\, d\mathbf{r}'\right\rangle \tag{4.45}$$

In general the flux **i** will depend on the frequency of the applied field. Suppose we separate the frequency and time dependence of **F** and **i** as follows:

$$\begin{aligned}\mathbf{F}(t) &= \mathbf{F}(\omega)\exp(i\omega t)\\ \mathbf{i}(t) &= \mathbf{i}(\omega)\exp(i\omega t)\end{aligned} \tag{4.46}$$

Then if μ and ν represent x, y, and z coordinates, eq. 4.45 becomes

$$\langle i_\mu(\omega)\rangle = \sum_\nu F_\nu(\omega) L^{\mu\nu}(\omega) \tag{4.47}$$

where the frequency dependent coefficient $L^{\mu\nu}(\omega)$ is given by

$$L^{\mu\nu}(\omega) = \frac{1}{kT}\int_0^\infty d\tau \exp(-i\omega\tau)\left\langle \int i_\mu(0) j_\nu(\mathbf{r}', -\tau)\, d\mathbf{r}'\right\rangle \tag{4.48}$$

Equation 4.47 is valid for a system in which only one disturbing force is responsible for the transport. Obviously, if a number of different driving forces $\mathbf{F}_1, \ldots \mathbf{F}_l$ are present it is possible to write

$$H_1 = \sum_{k=1}^{l} H_1(k) \tag{4.49}$$

for the disturbance to the total energy of the system. If the Liouville equation is then linearized in terms of $H_1(k)$ eq. 4.47 can be replaced by a linear sum of coefficients times the various forces corresponding to $H_1(k)$.

As an example of the application of eq. 4.47 consider a system in which the driving force is a chemical potential gradient. There is no frequency dependence in such a gradient, and the corresponding phenomenological coefficient to L is the diffusion coefficient. The appropriate flux, from the previous section, is the local velocity $\mathbf{j}$ so that eq. 4.48 is written

$$L^{\mu\nu} = \frac{1}{kT}\int_0^\infty d\tau \left\langle \int d\mathbf{s}\, j_\mu(0) j_\nu(-\tau) \right\rangle \tag{4.50}$$

If the formula defining the local velocity in terms of the molecular momentum, eq. 4.40, is introduced into this equation and use made of the fact that the fluid is spatially isotropic, the diffusion coefficient of the fluid is given by

$$D = \int_0^\infty C(t)\, dt = \tfrac{1}{3}\int_0^\infty \langle \mathbf{v}_1(0) \cdot \mathbf{v}_1(t) \rangle\, dt \tag{4.51}$$

where $C(t)$ is the *velocity autocorrelation function.* The average in eq. 4.51 is taken over the equilibrium state of the fluid. Although the equilibrium distribution function does not depend on time, the particles in the system are constantly in motion and so the function $C(t)$ can be calculated. A convenient method for following the motion of a particle in an equilibrium system is to solve Newton's equations of motion using the method of molecular dynamics discussed in Section 3.10.

There is an alternative expression for the diffusion coefficient that provides a more intuitive description of the diffusion process. Consider the mean square displacement of a particle, in the x direction, as a function of time

$$M_x^2(t) = \langle [x_1(t) - x_1(0)]^2 \rangle \tag{4.52}$$

where $x_1\,(t)$ is the x coordinate of particle 1 at time t. We know that the position at time t is given by

$$x_1(t) = x_1(0) + \int_0^t \dot{x}_1(t')\, dt'$$

where $\dot{x}_1(t)$ is the x component of the velocity vector $\mathbf{v}_1$ at time t. Introducing this result into eq. 4.52 gives

$$M_x^2(t) = \int_0^t dt' \int_0^t dt'' \langle \dot{x}_1(t') \dot{x}_1(t'') \rangle \tag{4.53}$$

where the time integrations can be taken outside the average over an equilibrium ensemble as the distribution function is time independent. The velocity autocorrelation function in eq. 4.53 only depends on the difference in times, $\tau = t'' - t'$, since the average over an equilibrium ensemble ensures that at any given time all values of the initial velocity have been considered. Consequently,

$$M_x^2(t) = \int_0^t dt' \int_0^t dt'' \langle \dot{x}_1(0) \dot{x}_1(\tau) \rangle \tag{4.54}$$

To proceed further, the variable τ is introduced as an integration variable and the order of integration changed, so that

$$\begin{aligned} \int_0^t dt' \int_0^t dt'' &= \int_0^t dt' \int_{-t'}^{t-t'} d\tau \\ &= \int_{-t}^{0} d\tau \int_{-\tau}^{t} dt' + \int_0^t d\tau \int_0^{t-\tau} dt' \\ &= 2\int_0^t d\tau \int_\tau^t dt' \end{aligned}$$

Using this result in eq. 4.54 gives

$$\begin{aligned} M_x^2(t) &= 2\int_0^t d\tau \, \langle \dot{x}_1(0) \dot{x}_1(\tau) \rangle \int_\tau^t dt' \\ &= 2\int_0^t (t-\tau) \, \langle \dot{x}_1(0) \dot{x}_1(\tau) \rangle \, d\tau \\ &= 2t \int_0^t \left(1 - \frac{\tau}{t}\right) \langle \dot{x}_1(0) \dot{x}_1(\tau) \rangle \, d\tau \end{aligned} \tag{4.55}$$

It is easy to extend this equation to include y and z displacements, to give

$$M^2(t) = 2t \int_0^t \left(1 - \frac{\tau}{t}\right) \langle \mathbf{v}_1(0) \cdot \mathbf{v}_1(\tau) \rangle \, d\tau \tag{4.56}$$

If we now let $t \to \infty$, and assume that $C(\tau)$ decays to zero sufficiently rapidly as $\tau \to \infty$, we obtain

$$\lim_{t \to \infty} \left(\frac{M^2(t)}{2t} \right) = \int_0^\infty \langle \mathbf{v}_1(0) \cdot \mathbf{v}_1(\tau) \rangle \, d\tau \tag{4.57}$$

Comparing this result with eq. 4.51 we see that the diffusion coefficient is also given by

$$D = \lim_{t\to\infty} \frac{\langle [\mathbf{r}_1(0) - \mathbf{r}_1(t)]^2 \rangle}{6t} \tag{4.58}$$

Thus the diffusion coefficient is related to the long time limit of the mean square displacement of a particle.

Returning to eq. 4.48, it can be seen that the zero frequency phenomenological coefficient is given as the time integral of a flux-flux correlation function

$$A(t) = \langle I(0)J(t) \rangle \tag{4.59}$$

where J and I are components of the appropriate fluxes. Expressions for the shear and bulk viscosities, thermal conductivity, and many other phenomenological coefficients can be obtained by considering the appropriate fluxes and thermodynamic forces. The equations for shear viscosity, η, bulk viscosity, ϕ, and thermal conductivity, λ, are obtained by using the result [44]

$$L = \int_0^\infty \langle J(0)J(t) \rangle \, dt \tag{4.60}$$

where the fluxes J are components of an appropriate tensor

$$J_\eta(xy) = \left(\frac{1}{VkT}\right)^{1/2} \sum_{i=1}^{N} [m\dot{x}_i\dot{y}_i + x_iF_i(y)] \tag{4.61}$$

$$J_\phi(xx) = \left(\frac{1}{VkT}\right)^{1/2} \sum_{i=1}^{N} \left[m\dot{x}_i\dot{x}_i + x_iF_i(x) - \frac{pV}{N}\right] \tag{4.62}$$

$$J_\lambda(x) = \left(\frac{1}{VkT^2}\right)^{1/2} \sum_{i=1}^{N} \frac{d}{dt}(x_iE_i') \tag{4.63}$$

In these equations $F_i(y)$ is the y component of the force on particle i due to interactions with all the other particles, and E_i' is the difference between the total energy of particle i and its average value $E_i' = E_i - \langle E_i \rangle$ with $E_i = m\dot{x}_i^2/2 + \Phi$. The bulk viscosity is obtained from the expression

$$\phi = -\tfrac{4}{3}\eta + \int_0^\infty \langle J_\phi(0)J_\phi(t) \rangle \, dt \tag{4.64}$$

where J_ϕ is given by eq. 4.62. It is worth noting that the equations for shear and bulk viscosity and thermal conductivity can also be written in terms of mean square displacements [44].

4.3 Theory of Brownian Motion

A whole class of approximate theories of transport in dense fluids has been constructed on the basis of Brownian motion theory [16, 20–22]. If macroscopic particles are suspended in a liquid they are subject to extremely rapid bombardment by the molecules forming the fluid. The result of this bombardment can be seen if the suspension is examined through a microscope, when it is discovered that the particles are in constant, erratic motion. This motion is not due to the effect of a single molecular impact but rather to the resultant force from very many collisions. A theory that gives an accurate description of the time evolution of the position of these macroscopic particles can be constructed, and a similar approximation for molecular sized particles derived. In this section we examine the derivation of results for the distribution in phase space of Brownian particles and then discuss why such theories cannot be extended into the realm of pure molecular systems.

Suppose that the force acting on a single particle can be written as the sum of two terms, the first representing a general slowing down due to frictional forces and being proportional to the velocity of the particle, and the second corresponding to a microscopic random fluctuating force. Then we may write the *Langevin equation* [13]

$$\frac{d\mathbf{v}}{dt} = -\zeta\mathbf{v} + \mathbf{X}(t) \tag{4.65}$$

where ζ, the friction constant, can be related to the viscosity by Stoke's law [45] so that $\zeta = 6\pi\eta a/m$, where a is the diameter of the particle and m is its mass, and $\mathbf{X}(t)$ represents the random fluctuating term. We assume that at time $t=0$ the particle has position $\mathbf{r}_0$ and velocity $\mathbf{v}_0$. At later times nothing can be said with certainty about the position, $\mathbf{r}$, and velocity, $\mathbf{v}$, of the particle as neither the magnitude nor direction of $\mathbf{X}(t)$ is known. However, it is possible to derive the probability density $f(\mathbf{r}, \mathbf{v}, t)$ that the center of the particle lies in the range $(\mathbf{r}, \mathbf{r}+d\mathbf{r})$ and its velocity is in the interval $(\mathbf{v}, \mathbf{v}+d\mathbf{v})$ at time t.

The form of $f(\mathbf{r}, \mathbf{v}, t)$ can be written down for very short and very long times. At the initial time we have by hypothesis

$$f(\mathbf{r}, \mathbf{v}, 0) = \delta(\mathbf{r}-\mathbf{r}_0)\,\delta(\mathbf{v}-\mathbf{v}_0) \tag{4.66}$$

where $\delta(\mathbf{r})$ is the Dirac delta function, introduced in eq. 4.39. At very long times the Brownian particle will have reached thermal equilibrium with the fluid and so will have the equilibrium single particle distribution given by statistical mechanics (eq. 3.34 for a single particle, before doing the momentum integration):

$$f(\mathbf{r}, \mathbf{v}, \infty) = A \exp \frac{-m\mathbf{v}^2}{2kT} \tag{4.67}$$

where A is proportional to the single particle position distribution function, or number density, of the Brownian particle.

Although the random fluctuating force, $\mathbf{X}(t)$, is not known we can discuss the probability density of the related quantity

$$\bar{\mathbf{X}}(\tau) = \int_t^{t+\tau} \mathbf{X}(t')\, dt' \tag{4.68}$$

which is given by $p(\bar{\mathbf{X}}, \tau)$, say. We assume when writing this expression that the interval τ is long compared with the period over which $\mathbf{X}(t)$ fluctuates so that $p(\bar{\mathbf{X}}, \tau)$ depends on the length of the interval, τ, but not on the time t at which the averaging began. Similarly, it is assumed that τ is short compared with the period over which the Brownian particle is studied, so that $\bar{\mathbf{X}}(\tau)$ represents the observable (e.g., through a microscope) effects of the random motions.

The change in velocity of a particle over some small period τ is obtained by integrating the Langevin equation (eq. 4.65), which results in the following:

$$\mathbf{u} = \mathbf{v} - \mathbf{v}_0 = -\zeta \mathbf{v}\tau + \bar{\mathbf{X}} \tag{4.69}$$

where it is assumed that τ is so small that $\mathbf{v}$ can be taken out of the integral. It follows that the probability that $\mathbf{u}$ should lie in the range ($\mathbf{u}$, $\mathbf{u}+d\mathbf{u}$) is just the probability that $\bar{\mathbf{X}}$ should lie in the range ($\bar{\mathbf{X}}$, $\bar{\mathbf{X}}+d\bar{\mathbf{X}}$), $p(\mathbf{u}+\zeta\mathbf{v}\tau, \tau)\, d\mathbf{u}$, using eq. 4.69 to relate $\bar{\mathbf{X}}$ and $\mathbf{u}$. The probability density that the particle is at point $\mathbf{r}$ with velocity $\mathbf{v}$ at time $t+\tau$, $f(\mathbf{r}, \mathbf{v}, t+\tau)$ is easily related to the probability of finding the particle at position $\mathbf{r}-\mathbf{v}\tau$, with velocity $\mathbf{v}-\mathbf{u}$, at time t, $f(\mathbf{r}-\mathbf{v}\tau, \mathbf{v}-\mathbf{u}, t)$. Multiplying $f(\mathbf{r}-\mathbf{v}\tau, \mathbf{v}-\mathbf{u}, t)$ by the probability that the velocity changes by an amount $\mathbf{u}$ in the interval τ, $p(\mathbf{u}+\zeta(\mathbf{v}-\mathbf{u})\tau, \tau)\, d\mathbf{u}$ and summing over all possible values of $\mathbf{u}$ gives

$$f(\mathbf{r}, \mathbf{v}, t+\tau) = \int f(\mathbf{r}-\mathbf{v}\tau, \mathbf{v}-\mathbf{u}, t)\, p(\mathbf{u}+\zeta(\mathbf{v}-\mathbf{u})\tau, \tau)\, d\mathbf{u} \tag{4.70}$$

This equation is valid provided we can make the assumption that the time interval τ is sufficiently small that changes in position and velocity are linear functions of the time.

The distribution function for $\bar{\mathbf{X}}(\tau)$, which is given by $p(\bar{\mathbf{X}}, \tau)$, can be obtained from eq. 4.70 by using the known long time expression for

$f(\mathbf{r}, \mathbf{v}, \infty)$, eq. 4.67. We have

$$\exp\frac{-m\mathbf{v}^2}{2kT} = \int \exp\frac{-m\mathbf{w}^2}{2kT}\, p(\mathbf{v}-\mathbf{w}+\zeta\mathbf{w}\tau, \tau)\, d\mathbf{w} \qquad (4.71)$$

where $\mathbf{w}=\mathbf{v}-\mathbf{u}$. The integral in eq. 4.71 is a convolution integral [46] and the equation can be solved using Fourier transforms. (We used a similar procedure in Section 3.8 to show that the Ornstein-Zernike equation was an acceptable definition of the direct correlation function.) The solution is

$$p(\bar{\mathbf{X}}, \tau) = \left(\frac{m}{4\pi\zeta kT\tau}\right)^{3/2} \exp\left(\frac{-m\bar{\mathbf{X}}^2}{4\zeta kT\tau}\right) \qquad (4.72)$$

That is, for a fluid at equilibrium the distribution of accelerations of a Brownian particle averaged over a period of time τ is Gaussian. It must be stressed that $\bar{\mathbf{X}}(t)$ is not the acceleration arising from an instantaneous force acting on the Brownian particle but is the result of a series of such forces taken over the period τ *and from its definition, eq.* 4.68, *has dimensions of a velocity.*

From eq. 4.72 it is easy to calculate the mean change in velocity over the period τ

$$\langle\mathbf{u}\rangle = \int \mathbf{u}p(\mathbf{u}+\zeta\mathbf{v}\tau, \tau)\, d\mathbf{u} = -\zeta\mathbf{v}\tau \qquad (4.73)$$

and the mean square deviation of $\mathbf{u}$ from the mean

$$\langle(\mathbf{u}+\zeta\mathbf{v}\tau)^2\rangle = \langle\bar{\mathbf{X}}^2\rangle = \frac{6\zeta\tau kT}{m} \qquad (4.74)$$

From these two results we see that the ratio of the mean square deviation of $\mathbf{u}$ from the mean, to the square of the mean is $6kT/m\zeta\mathbf{v}^2\tau$. Consequently, the root mean square deviation of $\mathbf{u}$ from the mean is much greater than the mean provided $\tau \ll 6kT/m\zeta\mathbf{v}^2$. Provided this condition is satisfied it is reasonable to assume that average changes in position and velocity are small over a period τ.

An equation for the distribution of particle position and velocity can be obtained from eq. 4.70 using eq. 4.72 for $p(\bar{\mathbf{X}}, \tau)$ together with the assumption that the interval τ is macroscopically short. We expand both sides of eq. 4.70 in a Taylor series in terms of τ about $\tau=0$, and truncate the series beyond the first power in τ.

After some careful algebra [13, 22] it is found that the distribution of position and velocity is determined by the *Fokker-Planck equation*

$$\frac{\partial f}{\partial t} + \mathbf{v}\cdot\nabla f = \zeta\frac{\partial}{\partial\mathbf{v}}\cdot\left(\mathbf{v}f + \frac{kT}{m}\frac{\partial f}{\partial\mathbf{v}}\right) \qquad (4.75)$$

This equation has to be solved subject to the initial condition eq. 4.66. If one makes the substitutions

$$\mathbf{V} = \mathbf{v}\exp(\zeta t) - \mathbf{v}_0$$

$$\mathbf{R} = \mathbf{r} - \mathbf{r}_0 + \frac{\mathbf{v} - \mathbf{v}_0}{\zeta}$$

$$\chi = \exp(-3\zeta t) f(\mathbf{r}, \mathbf{v}, t) \tag{4.76}$$

then the Fokker-Planck equation reduces to

$$\frac{\partial \chi}{\partial t} = \frac{\zeta kT}{m}\left(\exp(2\zeta t)\frac{\partial^2 \chi}{\partial \mathbf{V}^2} + \frac{2\exp(\zeta t)}{\zeta}\frac{\partial^2 \chi}{\partial \mathbf{V}\cdot\partial \mathbf{R}} + \frac{1}{\zeta^2}\frac{\partial^2 \chi}{\partial \mathbf{R}^2}\right)$$

and by using the trial solution

$$\chi = (2\pi\Delta^{1/2})^{-3}\exp\frac{-(a\mathbf{V}^2 + 2h\mathbf{R}\cdot\mathbf{V} + b\mathbf{R}^2)}{2\Delta}$$

where $\Delta = ab - h^2$, and where $a = 2tkT/\zeta m$, $b = kT(\exp(2\zeta t) - 1)/m$, and $h = -2kT(\exp(\zeta t) - 1)/\zeta m$, it can be shown that [13, 22]

$$f(\mathbf{r}, \mathbf{v}, t) = \left(\frac{1}{4\pi^2\Delta}\right)^{3/2}\exp\left[3\zeta t - \frac{(a\mathbf{V}^2 + 2h\mathbf{R}\cdot\mathbf{V} + b\mathbf{R}^2)}{2\Delta}\right] \tag{4.77}$$

where $\mathbf{r}$ and $\mathbf{v}$ are related to $\mathbf{R}$ and $\mathbf{V}$ through eq. 4.76.

Equation 4.77 gives the time dependent conditional single particle density function, the result being conditional on the initial position, $\mathbf{r}_0$, and velocity $\mathbf{v}_0$. We can use this function to compute various properties of the particle. For example, the mean velocity, mean displacement, mean square velocity, and mean square displacement are

$$\langle \mathbf{v} \rangle_1 = \int \mathbf{v}(t) f(\mathbf{r}, \mathbf{v}, t)\, d\mathbf{r}\, d\mathbf{v} = \mathbf{v}_0 \exp(-\zeta t) \tag{4.78}$$

$$\langle (\mathbf{r} - \mathbf{r}_0) \rangle_1 = \mathbf{v}_0 \frac{1 - \exp(-\zeta t)}{\zeta} \tag{4.79}$$

$$\langle \mathbf{v}\cdot\mathbf{v} \rangle_1 = \mathbf{v}_0^2 \exp(-2\zeta t) + 3kT\frac{1 - \exp(-2\zeta t)}{m} \tag{4.80}$$

$$\langle (\mathbf{r} - \mathbf{r}_0)^2 \rangle_1 = \mathbf{v}_0^2 \frac{[1 - \exp(-\zeta t)]^2}{\zeta^2} + 3kT\frac{2\zeta t - 3 + 4\exp(-\zeta t) - \exp(-2\zeta t)}{\zeta^2 m} \tag{4.81}$$

In these equations the notation $\langle\rangle_1$ implies that the averages are calculated for fixed initial conditions. These results can also be obtained by solving the Langevin equation. For example, to obtain eq. 4.78 multiply eq. 4.65 by the density function $f(\mathbf{r}, \mathbf{v}, \tau)$ and integrate over $\mathbf{r}$ and $\mathbf{v}$ to give

$$\left\langle \frac{d\mathbf{v}}{dt} \right\rangle_1 = \frac{d}{dt} \langle \mathbf{v} \rangle_1 = -\zeta \langle \mathbf{v} \rangle_1$$

provided we can commute the integration and differentiation on the left and that $\langle \mathbf{X} \rangle_1 = 0$. The solution to this equation is obviously given by eq. 4.78. Using similar arguments, the results in eqs. 4.79–4.81 can also be obtained.

At long times the results given in eqs. 4.78–4.81 reduce to

$$\langle \mathbf{v} \rangle_1 = 0; \qquad \langle (\mathbf{r} - \mathbf{r}_0) \rangle_1 = \frac{\mathbf{v}_0}{\zeta}$$

$$\langle \mathbf{v}^2 \rangle_1 = \frac{3kT}{m}; \qquad \langle (\mathbf{r} - \mathbf{r}_0)^2 \rangle_1 = \frac{6kTt}{\zeta m} \tag{4.82}$$

Note that the mean displacement is not zero but is related to the initial velocity. If we now average the results given in eq. 4.82 over a suitable distribution of initial positions and velocities the results alter. Suppose we choose to average over the equilibrium distribution, eq. 4.67. Then, for example, we find that $\langle (\mathbf{r} - \mathbf{r}_0) \rangle = 0$, so that not until the full ensemble average is taken does the expected displacement become zero.

If we examine the formula for the mean square displacement in eq. 4.82 we find that it is independent of the initial conditions. Taking an average over the initial conditions, therefore, gives

$$\langle (\mathbf{r} - \mathbf{r}_0)^2 \rangle = \frac{6kTt}{\zeta m}$$

for long times, or writing the equation as a limit

$$\lim_{t \to \infty} \frac{\langle (\mathbf{r} - \mathbf{r}_0)^2 \rangle}{6t} = \frac{kT}{\zeta m} \tag{4.83}$$

Comparing this result with that obtained in the previous section, eq. 4.58, we obtain the well known *Einstein equation*

$$D = \frac{kT}{\zeta m} \tag{4.84}$$

This result, together with eq. 4.83, was reported by Einstein in his study of Brownian motion [47].

Brownian motion theory also gives an expression for the velocity autocorrelation function occurring in eq. 4.51. The velocity autocorrelation function is given by

$$C(t)=\tfrac{1}{3}\langle \mathbf{v}(t)\cdot\mathbf{v}_0\rangle=\tfrac{1}{3}\int \mathbf{v}_0\cdot\mathbf{v}(t)f_0(\mathbf{v}_0)f(\mathbf{r},\mathbf{v},t)\,d\mathbf{v}_0\,d\mathbf{r}\,d\mathbf{v} \qquad (4.85)$$

where $f_0(\mathbf{v}_0)$ is the distribution of initial velocities. If $f_0(\mathbf{v}_0)$ is chosen to be the equilibrium distribution function and the result given in eq. 4.78 is used for the integration over $\mathbf{r}$ and $\mathbf{v}$ we have

$$C(t)=\tfrac{1}{3}\int \mathbf{v}_0\cdot\mathbf{v}_0\exp(-\zeta t)f_0(\mathbf{v}_0)\,d\mathbf{v}_0=\frac{kT}{m}\exp(-\zeta t) \qquad (4.86)$$

That is, the velocity autocorrelation function is a simple decreasing exponential. We can check that this function also gives the diffusion coefficient

$$\int_0^\infty C(t)\,dt=\int_0^\infty\left(\frac{kT}{m}\right)\exp(-\zeta t)\,dt=\frac{kT}{m\zeta}$$

and substituting for D using the Einstein relation (eq. 4.84) we see that

$$\int_0^\infty C(t)\,dt=D \qquad (4.87)$$

in agreement with eq. 4.51.

Brownian motion theory is valid only in cases where the equation of motion of a single particle is given by the Langevin equation. This equation holds when the fluctuating forces are random and small compared with the frictional drag term. In principle, such a situation can only arise if the mass of the diffusing particle is very much greater than the masses of the molecules forming the fluid in which it is moving. In practice, there have been several attempts to extend Brownian motion theory to one-component fluids, where the diffusing particles have the same mass as the molecules in the fluid. The best known of these is the Kirkwood theory [16], and it is instructive to examine the foundations of this theory in more detail.

We have seen that on the basis of the Langevin equation the instantaneous force acting on a particle is given as a sum of a term arising from friction between the particle and its surroundings and a random fluctuating term arising from collision processes. To obtain expressions for the time dependent properties of the fluid we assumed that the fluctuating acceleration $\mathbf{X}(t)$ at any time was independent of its value at all other times and that its average value was zero. Statistically, these assumptions can be written in the form

$$\langle \mathbf{X}(t)\rangle = 0 \qquad \langle \mathbf{X}(0)\cdot\mathbf{X}(t)\rangle = 3\phi(t) \tag{4.88}$$

where the function $\phi(t)$ is so sharply peaked that it is zero by the time $t=\tau$ where τ is the time interval discussed when writing down eq. 4.68. If these assumptions hold then Kirkwood [16] showed that the friction coefficient can be calculated using the equation

$$\zeta = \frac{1}{3kT}\int_0^{\tau} \langle \mathbf{F}_1(0)\cdot\mathbf{F}_1(t)\rangle\, dt \tag{4.89}$$

where $\mathbf{F}_1(t)$ is the instantaneous force acting on particle 1. The time τ is chosen so that the integral is independent of its value; in fact, if τ is allowed to go to infinity the integral vanishes [18], but provided $\phi(t)$ is sufficiently sharply peaked the integral rises rapidly to a plateau value and then falls very slowly to zero. Suddaby and Gray [18] analyzed the Kirkwood theory and showed that the force autocorrelation function appearing in eq. 4.89 takes the form

$$\tfrac{1}{3}\langle \mathbf{F}_1(0)\cdot\mathbf{F}_1(t)\rangle = m^2\phi(t) - m\zeta^2 kT \exp(-\zeta t) \tag{4.90}$$

Fisher and Watts [24] published a detailed analysis of the Kirkwood theory using the method of molecular dynamics together with an accurate pair potential for argon. They showed that the force autocorrelation function did not have a plateau value and that the term $\phi(t)$ dominated the exponential term in eq. 4.90 for all times at which $\langle \mathbf{F}_1(0)\cdot\mathbf{F}_1(t)\rangle$ was significant. Furthermore, the force autocorrelation function calculated

Table 4.1 Friction coefficients from the linear response theory (ζ_L) and Kirkwood theory (ζ_K) are compared as a function of the upper limit τ in eq. 4.89. Friction coefficients are in units of 10^{12} s^{-1}

	Gas phase		Liquid phase	
$\tau\times10^{12}$ s	ζ_L	ζ_K	ζ_L	ζ_K
0.22	—	1.10	—	3.39
0.25	—	0.91	—	1.91
0.31	—	0.77	—	0.57
4.08	0.99	−0.23	9.7×10^{-3}	−1.04
6.06	0.96	−0.27	10.7×10^{-3}	−0.81
9.84	0.94	−0.39	10.2×10^{-3}	−1.24

directly from the molecular dynamics calculations was in very poor agreement with eq. 4.90 if ζ was calculated from the self-diffusion coefficient. Fisher and Watts went on to calculate the friction coefficient both by integrating the force autocorrelation function and by calculating the self-diffusion coefficient from the velocity autocorrelation function, eq. 4.51, and then using the Einstein equation, eq. 4.84. Their results are given in Table 4.1 for various values of the upper limit τ in eq. 4.89. It can be seen that the agreement is very poor and that in the liquid state the Kirkwood theory predicts negative values for ζ, an unphysical result. A similar analysis of the friction coefficient theory has been given by Smedley and Woodcock [25].

4.4 Memory Function Formalism

The failure of the Brownian motion theory of transport appears to have rather unfortunate implications, for as we pointed out in the introduction attempts to extend the classical kinetic theory of dilute gases [1, 2] into the dense gas region have not been very successful [3]. Fortunately, it is possible to generalize the Langevin equation to give a good description of time dependent phenomena in dense fluids. The generalization involves including a memory function in the single particle equation of motion. Two approaches have been used to introduce this formalism into the transport theory of liquids. Zwanzig [48] and later Mori [49] rewrote the equations of motion of the system to obtain a generalized form of the Langevin equation containing a memory function. The equation was subsequently used by Akcasu and Daniels [50] in an analysis of correlations in simple liquids to obtain expressions for the correlation functions of a fluid. An alternative approach to that proposed by Zwanzig and Mori was reported by Kadanoff and Martin [51], who gave a systematic and general hydrodynamic description of fluids. Essentially, their method extends the normal hydrodynamic theory of fluids into shorter wavelength regions, where the standard equations break down. Chung and Yip [52] have used the Kadanoff-Martin theory to obtain rigorous expressions for the time correlation functions at certain frequencies and wavelengths. Unfortunately, both the Zwanzig-Mori theory and that attributed to Kadanoff and Martin are rather complicated and a description of them does not fall within the objectives of this book. However, the memory function approach has proven very useful in the analysis of computer simulation studies of dense fluids, and so we give a nonrigorous description of the theory.

The memory function theory is readily understood if we return for the moment to Brownian motion theory. In eq. 4.86 we showed that the velocity autocorrelation function is approximated by

$$C(t) = \frac{kT}{m} \exp(-\zeta t)$$

If we take the one-sided Fourier transform of this function we obtain the so-called *power spectrum*:

$$\tilde{C}(\omega) = \int_0^\infty \exp(-i\omega t) C(t)\, dt = \frac{kT}{m(\zeta + i\omega)} \tag{4.91}$$

This equation can also be obtained from the Langevin equation, but before doing this we introduce the memory function. The Langevin equation (eq. 4.65) assumes that in the absence of the fluctuating force, $\mathbf{X}(t)$, the diffusing particle continues in a straight line, with decreasing velocity. We can make this equation more general by assuming that the particle retains some memory of all its preceding movements, so that the friction coefficient is replaced by a function connecting the present rate of change of velocity with all previous velocities rather than with the immediate velocity. We write

$$\frac{d\mathbf{v}}{dt} = -\int_0^t M(\tau)\mathbf{v}(t-\tau)\, d\tau + \mathbf{X}(t) \tag{4.92}$$

where, as before $\mathbf{X}(t)$ is a random fluctuating term such that $\langle \mathbf{X} \rangle = 0$ and $M(t)$ is the memory function. This is the generalized Langevin equation. In his derivation of this equation Mori [49] related $M(t)$ to the intermolecular potential function. Unfortunately, the resulting expression is very complicated and the memory function cannot be readily calculated.

Rather than attempt to obtain the velocity autocorrelation function using a generalization of arguments given in the previous section, we solve the equation directly using simple, but definitely nonrigorous algebra. Multiply through by $\exp(-i\omega t)$ and integrate from $t=0$ to $t=\infty$ to give

$$\mathbf{v}_0 + i\omega\tilde{\mathbf{v}}(\omega) = -\int_0^\infty dt \exp(-i\omega t) \int_0^t d\tau\, M(\tau)\mathbf{v}(t-\tau) + \int_0^\infty dt \exp(-i\omega t)\mathbf{X}(t)$$

where $\mathbf{v}_0$ is the initial velocity, $\tilde{\mathbf{v}}(\omega)$ is the transform of $\mathbf{v}(t)$ and the result on the left hand side is obtained from an integration by parts. The term on the right hand side of the equation can be simplified by first reversing the order of the integrations and then by making the change of variable $q = t - \tau$:

$$\int_0^\infty dt\, \exp(-i\omega t) \int_0^t d\tau\, M(\tau)\mathbf{v}(t-\tau)$$

$$= \int_0^\infty d\tau\, M(\tau) \exp(-i\omega\tau) \int_\tau^\infty dt\, \exp[-i\omega(t-\tau)]\mathbf{v}(t-\tau)$$

$$= \int_0^\infty d\tau\, M(\tau) \exp(-i\omega\tau) \int_0^\infty dq\, \exp(-i\omega q)\mathbf{v}(q)$$

$$= \tilde{M}(\omega)\tilde{\mathbf{v}}(\omega)$$

so that

$$\mathbf{v}_0 + i\omega\tilde{\mathbf{v}}(\omega) = \tilde{M}(\omega)\tilde{\mathbf{v}}(\omega) + \int_0^\infty \exp(-i\omega t)\mathbf{X}(t)\, dt \qquad (4.93)$$

We now average over the distribution of $\mathbf{X}(t)$, take the inner product of all terms with the initial velocity, $\mathbf{v}_0$, and integrate over the distribution of initial velocities. The result is, after some rearrangement, an expression for the power spectrum analagous to eq. 4.91:

$$\tilde{C}(\omega) = \frac{kT}{m[\tilde{M}(\omega) + i\omega]} \qquad (4.94)$$

Thus we see that the generalized Langevin equation introduces a frequency dependent friction coefficient into Brownian motion theory.

The form of the memory function for liquids is still not fully understood, although its behavior at very low frequencies is known [52]. We can recover the original Langevin equation by making the assumption

$$M(t) = \zeta\delta(t) \qquad (4.95)$$

where $\delta(t)$ is the Dirac delta function. Other assumptions, such as the Gaussian Markov form $M(t) = M_0 \exp(-\lambda t)$, can also be used [53]. Finally, Mori [49] showed that under fairly general conditions the fluctuating term $\mathbf{X}(t)$ itself obeys a generalized Langevin equation, so that a hierarchy of equations can be established. We are not concerned with such details of transport theory, although in later chapters we show that semiempirical fits to $M(t)$ can be useful.

4.5 Time Dependent Distribution Functions

We have seen in Sections 4.2 and 4.3 that the transport properties of a system may be calculated in terms of averages of certain quantities over both space and time. It is reasonable to ask what form the distribution functions for the spatial and velocity coordinates take. The self-diffusion coefficient can be calculated in two ways. First, we can integrate the velocity autocorrelation function using eq. 4.51. In this case we need to know the distribution of velocities in the system. Second, we can obtain D using the long time limit of the mean square displacement of the particle, eq. 4.58. In this case we need to know the time dependent single particle position distribution function. It is worth noting that Brownian motion theory gives expressions for both of these functions. The general time dependent single particle velocity and position distribution is given by eq. 4.77, the velocity autocorrelation function is given by the negative exponential, eq. 4.86, and the single particle position distribution function is obtained by integrating the velocity variables out of eq. 4.77, to give a Gaussian. It is easy to generalize the concept of time dependent single particle distribution functions to more particles, and the expressions for transport coefficients such as the viscosity and thermal conductivity, eqs. 4.60–4.63, obviously require some knowledge of these functions. In this section and particularly in the following chapter we consider these time dependent functions in some detail.

The time dependent distribution function that we are most concerned with is the function $G(\mathbf{r}, t)$ which measures the probability of finding a particle at position $\mathbf{r}$ at time t given that there was a particle at the origin at time $t=0$. This function obviously bears some resemblance to the radial distribution function $g(r)$ discussed in the preceding chapter. $G(\mathbf{r}, t)$ is usually known as the van Hove correlation function after its originator [54]. The van Hove function can be measured experimentally using inelastic neutron scattering methods, as is discussed in the following chapter.

There are two ways in which we can have a particle at the origin at time $t=0$ and also a particle at position $\mathbf{r}$ at time t. First the particle initially at the origin can migrate through the fluid to position $\mathbf{r}$, and second some other particle can move to $\mathbf{r}$. This suggests that the van Hove function can be written as the sum of two terms

$$G(\mathbf{r}, t) = G_s(\mathbf{r}, t) + G_d(\mathbf{r}, t) \tag{4.96}$$

In this equation $G_s(\mathbf{r}, t)$ is known as the self-correlation function and $G_d(\mathbf{r}, t)$ is known as the distinct correlation function. Obviously, at time $t=0$, $G_s(\mathbf{r}, t)$ is given by a Dirac delta function

$$G_s(\mathbf{r}, 0) = \delta(\mathbf{r}) \tag{4.97}$$

We have mentioned earlier that the description of $G(\mathbf{r}, t)$ resembles that of the radial distribution function. In fact it can be shown [54,55] that in a uniform fluid the relation is such that

$$G_d(\mathbf{r}, 0) = \rho g(r) \tag{4.98}$$

In the next chapter we show that the van Hove functions are related to the so-called coherent and incoherent structure factors $S_c(q, \omega)$ and $S_i(q, \omega)$ [56]. These functions are related to the time and space transforms of the van Hove functions, and if we define the intermediate scattering functions

$$F(q, t) = \int d\mathbf{r}\; G(\mathbf{r}, t) \exp(-i\mathbf{q}\cdot\mathbf{r}) \tag{4.99}$$

$$F_s(q, t) = \int d\mathbf{r}\; G_s(\mathbf{r}, t) \exp(-i\mathbf{q}\cdot\mathbf{r}) \tag{4.100}$$

we have

$$S_c(q, \omega) = \int_0^\infty F(q, t) \exp(-i\omega t)\, dt \tag{4.101}$$

$$S_i(q, \omega) = \int_0^\infty F_s(q, t) \exp(-i\omega t)\, dt \tag{4.102}$$

As we mentioned earlier, Brownian motion theory gives an expression for $G_s(\mathbf{r}, t)$, from eq. 4.77, and so we might expect the memory function formalism outlined in the preceding section to be useful for analyzing the van Hove functions. This is so, and various authors have shown that [57]

$$S_i(q, \omega) = \frac{1}{\tilde{M}_i(q, \omega) + i\omega} \tag{4.103}$$

We show in the next chapter that $S_i(q, \omega)$ can be described by semiempirical models of $\tilde{M}_i(q, \omega)$.

References

1. L. Boltzmann, 1872–1875, *Collected Works*, Chelsea Publishers, New York.
2. S. Chapman and T. G. Cowling, 1952, *The Mathematical Theory of Non-Uniform Gases*, Cambridge University Press.
3. M. H. Ernst, L. K. Haines, and J. R. Dorfman, 1969, *Rev. Mod. Phys.*, **41,** 296.
4. C. S. Wang Chang and G. E. Uhlenbeck, 1951, "Transport Phenomena in Polyatomic Gases," Michigan University Engineering Research Institute Report CM-681.
5. N. Taxman, 1958, *Phys. Rev.*, **110,** 1235.
6. R. S. C. She and N. F. Sather, 1967, *J. Chem. Phys.*, **47,** 4978.
7. C. S. Wang Chang, G. E. Uhlenbeck, and J. De Boer, 1964, in *Studies in Statistical Mechanics*, J. De Boer, G. E. Uhlenbeck, Eds., Vol. 3, Part c, North Holland, Amsterdam.
8. D. J. Evans and R. O. Watts, 1976, *Mol. Phys.*, (to be published).

9. A. G. Clarke and E. B. Smith, 1970, *J. Chem. Phys.*, **53,** 1235; J. C. Parker, 1959, *Phys. Fluids*, **7,** 449.
10. "Molecular Beam Scattering," 1973, *Discuss. Faraday Soc.*, **55.**
11. E. A. Mason and L. Monchick, 1962, *J. Chem. Phys.*, **36,** 1622.
12. C. F. Curtiss, 1956, *J. Chem. Phys.*, **24,** 225.
13. H. S. Green, 1952, *Molecular Theory of Fluids*, North Holland, Amsterdam.
14. N. N. Boguliubov, 1946, *J. Phys.* (*USSR*), **10,** 265.
15. M. Born and H. S. Green, 1949, *A General Kinetic Theory of Liquids*, Cambridge University Press.
16. J. G. Kirkwood, 1946, *J. Chem. Phys.*, **14,** 180.
17. J. Yvon, 1935, *Actualities Scientifiques et Industriel*, Herman et Cie, Paris.
18. A. Suddaby and P. Gray, 1960, *Proc. Phys. Soc.* (*London*), **75,** 109.
19. D. Enskog, 1922, *Arkiv. Mat. Astronomi Fys.*, **16,** 16.
20. S. A. Rice and A. R. Allnatt, 1961, *J. Chem. Phys.*, **34,** 2144.
21. A. R. Allnatt and S. A. Rice, 1961, *J. Chem. Phys.*, **34,** 2156.
22. S. A. Rice and P. Gray, 1965, *The Statistical Mechanics of Simple Liquids*, Interscience, New York.
23. A. F. Collings, R. O. Watts, and L. A. Woolf, 1971, *Mol. Phys.*, **20,** 1121.
24. R. A. Fisher and R. O. Watts, 1972, *Aust. J. Phys.*, **25,** 21.
25. S. I. Smedley and L. V. Woodcock, 1974, *Trans. Faraday Soc. II*, **70,** 955.
26. J. H. Dymond, 1972, *Trans. Faraday Soc. II*, **68,** 1789; B. J. Alder, W. E. Alley, and J. H. Dymond, 1974, *J. Chem. Phys.*, **61,** 1415.
27. K. Kim and D. Chandler, 1973, *J. Chem. Phys.*, **59,** 5215.
28. L. A. Woolf, 1973, *J. Chem. Phys.*, **57,** 3013.
29. M. S. Green, 1952, *J. Chem. Phys.*, **20,** 1281.
30. M. S. Green, 1954, *J. Chem. Phys.*, **22,** 398.
31. R. Kubo, 1958, in *Lectures in Theoretical Physics*, W. E. Britten, L. G. Dunham, Eds., Vol. 1, p. 120, Interscience, New York; R. Kubo, M. Yokota, and S. Nakajima, 1957, *J. Phys. Soc. Japan*, **12,** 1203.
32. R. Zwanzig, 1964, *J. Chem. Phys.*, **40,** 2527.
33. H. Mori, 1958, *Phys. Rev.*, **112,** 1829; J. M. Luttinger, 1964, *ibid.*, **135,** A1505; J. A. McLennan, Jr., 1960, *Phys. Fluids*, **3,** 493.
34. L. Onsager, 1931, *Phys. Rev.*, **37,** 405.
35. L. Onsager, 1931, *Phys. Rev.*, **37,** 2265.
36. A. B. Pippard, 1957, *Elements of Classical Thermodynamics*, Cambridge University Press.
37. D. D. Fitts, 1962, *Nonequilibrium Thermodynamics*, McGraw-Hill, New York.
38. S. R. De Groot and P. Mazur, 1962, *Nonequilibrium Thermodynamics*, North Holland, Amsterdam.
39. P. Curie, 1894, *J. Phys.* (*Paris*), **3,** 393.
40. L. A. Woolf, D. G. Miller, and L. J. Gosting, 1962, *J. Am. Chem. Soc.*, **84,** 317; D. G. Miller, 1969, Chapter 11 in *Transport Phenomena in Fluids*, H. J. M. Hanley, Ed., Dekker, New York.
41. J. G. Kirkwood and D. D. Fitts, 1960, *J. Chem. Phys.*, **33,** 1317; M-K. Ahn, S. J. K. Jensen, and D. Kivelson, 1972, *ibid.*, **57,** 2940.
42. T. L. Hill, 1956, *Statistical Mechanics*, McGraw-Hill, New York.
43. H. Goldstein, 1953, *Classical Mechanics*, Addison-Wesley, Cambridge, Mass.
44. E. Helfand, 1960, *Phys. Rev.*, **119,** 1.
45. P. M. Morse and H. Feshbach, 1953, *Methods of Theoretical Physics*, p. 1807, McGraw-Hill, New York.

46. I. N. Sneddon, 1951, *Fourier Transforms*, McGraw-Hill, New York.
47. A. Einstein, 1905, *Ann. Phys.* (*Leipzig*), **17,** 549; 1906, *ibid.*, **19,** 371.
48. R. W. Zwanzig, 1960, in *Lectures in Theoretical Physics*, W. E. Britten, B. W. Downs, and J. Downs, Eds., Vol. 3, p. 106, Interscience, New York.
49. H. Mori, 1965, *Progr. Theor. Phys.* (*Kyoto*), **33,** 423; 1965, *ibid.*, **34,** 399.
50. A. Z. Akcasu and E. Daniels, 1970, *Phys. Rev.*, A**2,** 962.
51. L. P. Kadanoff and P. C. Martin, 1963, *Ann. Phys.* (*N. Y.*), **24,** 419.
52. C. H. Chung and S. Yip, 1969, *Phys. Rev.*, **182,** 323.
53. G. D. Harp and B. J. Berne, 1970, *Phys. Rev.*, A**2,** 975; B. J. Berne and G. D. Harp, 1970, *Adv. Chem. Phys.*, **17,** 63.
54. L. van Hove, 1953, *Phys. Rev.*, **89,** 1189.
55. P. A. Egelstaff, 1967, *Introduction to the Liquid State*, Academic, New York.
56. K. E. Larsson, U. Dahlborg, and K. Sköld, 1968, in *Simple Dense Fluids*, H. L. Frisch and Z. W. Salsburg, Eds., Academic, New York.
57. D. Levesque and L. Verlet, 1970, *Phys. Rev.*, A**2,** 2514.

Five

Structure of Liquids

The structure of all matter is changing continuously with time because of the motions of the particles making up the particular phase being considered. In the solid state the mean square displacements of atoms and molecules is usually small, and it is an excellent approximation for most studies to assume that the particles are localized at specific positions called lattice sites. At the other extreme of low densities, it is well known that the particles move over such large distances that no lattice sites exist and the term "structure" is seldom used in dilute gas studies. Liquids have densities similar to those found in the solid state but many of their properties are qualitatively similar to those expected of a gas, the most obvious being the lack of shear strength. As densities are high in liquids we can expect some resemblance between the packing of particles in the solid and liquid states, and this has resulted in the use of the term "structure of a liquid." This term has a great deal of meaning since the relative packing of particles in the liquid state shows considerable order. However, the large ratio of the self diffusion coefficients for a system just above and just below its melting point ($\sim 10^5$) together with other differences tell us that the structure of the liquid must be constantly altering and that it is not a good approximation to assume that particles in the liquid state are localized to specific lattice sites. It is particularly important that the difference between "liquid structure" and "solid structure" is understood before the discussion of the structure of liquid water in Chapter 9. A great deal of confusion has been caused in the past by workers who have taken this term too literally and have constructed theories of liquid water (and other liquids) based on mixtures of solidlike structures.

5.1 Experimental Determination of the Structure

As the particles in the liquid are constantly moving over considerable distances, the most appropriate quantity for measuring the static structure is a probability density function $f^{(s)}(\mathbf{r}_1, \ldots \mathbf{r}_s)$ measuring the probability of finding particle 1 at $\mathbf{r}_1$, particle 2 at $\mathbf{r}_2 \ldots$, and particle s at $\mathbf{r}_s$. These quantities are defined in Chapter 3 in terms of the intermolecular potential (Section 3.5) and so in principle we can calculate them. Experimentally, it has not been possible to determine the higher order density functions directly, and we restrict ourselves to measurements and calculations of the two-body radial distribution function

$$g(\mathbf{r}_1, \mathbf{r}_2) = f^{(2)}(\mathbf{r}_1, \mathbf{r}_2)/f^{(1)}(\mathbf{r}_1)f^{(1)}(\mathbf{r}_2) \tag{5.1}$$

As we discuss in Section 3.5, the radial distribution function measures the probability of finding a particle at $\mathbf{r}_2$ given that there is a particle at $\mathbf{r}_1$. For a uniform fluid it can be written as a function of the distance between the particles alone, $g(r)$. The single particle distribution function for a uniform fluid is the average number density, $N/V = \rho$. Experimentally, it is possible to measure the radial distribution function using methods commonly found in solid state studies of structure, primarily X-ray [1] and neutron diffraction [2, 3]. Although studies in both solid and liquid phases measure the intensity of scattering as a function of angle, the analysis of the data is rather different.

Scattering of electromagnetic radiation may occur with or without a change in frequency of the scattered radiation [4]. If there is no change in frequency the scattered radiation has no change in energy and the phenomenon is termed Rayleigh scattering. When there is a change in frequency of the incident and scattered radiation there is a change in energy of the radiation, and this inelastic process is called Raman or Brillouin scattering. In this section we begin by discussing Rayleigh scattering of electromagnetic radiation, primarily X-rays, as this measures the radial distribution function. Later in the section we discuss neutron scattering experiments in which frequency changes occur, and it is shown that such experiments can be used to measure time dependent two-body distribution functions. The theory developed here is general for all electromagnetic wave scattering measurements and can be used, for example, to discuss some aspects of light scattering from liquids.

Consider a beam of electromagnetic radiation with wave vector $\mathbf{k}_0$ entering a system of N particles which scatter weakly so that no photon is deflected more than once [4]. Let the intensity of the scattered radiation with wave vector $\mathbf{k}_s$ be measured and let there be no change in frequency so that $|\mathbf{k}_s| = |\mathbf{k}_0| = 2\pi/\lambda$ where λ is the wavelength of the radiation. Suppose the radiation is incident on two particles at positions $\mathbf{r}_i$ and $\mathbf{r}_j$, then the difference in phase between the two scattered waves is related to the difference in path lengths of the radiation scattered from each particle. If $\mathbf{r}_{ij} = \mathbf{r}_j - \mathbf{r}_i$ is the vector distance between i and j, $s = r_{ij} \cos \theta_0$ and $t = r_{ij} \cos(\pi - \theta_s) = -r_{ij} \cos \theta_s$, where θ_0 is the angle between the vector $\mathbf{k}_0$ and $\mathbf{r}_{ij}$ and θ_s is the angle between $\mathbf{k}_s$ and $\mathbf{r}_{ij}$, the phase difference between the two waves is given by

$$\frac{2\pi}{\lambda}(s+t) = -\mathbf{r}_{ij} \cdot \mathbf{q} \tag{5.2}$$

where $\mathbf{q} = \mathbf{k}_s - \mathbf{k}_0$. Suppose the amplitude of the scattered wave is A, then the magnitude of the electric field vector at some point $\mathbf{r}$ due to scattering from particle j is

$$\frac{A \exp(i\eta_j)}{|\mathbf{r}-\mathbf{r}_j|} = \frac{A \exp(i\eta_i)}{|\mathbf{r}-\mathbf{r}_j|} \exp[-i\mathbf{q} \cdot \mathbf{r}_{ij}] \tag{5.3}$$

where η_j, the phase of the wave scattered from j, is equal to the sum of the phase η_i and the phase difference given in eq. 5.2. The total magnitude of the electric field vector when $|\mathbf{r}|$ is very large is the sum of the result given in eq. 5.3 over all particles j:

$$E_s = \frac{A}{r} \exp(i\eta_i) \exp[i\mathbf{q} \cdot \mathbf{r}_i] \sum_{j=1}^{N} \exp[-i\mathbf{q} \cdot \mathbf{r}_j] \tag{5.4}$$

The intensity of the scattered wave is then given by $I_s = E_s E_s^*/4\pi$ or,

$$I_s = \frac{A^2}{4\pi r^2}\left[N + \sum_{i \neq j} \exp(-i\mathbf{q} \cdot \mathbf{r}_{ij})\right] \tag{5.5}$$

Now the probability of finding a particle at $\mathbf{r}_1$ and a second one at $\mathbf{r}_2$ is given by the two-body density function $f^{(2)}(\mathbf{r}_1, \mathbf{r}_2) = \rho^2 g(r_{12})$ for a uniform fluid where $g(r_{12})$ is the radial distribution function. Consequently, we can multiply the term under the summation by $\rho^2\, g(r_{12})$ and replace the sums of particle pairs by integrals over $\mathbf{r}_1$ and $\mathbf{r}_2$. This gives the required expression for the scattered intensity

$$I_s = \frac{A^2}{4\pi r^2}\left[N + \rho^2 \int\int \exp(-i\mathbf{q} \cdot \mathbf{r}_{12})\, d\mathbf{r}_1\, d\mathbf{r}_2 + \rho^2 \int\int \exp(-i\mathbf{q} \cdot \mathbf{r}_{12})[g(\mathbf{r}_{12}) - 1]\, d\mathbf{r}_1\, d\mathbf{r}_2\right] \tag{5.6}$$

or $I_s = I_m + I_t + I_i$ where I_m is the intensity of radiation scattered by a single particle, I_t is that scattered when $g(r) = 1$ at the surface, and I_i is the intensity measuring interference effects between pairs of particles. Given a large system and ignoring surface scattering we obtain the required expression

$$I_s = I_m\left[1 + \frac{4\pi\rho}{q}\int_0^\infty r \sin(qr)[g(r)-1]\,dr\right] \tag{5.7}$$

where $q = |\mathbf{k}_0 - \mathbf{k}_s| = (4\pi/\lambda)\sin\theta$, the angle θ being half the angle between the incident and scattered wave vectors. The last step is to define the *structure factor,* $S(q)$, as the ratio $I_s(q)/I_m(q)$ and invert the Fourier transform given in eq. 5.7 which leads to

$$g(r) = 1 + \frac{1}{2\pi^2\rho r}\int_0^\infty q \sin(qr)[S(q)-1]\,dq \tag{5.8}$$

Experimentally, the problem has now been resolved into measuring the ratio of the total scattering intensity, I_s, to the molecular scattering intensity, I_m, as a function of scattering angle.

In Section 3.8 it was shown that at low densities the radial distribution function is of the form

$$g(r) = \exp\frac{-\phi(r)}{kT}(1 + g_1(r)\rho + \cdots)$$

where $\phi(r)$ is the pair potential and the coefficient $g_1(r)$ is independent of density. The molecular scattering intensity does not depend on the density of the system, and consequently, at low densities eq. 5.7 is linear in density, of the form

$$I_s(q) = I_m(q) + \rho G_1(q) + \rho^2 G_2(q) + \cdots$$

where G_1, G_2 are coefficients independent of ρ. Consequently, $I_m(q)$ can be determined from scattering measurements made at very low densities. In practice this is not an easy measurement to make; the molecular scattering intensity is calculated quantum mechanically from the molecular electron densities, and $I_m(q)$ is obtained by fitting the calculated result to experiment.

Although the theory developed so far in this chapter is specifically for the scattering of electromagnetic radiation, similar results follow for electron scattering, even though the quantities have a slightly different interpretation. In particular, the electromagnetic radiation is scattered by the electrons around an atom or molecule whereas an electron beam is scattered by the total charge density. Consequently, the molecular scattering intensity I_m is determined by different quantities. A neutron beam is scattered by the nuclei in the system and in this case the single particle scattering $I_m(q)$ may be treated more easily.

The energies of neutron beams used to measure particle distribution functions is such that no intranuclear energy levels are excited [5]. Consequently, we can treat the problem using particle-particle scattering theory rather similar to the atom-atom problem dealt with in Sections 2.4–2.6. Some approximation must be introduced for the neutron-nucleus interaction, and, given the absence of more information, it is usual to assume a hard sphere interaction with radius, or *scattering length*, a, that is much less than the wavelength of the neutrons. In general a nucleus exists in a nonzero spin state, J, and this must be taken into account. Under the conditions of the neutron diffraction experiment only the zero angular momentum, or s wave scattering, contributes and, consequently, the important spin states in the problem are $J-\frac{1}{2}$ and $J+\frac{1}{2}$ carrying weights $J/(2J+1)$ and $(J+1)/(2J+1)$, respectively [5]. It is readily shown that a hard sphere potential leads to a constant collision cross-section $\sigma = 4\pi a^2$, and so it follows that the free atom cross-section is

$$\sigma_f = 4\pi\left(\frac{J+1}{2J+1}\,a_+^2 + \frac{J}{2J+1}\,a_-^2\right) \tag{5.9}$$

where a_+ and a_- are the scattering lengths corresponding to the $J+\frac{1}{2}$ and $J-\frac{1}{2}$ spin states, respectively. We are interested in scattering such that the waves from different nuclei interfere, leading to diffraction effects. For such scattering, called *coherent scattering*, the scattering length is the weighted mean of the two possible scattering lengths

$$a_{\text{coh}} = \frac{J+1}{2J+1}\,a_+ + \frac{J}{2J+1}\,a_-$$

and the coherent scattering cross-section is

$$\sigma_{\text{coh}} = 4\pi a_{\text{coh}}^2 \tag{5.10}$$

The *incoherent scattering cross-section*, given by the difference between the free atom scattering and coherent scattering cross-sections, is

$$\sigma_{\text{inc}} = 4\pi\frac{J(J+1)}{(2J+1)^2}(a_+ - a_-)^2 \tag{5.11}$$

so that

$$a_{\text{inc}} = \frac{\sqrt{J(J+1)}}{2J+1}(a_+ - a_-)$$

We have separated the scattering into two terms, one that arises from the interference of waves from different atoms and another that contains no interference phenomena. The second type of scattering is expected to be isotropic and provides a uniform background to the coherent scattering. Both the coherent and incoherent scattering can undergo Brillouin scattering and, in fact, the incoherent scattering cross-section contains the information necessary to measure the time dependent, single particle distribution function $G_s(r, t)$ referred to at the end of the preceding chapter. Once the uniform background is removed from the neutron scattering measurements the intensity of coherent scattering at a particular angle, $I_{coh}(q)$, gives the radial distribution function from the equation [2, 5]

$$I_{coh}(q) = b_{coh}^2 \left[1 + \frac{4\pi\rho}{q} \int_0^\infty r \sin(qr)[g(r) - 1]\, dr \right] \tag{5.12}$$

where $b_{coh} = (1 + m/M)\, a_{coh}$, M being the mass of the atom and m the mass of a neutron. Comparing this with eq. 5.7 we see that the ratio I_s/I_m is not required for neutrons, although the scattering length a_{coh} has to be determined. The scattering lengths are quantities particular to a given atom, do not depend on the system in which the atom is found, and can be determined separately. It is worth noting that the relative simplicity of neutron scattering is due to the radiation not exciting nuclear energy levels. X-rays excite a large number of intraatomic levels, giving rise to Compton scattering, and it is this phenomenon that is reflected in the function $I_m(q)$ in eq. 5.7.

When we consider particle-particle scattering in Section 2.6 we see that the scattering cross-section can also be measured in terms of energy change during the collision, the so-called differential scattering cross-section $I(q, \theta)$. Similarly, we can measure the differential scattering cross-section for the neutron wave, the energy change being measured by the frequency shift in the scattered wave. From eq. 2.90 we see that the energy of a wave of wavenumber $\mathbf{q}$ is $E = \hbar\omega = \hbar^2 q^2/2m$ where in this case m is the mass of the neutron. If follows that the energy exchange between an incident neutron beam of wave vector $\mathbf{k}_0$ and the fluid is given by $\Delta E = \hbar^2(\mathbf{k}_0^2 - \mathbf{k}_s^2)/2m$. In neutron spectroscopy it is possible to measure two differential intensity functions, the coherent scattering function $I_c(q, \omega)$ and the incoherent scattering function $I_i(q, \omega)$. As with the corresponding quantities for total scattering, we can now define frequency dependent scattering factors so that by analogy with eqs. 5.7 and 5.12 we have [5, 6]

$$I_c(q, \omega) = b_{coh}^2 S_c(q, \omega) \tag{5.13}$$

$$I_i(q, \omega) = b_{inc}^2 S_i(q, \omega) \tag{5.14}$$

The frequency dependent structure factors are themselves related to the time dependent van Hove distribution functions [7], $G_s(r, t)$ and $G_d(r, t)$ introduced in Section 4.5.

In that section we state that one of the simplest time dependent distribution functions, $G(r, t)$, measures the probability of finding a particle at position $\mathbf{r}$ at time t given that there is a particle at the origin at time $t=0$. By arguing that the particle at $\mathbf{r}$ could be either the original particle or one of the other particles in the system we show that the van Hove function can be written for a uniform fluid as

$$G(r, t) = G_s(r, t) + G_d(r, t) \tag{5.15}$$

where $G_s(r, t)$ is the self correlation function and $G_d(r, t)$ is the distinct correlation function. We introduce the radial distribution function in eq. 5.6 by replacing the sum over pairs in eq. 5.5 by an integration over the two-body density function $f^{(2)}(\mathbf{r}_1, \mathbf{r}_2) = \rho^2 g(r)$. If the analysis leading up to these equations is repeated for the inelastic scattering problem [5–7] it is found that the time dependent van Hove distribution function gives the appropriate generalization. In particular, it is found that the $S(q, \omega)$ are given by the time and space transforms

$$S_c(q, \omega) = \iint \exp\left[-i(\mathbf{q}\cdot\mathbf{r} + \omega t)\right] G(r, t)\, d\mathbf{r}\, dt \tag{5.16}$$

$$S_i(q, \omega) = \iint \exp\left[-i(\mathbf{q}\cdot\mathbf{r} + \omega t)\right] G_s(r, t)\, d\mathbf{r}\, dt \tag{5.17}$$

where, as before, $\mathbf{q}$ is the difference between the wave vectors of the scattered and incident beams. The relation between the self-correlation function and the incoherent scattering factor is self-evident in that both functions are related to the properties of a single particle. Perhaps less evident is the conclusion that if energy exchange is allowed between the neutrons and a single particle, it is also possible to get interference effects, and hence coherent scattering between waves scattered from the same particle. Thus in general $S_c(q, \omega)$ is related to the total van Hove function. There are some substances that show only one type of scattering; for example, if the scattering lengths for the $J+\frac{1}{2}$ and $J-\frac{1}{2}$ spin states are equal, that is $a_+ = a_-$ in eq. 5.11, the incoherent scattering cross-section is zero. In fact, most substances show both coherent and incoherent scattering, although the cross-section for hydrogen is almost completely incoherent and that for helium is completely coherent.

5.2 Theoretical Determination of Static Structure

A number of theoretical methods exist for obtaining the radial distribution function predicted by a particular interaction potential. In this section the more important of these are discussed and the following sections are used to compare theoretical and experimental results. It becomes apparent that the radial distribution function is a poor test of any potential function, being rather insensitive to details of the attractive forces. This observation is particularly important as it is the reason for the success of a number of perturbation theories in predicting bulk properties of fluids—a topic dealt with in the next chapter.

For a given interaction potential the most accurate procedures available for calculating distribution functions are the two machine simulation methods described in Section 3.10. The Monte Carlo (MC) method [8] uses the theory of Markov processes to generate a series of distributions of the particles such that the probability of finding a particular distribution is proportional to its Boltzmann weighting factor $\exp[-\Phi(\mathbf{r}_1 \ldots \mathbf{r}_N)/kT]$. The molecular dynamics (MD) method [9] is based on numerical integration of the equations of motion of the system. If it is assumed that the numerical errors in both methods are low, then we can expect good agreement between distribution functions obtained from the ensemble average (MC method) and those obtained from the time averages generated by the MD method. As we see, the agreement is excellent and this is a useful numerical test of the validity of the postulates on which equilibrium statistical mechanics is based. Equation 3.143 gives the connection between the radial distribution function and the average number of particles in a spherical shell of radius r and thickness dr around a given particle

$$\langle N(r, r+dr)\rangle = 4\pi\rho r^2 g(r)\, dr \tag{5.18}$$

and the machine simulation methods use this result to calculate $g(r)$. In the case of the MC method a distribution function that is accurate to a few percent can be obtained from a system of about 100 particles if around 10^6 configurations are generated. Similar accuracy is obtained from the MD procedure if the time average extends over about 5×10^{-11} s.

It is usually less expensive, and less accurate, to use an approximate method to calculate $g(r)$, and the methods most often used are the integral equation approximations. Section 3.8 gives the necessary theory for several of these, including the Born-Green, Kirkwood, Percus-Yevick (PY), and Hypernetted Chain (HNC) approximations [10]. Historically the Born-Green and Kirkwood approximations were the earliest integral equations used to obtain $g(r)$ [11]. However, the results were not particularly good except possibly for liquid metals [12]. More recently the Born-Green-Yvon hierarchy has been truncated at the stage where it relates $g^{(4)}$ to $g^{(3)}$ (see Section 3.8) and the resulting $g^{(3)}$ used to calculate the two-body radial distribution function for hard spheres [13]. There was a substantial increase in accuracy using this method but this was accompanied by a very substantial increase in computer time needed to solve the equations and thus the method is no less expensive than the simulation calculations. The PY and HNC approximations, together with a number of related results, are best considered as closure relations for the Ornstein-Zernike equation. As these methods have been widely used and have proven to be reasonably successful it is worth giving a brief outline of one numerical procedure for solving the equations. It should be recalled that the closure relations are of the form (see eqs. 3.102 and 3.104)

$$\begin{aligned} c(r) &= g(r) - 1 - \log\left[g(r)\exp\frac{\phi(r)}{kT}\right] \qquad \text{(HNC)} \\ c(r) &= g(r)\left[1 - \exp\frac{\phi(r)}{kT}\right] \qquad \text{(PY)} \end{aligned} \tag{5.19}$$

where $c(r)$, the direct correlation function, is introduced using the Ornstein-Zernike equation

$$h(\mathbf{r}) = c(\mathbf{r}) + \rho\int c(\mathbf{s})h(\mathbf{r}-\mathbf{s})\,d\mathbf{s} \tag{5.20}$$

with $h(r) = g(r) - 1$, the total correlation function.

Several numerical methods exist for solving eq. 5.20 together with either the PY or HNC closure; they have been discussed in an article published by one of the authors [10]. The simplest method to discuss is probably that based on the fact that eq. 5.20 contains a convolution integral. Taking the Fourier transform of both sides of the Ornstein-Zernike equation gives [14] (see Section 3.8)

$$\tilde{h}(q) = \tilde{c}(q) + \rho\tilde{c}(q)\tilde{h}(q)$$

which can be solved for $\tilde{h}(q)$ as follows:

$$\tilde{h}(q) = \frac{\tilde{c}(q)}{1 - \rho\tilde{c}(q)} \tag{5.21}$$

The approximate integral equations obtained by combining eqs. 5.19 and 5.20 can be solved by the method of direct iteration. Suppose we begin our solution with an initial guess for the direct correlation function,

$c^{(0)}(r)$, and calculate its three dimensional Fourier transform as follows:

$$\tilde{c}^{(0)}(q) = \frac{4\pi}{q} \int_0^\infty r \sin(qr) c^{(0)}(r)\, dr \tag{5.22}$$

We can obtain the Fourier transform of the total correlation function from eq. 5.21 and then obtain an initial guess for $h(r)$

$$h^{(0)}(r) = \frac{1}{2\pi^2 r} \int_0^\infty q \sin(qr) \tilde{h}^{(0)}(q)\, dq \tag{5.23}$$

Substituting this guess into the appropriate member of eq. 5.19 gives a new guess for the direct correlation function, $c^{(1)}(r)$. Continuing the iterations around the cycle $\rightarrow c^{(i)} \rightarrow \tilde{c}^{(i)} \rightarrow \tilde{h}^{(i)} \rightarrow h^{(i)} \rightarrow c^{(i+1)} \rightarrow$ eventually gives several consecutive iterates that are in good agreement with each other, so that the method has converged. In practice, a number of devices are employed to ensure that the iterations do converge, but these are discussed in the article referred to earlier [10] and are not relevant here. It is possible using any one of several such numerical methods to obtain solutions to the integral equations that are accurate to better than 1%. Although the integral equation approximations give distribution functions much more quickly than do the simulation methods they are not as accurate.

In Section 3.9 a number of perturbation theories were used to obtain expressions for the free energy of a system in terms of the free energy and other properties of a reference system, usually a hard sphere fluid [15–18]. Perturbation theory can also be used to obtain the radial distribution function, the starting point being the expression for $g(r)$ in terms of the average number of particles in a shell of radius r, thickness dr, about a given particle—eq. 5.18. Let $N(r, r+dr)$ in eq. 5.18 be denoted by N_i for the number of particles in the ith shell. It can be shown [15] that if the potential is written as a combination of reference potential and perturbing potential, $\phi(r) = \phi_0(r) + \phi_1(r)$, the zero and first order terms in the expansion of $\langle N_i \rangle$ give

$$\langle N_i \rangle = \langle N_i \rangle_0 - \frac{1}{kT} \sum_j [\langle N_i N_j \rangle_0 - \langle N_i \rangle_0 \langle N_j \rangle_0] \phi(r_j) \tag{5.24}$$

where in terms of the Barker-Henderson theory [16] $\phi(r_j)$ is the potential energy of two particles separated by the distance r_j. Using eq. 5.18 we can obtain an expression for the radial distribution function

$$g(r) = g_0(r) - \frac{1}{2\pi N\rho kTr^2\, dr} \sum_j [\langle N_i N_j \rangle_0 - \langle N_i \rangle_0 \langle N_j \rangle_0] \phi(r_j) \tag{5.25}$$

The covariance term in eqs. 5.24 and 5.25 can be written in terms of the three-body and four-body distribution functions of the reference system. It is difficult to calculate such functions, and Barker and Henderson overcame the problem by making very extensive calculations of the covariance term using Monte Carlo results for their reference system, the hard sphere fluid [17].

Many other theories of the radial distribution function have been proposed but generally they are not sufficiently accurate to warrant detailed consideration here. Methods based on lattice theories of the liquid state have been reviewed by Barker [19], but, although later developments in that field were reasonably successful [20], the methods have fallen into disuse in liquid state physics. Before the advent of simulation methods considerable efforts were made to determine the packing of hard spheres together with their radial distribution function by experimenting with ball bearings and other spherical objects [21]. Although these methods were able to generate some random packing configurations and gave a qualitative understanding of liquid state structure, it was not possible to obtain distribution functions that were quantitatively accurate.

5.3 The Hard Sphere Fluid

Although it is an idealized model of a real liquid, the hard sphere fluid has several attributes that make it worth examining in some detail. The first advantage is that it is probably the simplest three-dimensional system to study, and it is possible to calculate many of its properties relatively easily. As an example Wertheim [22] and Thiele [23], and later Baxter [24], obtained analytic expressions for $g(r)$ and thermodynamic properties of hard spheres in terms of the Percus-Yevick approximation. In addition to its computational simplicity the qualitative agreement between hard sphere radial distribution functions and those of simple fluids, such as the liquefied noble gases, shows that the structure of dense fluids is largely determined by the strongly repulsive short range intermolecular forces. This observation has been exploited in various perturbation theories [15–18], and to a very large extent the success of such theories depends on this qualitative similarity between hard sphere fluids and more realistic systems.

We recalled in the preceding section that the PY approximation can be considered as a closure relation for the Ornstein-Zernike equation. Although details of the solution of the resulting integral equation are too long to reproduce here, it is worth recording the result that leads eventually to the analytic expression for the distribution function. From the PY approximation given in eq. 5.18 it follows that when the interaction potential is zero the direct correlation function is also zero; the radial distribution function measures the probability of finding two particles a distance r apart and vanishes when $\phi(r)$ is infinite. We see that if distances are measured in terms of the position of the first zero in a potential, σ, then the hard sphere potential $g(r)$ is zero in the range $(0, 1)$ and $c(r)$ is zero in the range $(1, \infty)$. Consequently, we can expect (at least intuitively) that solving the PY equation for hard spheres will be easier than solving the equation for a more realistic potential, when neither function will be known. Using this information Wertheim and Thiele independently obtained solutions to the PY equation for hard spheres and showed that the direct correlation function is given by

$$\begin{aligned} c(r) &= -\lambda_1 - \frac{6\lambda_2 \eta r}{\sigma} - \frac{\eta \lambda_1 r^3}{2\sigma^3}, \qquad r \leqslant \sigma \\ &= 0, \qquad r > \sigma \end{aligned} \tag{5.26}$$

where σ is the hard sphere diameter, $\eta = \pi\rho\sigma^3/6$ and

$$\lambda_1 = \frac{(1+2\eta)^2}{(1-\eta)^4}$$

$$\lambda_2 = -\frac{(1+\eta/2)^2}{(1-\eta)^4}$$

The radial distribution function can be obtained by substituting eq. 5.26 into the Ornstein-Zernike equation, in the form of eq. 5.21, and calculating the inverse Fourier transform. Expressions for $g(r)$ in the intervals $(\sigma, 2\sigma)$, $(2\sigma, 3\sigma)$, . . . , and so on can be written down in terms of Laplace transforms [25], but they are lengthy and are not pertinent to subsequent discussion.

It is relatively easy to obtain expressions for the thermodynamic properties of hard spheres in terms of the PY approximation; for completeness we derive them here, leaving detailed discussion of the results until the following chapter. Equation 3.63 gives an expression for the isothermal compressibility of the system

$$kT\left(\frac{\partial \rho}{\partial p}\right)_T = 1 + 4\pi\rho \int_0^\infty r^2 h(r)\, dr \tag{5.27}$$

where $h(r) = g(r) - 1$. If we examine eq. 5.22, which gives the Fourier transform of the function $c(r)$, we see that if $c(r)$ is replaced by $h(r)$ we can write eq. 5.27 in the form

$$kT\left(\frac{\partial \rho}{\partial p}\right)_T = 1 + \rho\tilde{h}(0) \tag{5.28}$$

where we have used the fact that $\lim_{q\to 0} \sin(qr)/q = r$. Using the Ornstein-Zernike equation in its Fourier transform representation, eq. 5.21, we can rewrite eq. 5.28 in terms of the direct correlation function as

$$\frac{1}{kT}\left(\frac{\partial p}{\partial \rho}\right)_T = 1 - \rho\tilde{c}(0) = 1 - 4\pi\rho\int_0^\infty r^2 c(r)\,dr \tag{5.29}$$

Thus we see that the isothermal compressibility can be written in terms of the integral over the direct correlation function. Substituting eq. 5.26 for $c(r)$ and integrating over r and ρ gives an expression for the pressure of the hard sphere fluid

$$\frac{p}{\rho kT} = \frac{1 + \eta + \eta^2}{(1-\eta)^3} \tag{5.30}$$

a result that we call the "compressibility equation" pressure for a reason that becomes obvious shortly.

We have also given an equation for the pressure of a fluid in terms of the derivative of the potential energy (eq. 3.54)

$$\frac{p}{\rho kT} = 1 - \frac{2\pi\rho}{3kT}\int_0^\infty r^3 \frac{d\phi}{dr} g(r)\,dr$$

If we note that for a hard sphere potential $d\phi/dr$ is the Dirac delta function, $-\delta(r-\sigma)$, it follows that

$$\frac{p}{\rho kT} = 1 + \frac{2\pi\rho\sigma^3}{3} g(\sigma) \tag{5.31}$$

We observed in eq. 3.93 that the radial distribution function can be written $g(r) = \exp[-\phi(r)/kT]y(r)$ where $y(r)$ is given as a density expansion, each term of which involves integrals over the function $\exp[-\phi(r)/kT] - 1$. It follows that even if $\phi(r)$ is discontinuous, as it is for the hard sphere fluid, $y(r)$ is continuous. Examining the PY approximation again we see that for a hard sphere fluid

$$c(r) = -y(r), \qquad r \leqslant \sigma$$

and

$$g(r) = y(r), \qquad r \geqslant \sigma \tag{5.32}$$

Consequently, the equation of state given in eq. 5.31 can also be written in terms of $c(\sigma)$. The result is

$$\frac{p}{\rho kT} = \frac{1+2\eta+3\eta^2}{(1-\eta)^2} \tag{5.33}$$

from which we see that the equation of state obtained from the virial equation differs from that obtained using the compressibility equation. This discrepancy occurs for almost all integral equation approximations and is used in the following chapter as one method of discriminating among them.

Radial distribution functions for hard spheres calculated from Monte Carlo studies [17], the PY [25], and the HNC [26] approximations are shown in Fig. 5.1, both at a low and high density. Since $g(r)=0$ for $r<\sigma$, this region has not been included in the figure. At the low density $\rho^* = \rho\sigma^3 = 0.4$, $g(r)$ falls monotonically until about 1.8σ. There is a small second peak at about 2.2σ and thereafter the function is very close to unity. We can associate the large peak with the nearest neighbor distance in the fluid, and it is evident that at this fairly low density the particle distribution rapidly becomes random. Also in Fig. 5.1 are the corresponding distribution functions at a fairly high density, $\rho^* = 0.9$, typical of the number densities found in simple liquids. It can be seen that there is more structure in the fluid at this density, with strong second and third neighbor maxima. The minima in the distribution function represent regions where there is a relative deficiency of particles. It can be seen that near $r=\sigma$ the PY approximation predicts a rather low value of $g(r)$ and the HNC approximation a rather high result. At longer distances the PY approximation gives rather stronger maxima than the Monte Carlo results and it also oscillates slightly out of phase.

A number of workers have used the differences between the PY and HNC approximations to construct new distribution function theories of fluids [27, 28]. We have observed that the PY approximation gives different results for the equation of state of hard spheres from the compressibility and virial equations, and it is seen in the next chapter that similar results are found for the HNC approximation. It was shown by eq. 5.31 that for hard spheres the virial equation of state can be obtained from $g(\sigma)$, and if we examine Fig. 5.1 again it is obvious that the PY virial pressure is lower than the exact result and that the HNC approximation gives a higher result. It is possible to improve considerably the approximation to the equation of state and $g(r)$ if use is made of these observations. We write a new approximation for the radial distribution function [28] of the form

$$g(r) = [1-\alpha(\rho)]g_{\mathrm{PY}}(r) + \alpha(\rho)g_{\mathrm{HNC}}(r) \tag{5.34}$$

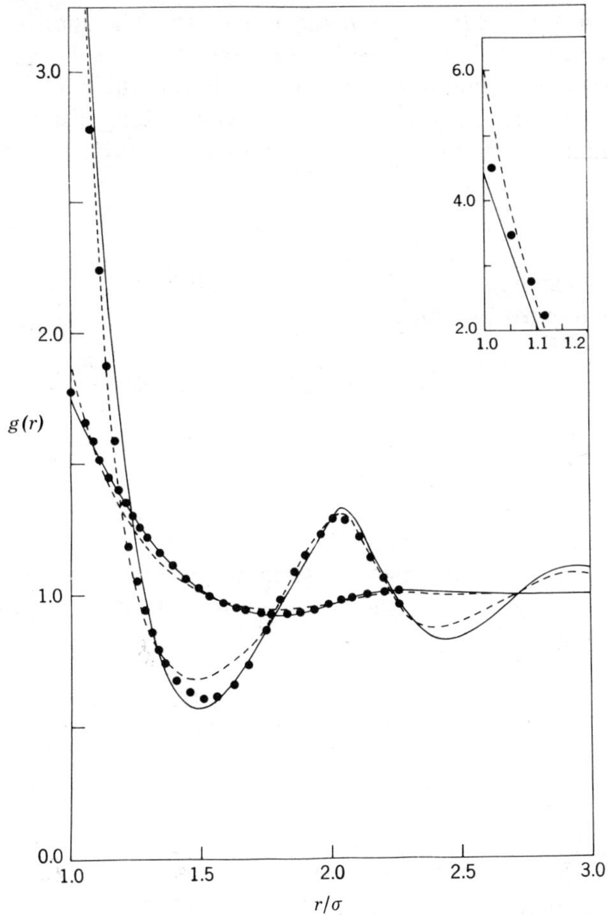

Figure 5.1 Radial distribution functions from the hard sphere fluid. ● simulation results; ——— PY results; ---- HNC results. The curves with the lower intercept at $r=\sigma$ represent $\rho\sigma^3=0.4$ and the others represent $\rho\sigma^3=0.9$.

where $\alpha(\rho)$ is a density dependent function chosen so that at every density the equation of state calculated from the compressibility equation is identical to that calculated from the virial equation. The resulting distribution function is such that the agreement with the exact results is much improved, particularly at lower densities. It is possible to apply these so-called *self-consistent approximations* to other fluids providing the fluid is not below its critical temperature and in the liquid region. As the equation of state is obtained from the isothermal compressibility by integrating over densities at fixed temperature there is no consistent method of integrating through the two-phase region.

Verlet and Weiss [29] have made use of the PY approximation to obtain an empirical fit to the machine simulation results for the hard sphere radial distribution function at all densities in the fluid region. They began by correcting the out-of-phase behavior at long distances, using a PY distribution function for a different density in this region as follows:

$$g(r/\sigma;\rho)=g_p(r/\sigma_p;\rho_p) \tag{5.35}$$

where the diameter σ_p used in the PY distribution function, g_p, was chosen to minimize the integral

$$\int_{r_1}^{3\sigma} |g(r/\sigma;\rho)-g_p(r/\sigma_p;\rho_p)|\, r^2\, dr \tag{5.36}$$

In this expression $g(r/\sigma;\rho)$ is the exact distribution function obtained from machine simulation calculations $\rho_p=\rho(\sigma_p/\sigma)^3$, and $r_1=1.6\sigma$ is chosen large enough for the region close to contact to be ignored. This adjustment fitted the long range behavior of $g(r)$ but did not improve matters in the region of the nearest neighbor peak. At short distances the fit was improved by adding a short range form that was essentially zero by $r=1.6\sigma$ so that

$$g(r/\sigma;\rho)=\begin{cases} 0 & r<\sigma \\ g(r/\sigma;\rho_p)+g_1(r) & r\geqslant\sigma \end{cases} \tag{5.37}$$

where

$$g_1(r)=\frac{A}{r}\exp\left[-\mu(r-\sigma)\right]\cos\mu(r-\sigma) \tag{5.38}$$

The coefficient A was chosen so that the correct contact value of the radial distribution function was obtained, and μ was determined by requiring the fitted distribution function to give the correct isothermal compressibility. Parameters A and μ in eq. 5.38 are given by

$$\frac{A}{\sigma}=\frac{3}{4}\,\frac{\eta^2(1-0.7117\eta-0.114\eta^2)}{(1-\eta)^4}$$

$$\mu\sigma=\frac{24A/\sigma}{\eta g_1(1;\rho_p)}$$

where $\eta=\pi\rho\sigma_p^3/6$. Verlet and Weis found that this method gives hard sphere distribution functions that agree with the exact results to within 1%. An analytic model similar in form to the Verlet and Weiss model has been given by Grundke and Henderson [30] for mixtures of hard spheres, but, as we are considering one-component systems only, it is not reported here. The analytic fit to the hard sphere distribution function is particularly useful in perturbation theory where it can be used to generate rapidly the properties of the hard sphere reference system.

5.4 Structure of the Noble Gas Fluids

Although several more realistic models of the noble gas interactions have been proposed (see Chapter 7), most of the theoretical work on liquid structure has been based on the Lennard-Jones 12-6 potential function (LJ). Consequently, we begin by examining the ability of several approximate theories to predict the radial distribution function of this system. To do this we use the fact that computer simulation methods, both Monte Carlo and molecular dynamics, give distribution functions that contain errors of the order of a few percent. Comparing predictions from approximate theories with these results enables errors occurring from the nature of the various approximations to be examined; we do not have to worry about inaccuracies in the potential function. Moreover, the LJ potential is a good effective pair potential for the noble gases and some qualitative conclusions can be drawn by comparing both approximate theories and machine simulation results for this potential with experimental results for argon.

The results in Fig. 5.2 are obtained from the PY approximation for the LJ fluid [31] and are compared with experimental measurements by Mikolaj and Pings [32] of the radial distribution function of argon. Comparisons are given as a function of both density and temperature around the critical point and this figure shows several qualitative features. The first comes from examining the behavior as a function of density and is the most important. At very low densities the first peak in the distribution function is fairly wide and is also lower than that for the other densities. There is a shallow well at around 1.75σ and a small second maximum at around 2.2σ, with no structure apparent at longer distances. At short distances the strongly repulsive core of a particle effectively excludes all other particles from within a distance of 1.0σ. Beyond this distance, the pronounced first peak can be attributed to the nearest neighbors in the fluid. As we show shortly, these peaks become much stronger in the liquid region and can be associated with the various shells of particles surrounding the central particle. It is also apparent from Fig. 5.2 that $g(r)$ is less sensitive to changes in temperature than to changes in density. This behavior, which is not entirely unexpected, arises from the fact that the structure of liquids is largely determined by the shape and

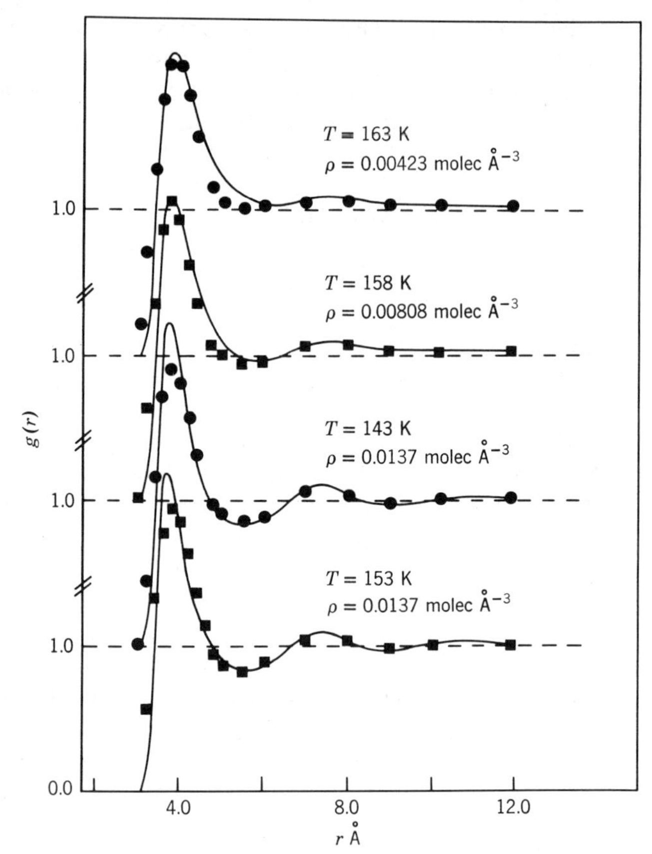

Figure 5.2 Radial distribution function for the LJ fluid at medium densities compared with experimental data for argon. (Adapted with permission from *J. Chem. Phys.*, **50,** 984, 1969.)

nature of the repulsive forces and is responsible for the success of perturbation theories of fluids composed of spherical particles. We can also see from Fig. 5.2 that despite their qualitative similarities there are several differences between the Percus-Yevick results and the experimental ones, particularly at low densities. It is not possible on the basis of this comparison to say if the differences are due to inadequacies in the theory or in the LJ potential used to model the argon interactions. To resolve this problem we need to compare the PY results with those obtained from machine simulations and then compare the latter with experiment.

Figure 5.3 is used to compare several approximations with the results of molecular dynamics calculations for the LJ potential reported by Verlet [33]. The approximations considered in the figure [10] are the PY theory, the HNC theory, and the Mean Spherical Model (MSM) theory

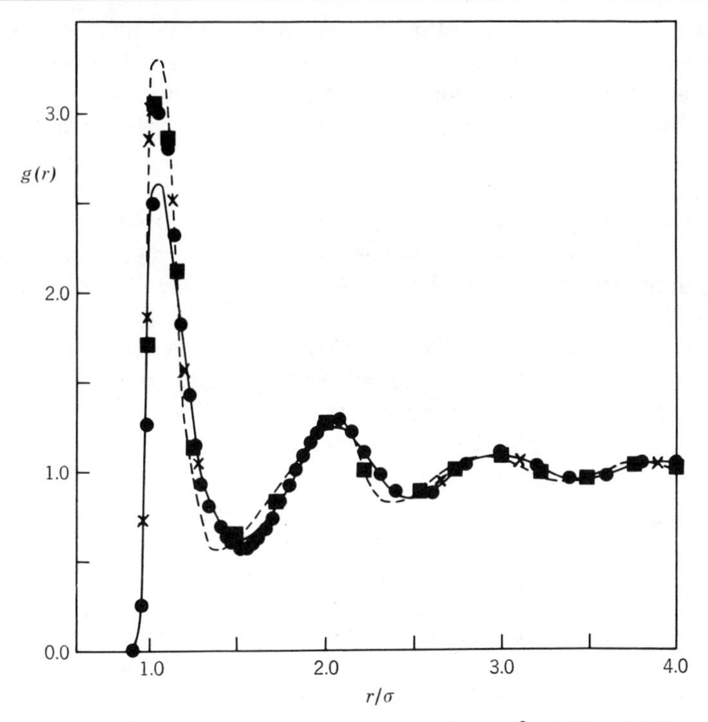

Figure 5.3 Radial distribution function for the LJ fluid at $\rho\sigma^3 = 0.85$ and $kT/\varepsilon = 0.72$. ● simulation results; ---- PY results; —— MSM results; ■ HNC results; × experimental results for argon.

[34, 35], a third integral equation approximation. The temperature and density are given in reduced units $kT/\varepsilon = 0.72$ and $\rho\sigma^3 = 0.85$ and correspond to liquid argon close to the triple point. Also given in the figure are the experimental results of Yarnell et al. [36] who have reported $g(r)$ for argon near the triple point. It is at once obvious that the liquid state results contain much more structure than do those obtained in the region of the critical point. Scaling the distances by σ, as has been done in this diagram, also gives an excellent idea of the meaning we can attribute to the adjective "structure" when applied to a liquid. There are several maxima in $g(r)$, positioned at approximately 1.0σ, 2.0σ, 3.0σ, and 4.0σ in Fig. 5.3. In the preceding paragraph, the first maximum was associated with the nearest neighbor particles in the liquid, found at about one diameter from the central atom, and it is reasonable to associate the other maxima with second, third, and fourth neighbor shells. The figure also shows that these maxima broaden with increasing r. For example, whereas the probability of finding a particle at 1.5σ is very much less than that of

finding a particle at either 1.0σ or 2.0σ, there is little difference between the probabilities of finding particles at 3.5σ and 4.0σ. We can conclude from the qualitative behavior of $g(r)$ that at least for spherical particles there is no long range ordering in the liquid and that the term "structure of a liquid" must be used with care.

The differences between the radial distribution functions shown in Fig. 5.3 are also interesting and should be discussed on the basis of the agreement between each approximation and the molecular dynamics calculations. In the region of the first maximum the PY approximation gives a higher peak than does the exact result and the MSM model gives a lower peak. However, the HNC approximation gives values closer to the molecular dynamics results, being only slightly higher at the first peak and in very good agreement at longer distances. In addition the PY approximation has a sharper first maximum than does the simulation result and it begins to oscillate out-of-phase at the first minimum. The MSM approximation is generally in very good agreement with the exact result, particularly beyond about 1.7σ. On the basis of this comparison, we would conclude that the MSM approximation is a good integral equation approximation to the radial distribution function. However, Watts et al. also examined the thermodynamic properties of the MSM approximation [34] and showed that the errors in the first peak in $g(r)$ are reflected in relatively poor predictions of the bulk data. No detailed study of the HNC approximation in the liquid region has been reported. Finally, it can be seen that the machine simulation results are in excellent agreement with the experimental data. This is a good indication of the usefulness of the LJ potential as an effective pair potential for argon. Other interaction potentials also give good $g(r)$ for the liquid, and in particular the very accurate pair potential proposed by Barker, Fisher, and Watts [37] is excellent [36, 38].

The Barker-Henderson (BH) [16] and Weeks, Chandler and Andersen (WCA) [18] perturbation theories also reproduce the radial distribution function of the LJ fluid with considerable accuracy [39]. It is found that the first order BH result, which is the hard sphere distribution function with the diameter determined by the expression (see eq. 3.120)

$$\sigma_{\mathrm{HS}} = -\int_0^{\sigma_{\mathrm{LJ}}} \left(\exp \frac{-\phi_{\mathrm{LJ}}(r)}{kT} - 1 \right) dr \tag{5.39}$$

is in reasonable agreement with the exact result. This confirms the statement made earlier that the repulsive interactions largely determine the structure of the liquid. The second order BH result is in very good agreement with the molecular dynamics results, and we can expect second order BH perturbation theory to give good results for thermodynamic properties. In the WCA perturbation theory the first order $g(r)$ is given by eq. 3.125

$$g_0(r) = \exp\left\{-\frac{u_0(r)}{kT}\right\} y_{HS}(r)$$

where $y(r)$ is the function $\exp\,[\phi(r)/kT]g(r)$, and where $u_0(r)$ is the reference potential in the WCA theory (eq. 3.121). The WCA perturbation theory is also in excellent agreement with the computer simulation results, and again this theory should give good predictions of the thermodynamic properties.

5.5 Time Dependent Distribution Functions

As we discussed in Section 5.1, the time dependent van Hove distribution functions can be measured experimentally using inelastic neutron scattering methods. Several authors have reviewed experimental results, both for liquid metals [2] and the inert gases [6]. Among the noble gases, the time dependent distribution functions for liquid helium [40] must be considered separately. As we discuss in Chapter 7, liquid state calculations for this substance have to be carried out quantum mechanically and so they fall largely outside the scope of this book. The interested reader is referred to papers by Cohen and Feynman [41] for a description of this work. Fortunately, other simple gases are reasonably well described by classical statistical mechanics and so are of interest here. Experimental results exist for argon and methane [6, 42], and it is found that the agreement of theoretical calculations with this work is good.

The most important method of determining the time dependent distribution functions theoretically is from molecular dynamics calculations [43–45]. Probably the first worker to use this method to make a detailed study of the van Hove functions was Rahman [43], who based his analysis in part on the Brownian motion theory of $G_s(r, t)$. If the velocity coordinates are integrated from eq. 4.77 it is found, after some algebra, that the approximate expression for $G_s(r, t)$ is

$$G_s(r, t) = \left(\frac{1}{4\pi Dt}\right)^{3/2} \exp -\frac{r^2}{4Dt} \tag{5.40}$$

where D, the self-diffusion coefficient, can be introduced into the equation by using the Einstein relation, eq. 4.84, to eliminate the friction constant. Rahman found that if the value of D calculated from his molecular dynamics study was used in eq. 5.40, he obtained a reasonable representation of the molecular dynamics result for $G_s(r, t)$. He went on to examine the departure from Gaussian shape by expanding the van Hove function in terms of the Hermite polynomials [46], $H_{2n}(r)$. The coefficients in this expansion are related to the even moments of the displacement of the particle from its initial position. Rahman showed that a reasonable representation of $G_s(r, t)$ could be obtained using the first few Hermite polynomials. Fisher [38] obtained similar results to Rahman in a study of the Barker-Bobetic potential for argon [47], and his results are compared with the Gaussian approximation of eq. 5.40 in Fig. 5.4. Fisher's values for the Gaussian approximation were obtained using his estimate of the diffusion coefficient [48] from the equation (eq. 4.58)

$$D = \lim_{t\to\infty} \frac{\langle[\mathbf{r}_1(t) - \mathbf{r}_1(0)]^2\rangle}{6t}$$

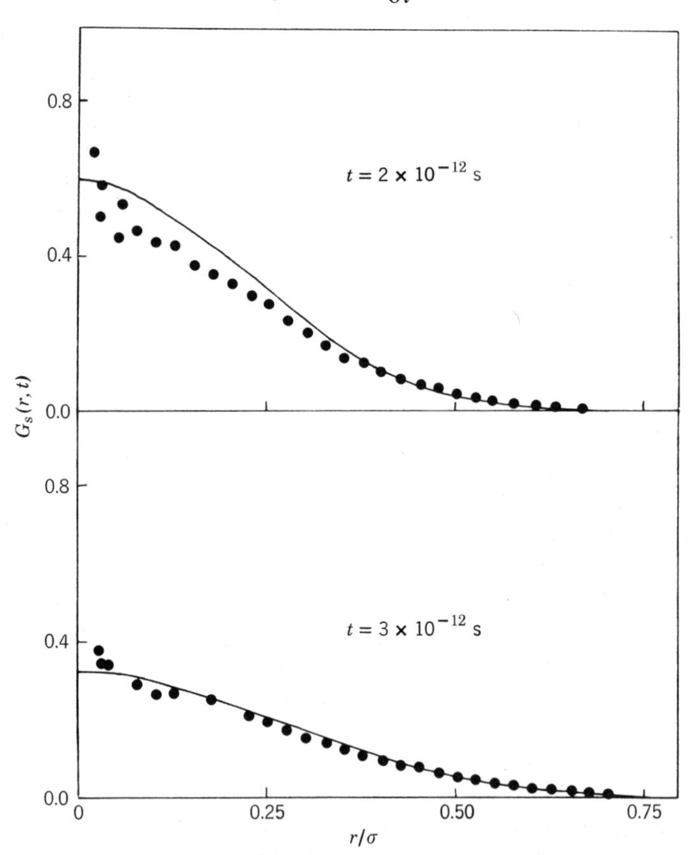

Figure 5.4 $G_s(r, t)$ for the Barker-Bobetic potential at $T = 94$ K, $\rho = 0.0223$ molec Å^{-3} from a simulation study (· · · ·) compared with the Gaussian approximation (———).

If D in eq. 5.40 is replaced by the time dependent function

$$D(t)=\frac{\langle[\mathbf{r}_1(t)-\mathbf{r}_1(0)]^2\rangle}{6t}$$

as was done by Rahman [43], the agreement between the Gaussian approximation and the simulation result is improved.

We state in Section 4.5 that in terms of the generalized Langevin equation [49] the frequency dependent incoherent structure factor $S_i(q, \omega)$ is given by a memory function, so that

$$S_i(q, \omega)=\frac{1}{\tilde{M}_i(q, \omega)+i\omega} \tag{5.41}$$

Although the exact form of the memory function is unknown, various approximations have been suggested for the spatially transformed memory function $\tilde{M}_i(q, t)$, one of which is the exponential approximation [50]

$$\tilde{M}_i(q, t)=\frac{kT}{m}q^2 \exp -\frac{t}{\tau_i(q)} \tag{5.42}$$

where $\tau_i(q)$ is a function such that $\tau_i(0)=mD/kT$. This approximation was considered in detail by Levesque and Verlet [45] who showed that it is only successful if $\tau_i(q)$ is treated as an empirically determined function. Levesque and Verlet also considered various approximations to the memory function for the velocity autocorrelation function, and they showed that these approximations could also be used to estimate $M_i(r, t)$. We return to these approximations shortly.

The van Hove function $G_d(r, t)$ is related to the radial distribution function and in particular

$$\rho g(r)=\lim_{t\to 0} G_d(r, t) \tag{5.43}$$

It would be useful to obtain an expression for $G_d(r, t)$ in terms of $g(r)$, and one approximation along these lines has been proposed by Vineyard [51]. He suggested that if the function $H(r, t)$ was defined by the equation

$$G_d(r, t)=\int d\mathbf{s}\, g(\mathbf{s})H(\mathbf{r}-\mathbf{s}, t) \tag{5.44}$$

a useful first approximation would be $H=G_s$. Rahman [43] showed that this approximation gives too rapid a decay of $G_d(r, t)$ and went on to suggest that if the convolution is delayed, by considering $H(r, t')$, $t'<t$ in eq. 5.44, the agreement can be improved. Others workers have proposed theories relating $G_d(r, t)$, and $G_s(r, t)$, to $g(r)$ with reasonable success

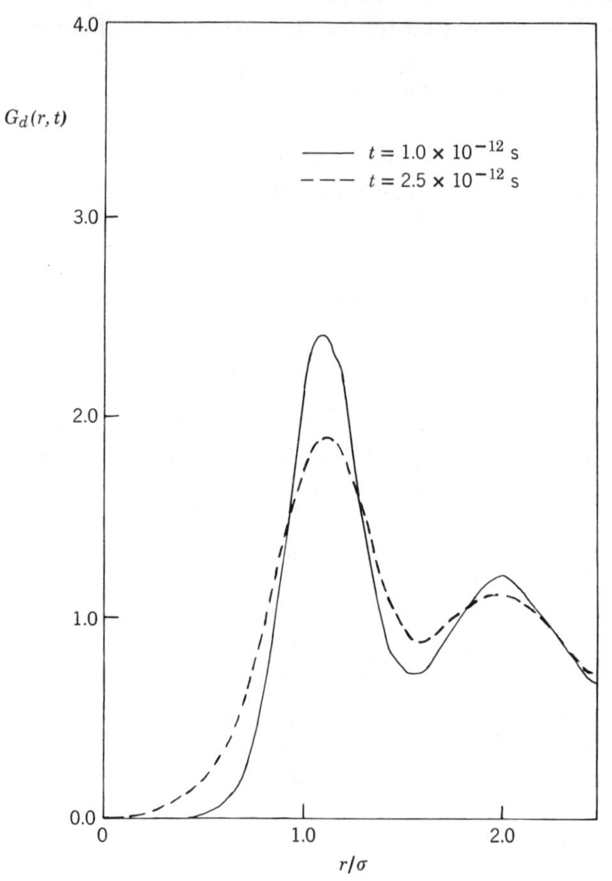

Figure 5.5 $G_d(r, t)$ for the Barker-Bobetic potential from a simulation study. Conditions are as for Fig. 5.4.

[52]. The behavior of $G_d(r, t)$ at various times is shown in Fig. 5.5 for the Barker-Bobetic potential and is taken from the work of Fisher [38]. It can be seen that as t increases the initial form $G_d(r, 0) = \rho g(r)$ relaxes, the various peaks becoming less pronounced. At large times $G_d(r, t)$ reaches its limiting value of ρ.

The third time distribution function with which we are concerned is the velocity autocorrelation function (eq. 4.51)

$$C(t) = \tfrac{1}{3}\langle \mathbf{v}_1(0) \cdot \mathbf{v}_1(t) \rangle \tag{5.45}$$

In Section 4.3 we showed that Brownian motion theory gave an expression for $C(t)$ in the form of a negative exponential. If we examine this expression (eq. 4.86) and replace the friction coefficient by the self-diffusion coefficient using the Einstein relation (eq. 4.84), we find the approximation

$$C(t) = \frac{kT}{m} \exp - \frac{kTt}{mD} \tag{5.46}$$

Figure 5.6 compares this approximation to $C(t)$ with the molecular dynamics results for argon of Fisher and Watts [48]. Calculations are given at two densities and temperatures, one in the liquid region and the other for the dense fluid near the critical point. We see that the negative exponential is a reasonably good approximation at the lower density but that it is poor in the liquid region. Of particular interest is the negative region in $C(t)$ at the liquid density as this is a qualitative feature not evident in the Brownian motion approximation. The negative region can be interpreted in terms of a cagelike structure in the dense fluid. At short times a given particle is moving in a "cage" formed by its neighbors. After a time that can be roughly correlated with the time between collisions—a concept that is not really valid at high densities—the particle collides with other particles at the end of the "cage" and its velocity is reversed. Thereafter there appears to be a weakly negative tail in the velocity autocorrelation function. We discuss the long time behavior of $C(t)$ shortly, but before doing so we consider the memory function for this correlation function.

We showed in Section 4.4 that the power spectrum associated with $C(t)$ is given by

$$\tilde{C}(\omega) = \frac{kT}{m[\tilde{M}(\omega) + i\omega]} \tag{5.47}$$

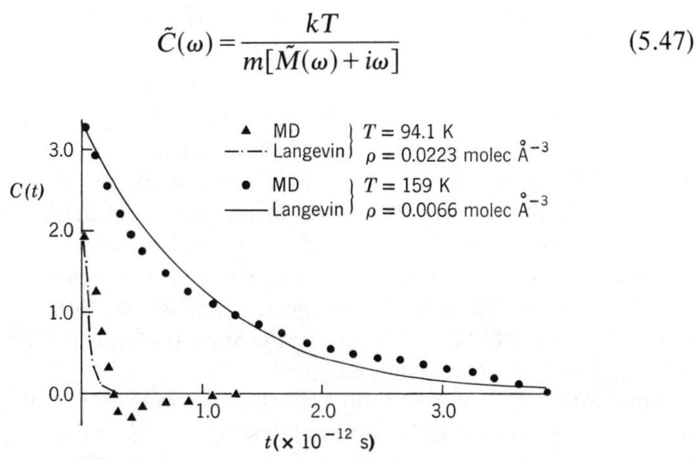

Figure 5.6 Velocity autocorrelation function in $10^4\ m^2\ s^{-2}$ for the Barker-Bobetic fluid. Molecular dynamics results are compared with the Langevin approximation at a liquid density and at a density close to the critical point.

with $\tilde{M}(\omega) = kT/mD$ in the Langevin approximation. This memory function can be related to that introduced for the van Hove function $G_s(r, t)$ in eq. 5.41. If the self current-current correlation function is defined through

$$\tilde{C}_s(q, t) = -\frac{\partial^2}{\partial t^2} F_s(q, t) \tag{5.48}$$

where F_s is the spatial Fourier transform of $G_s(r, t)$, then as Levesque and Verlet point out [45]

$$\lim_{q \to 0} C_s(q, t) = C(t) \tag{5.49}$$

A memory function can now be introduced for $\tilde{C}_s(q, \omega)$, and obviously this relates $\tilde{M}_i(q, \omega)$ in eq. 5.41 and $\tilde{M}(\omega)$ in eq. 5.47 through eqs. 5.48 and 5.49. Levesque and Verlet [45] explored this relation in some detail and used it to suggest approximations for both $\tilde{M}(\omega)$, and hence $\tilde{M}_i(q, \omega)$. They found that it was possible to construct semiempirical models that reproduced their machine calculations for the LJ potential.

When discussing the form of the velocity autocorrelation function in Fig. 5.6 we mention that $C(t)$ for the liquid has a weakly negative tail. Alder and Wainwright [53] have made a detailed study of this function at long times, both for hard spheres and hard discs, and have shown that $C(t)$ does not decay exponentially as suggested by the Langevin approximation. Rather, it appears that at very long times $C(t)$ for hard spheres has a significant *positive* tail for all densities that decays as $t^{-3/2}$. The corresponding function for hard discs decays as t^{-1}. Alder and Wainwright related this behavior to the existence of a so-called vortex motion of particles about a given particle. If the velocities of all the particles in a fluid are measured with respect to a given particle, so that $\mathbf{v}_i \to \mathbf{v}_i - \mathbf{v}_1$ if $\mathbf{v}_1$ is the velocity of the given particle, hydrodynamic flow lines can be observed. In the immediate vicinity of particle 1 the neighboring particles tend to be moving in the direction $\mathbf{v}_1$, whereas at longer distances this is not so. The result is a vortex about particle 1 which moves in such a direction that it tends to encourage this particle to diffuse. Thus, as the self-diffusion coefficient is given by the time integral of $C(t)$ (see Section 4.2) the result is that D is enhanced.

Since Alder and Wainwright reported the long time tail, a number of workers have given a semiempirical description of its nature [54]. Most of these studies have been based either on extending hydrodynamic descriptions of fluid flow into the microscopic region or on extending kinetic theory treatments of dilute gases. The various theories give an analytic form for the velocity autocorrelation function at large times

$$C(t) = \frac{2kT}{\rho m} \frac{1}{[4\pi(\eta/\rho m + D)t]^{3/2}} \tag{5.50}$$

where η is the shear viscosity and ρ the number density. This equation can be used to correct calculated self-diffusion coefficients [55] and is discussed at greater length in the following chapter. It is of interest to note that there is now experimental evidence that seems to support the existence of the long time tail in real fluids [56].

A limited amount of work has been reported on the autocorrelation functions for shear and bulk viscosity and thermal conductivity (discussed in Section 4.2). Alder et al. [57] have examined the correlation functions for shear and bulk viscosity for a fluid of hard spheres but do not discuss the long time behavior of these functions. They showed that both functions departed significantly from the exponential decay predicted by the Enskog theory [58]. Levesque et al. [59] examined these functions in detail for the LJ potential and showed that a long time tail exists. They used the hydrodynamic theory of Zwanzig and Bixon [54] to show that the long time tail in the viscosity autocorrelation function offers an explanation of the negative region in the liquid state velocity autocorrelation function. Levesque et al. also reported values for the shear and bulk viscosity and thermal conductivity; these are discussed in the following chapter.

References

1. C. J. Pings, 1968, Chapter 10 in *Physics of Simple Liquids*, H. N. V. Temperley, J. S. Rowlinson, and G. S. Rushbrooke, Eds., North Holland, Amsterdam.
2. J. E. Enderby, 1968, Chapter 14 in *Physics of Simple Liquids*, H. N. V. Temperley, J. S. Rowlinson, and G. S. Rushbrooke, Eds., North Holland, Amsterdam.
3. P. A. Egelstaff, Ed., 1965, *Thermal Neutron Scattering*, Academic, London.
4. J. O. Hirschfelder, C. F. Curtiss, and R. B. Bird, 1954, Chapter 12 in *Molecular Theory of Gases and Liquids*, Wiley, New York.
5. P. A. Egelstaff, 1967, *An Introduction to the Liquid State*, Academic, London.
6. K. E. Larsson, U. Dahlborg, and K. Sköld, 1968, in *Simple Dense Fluids*, H. L. Frisch and Z. W. Salsburg, Eds., Academic, New York.
7. L. van Hove, 1953, *Phys. Rev.*, **89,** 1189.
8. I. R. McDonald and K. Singer, 1970, *Quart. Revs.*, **24,** 238; W. W. Wood, 1968, Chapter 5 in *Physics of Simple Liquids*, H. N. V. Temperley, J. S. Rowlinson, and G. S. Rushbrooke, Eds., North Holland, Amsterdam; N. Metropolis, A. W. Rosenbluth, M. N. Rosenbluth, E. Teller, and A. H. Teller, 1953, *J. Chem. Phys.*, **21,** 1087.
9. T. E. Wainwright and B. J. Alder, 1958, *Nuovo Cim. Suppl.*, **9,** 117; B. J. Alder and T. E. Wainwright, 1960, *J. Chem. Phys.*, **33,** 1439; 1959, *ibid.*, **31,** 459, B. J. Alder and W. G. Hoover, Chapter 4 in *Physics of Simple Liquids*, H. N. V. Temperley, J. S. Rowlinson, and G. S. Rushbrooke, Eds., North Holland, Amsterdam.

10. R. O. Watts, 1973, "Integral Equation Approximations in the Theory of Fluids," Chapter 1 in *Specialist Periodical Reports of the Chemical Society: Statistical Mechanics*, K. Singer, Ed., Vol. 1, Chemical Society, London.
11. J. G. Kirkwood, V. A. Lewinson, and B. J. Alder, 1952, *J. Chem. Phys.*, **20,** 929.
12. J. E. Enderby, 1972, Chapter 14 in *Liquid Metals Chemistry and Physics*, S. Z. Beer, Ed., Dekker, New York.
13. Y. T. Lee, F. H. Ree, and T. Ree, 1968, *J. Chem. Phys.*, **48,** 3506.
14. I. N. Sneddon, 1951, *Fourier Transforms*, McGraw-Hill, New York.
15. W. R. Smith, 1973, Chapter 2 in *Specialist Periodical Reports of the Chemical Society: Statistical Mechanics*, K. Singer, Ed., Vol. 1, Chemical Society, London; G. A. Mansoori and F. B. Canfield, 1970, *Ind. Eng. Chem. Fundam.*, **62,** 12.
16. J. A. Barker and D. Henderson, 1967, *J. Chem. Phys.*, **47,** 2856; 1967, *ibid.*, **47,** 4714.
17. J. A. Barker and D. Henderson, 1971, *Mol. Phys.*, **21,** 187.
18. J. D. Weeks, D. Chandler, and H. C. Andersen, 1971, *J. Chem. Phys.*, **54,** 5237; 1971, *ibid.*, **55,** 5422; H. C. Andersen, D. Chandler, and J. D. Weeks, 1971, *Phys. Rev.*, A**4,** 1597; 1972, *J. Chem. Phys.*, **56,** 3812.
19. J. A. Barker, 1963, *Lattice Theories of the Liquid State*, Pergamon, London.
20. J. A. Barker, 1962, *J. Chem. Phys.*, **37,** 1061.
21. J. D. Bernal and S. V. King, 1968, Chapter 6 in *Physics of Simple Liquids*, H. N. V. Temperley, J. S. Rowlinson, and G. S. Rushbrooke, Eds., North Holland, Amsterdam.
22. M. S. Wertheim, 1963, *Phys. Rev. Lett.*, **10,** 321; 1964, *J. Math. Phys.*, **5,** 643.
23. E. Thiele, 1963, *J. Chem. Phys.*, **39,** 474.
24. R. J. Baxter, 1967, *Phys. Rev.*, **154,** 170.
25. G. J. Throop and R. J. Bearman, 1965, *J. Chem. Phys.*, **42,** 2408.
26. M. Klein, 1963, *J. Chem. Phys.*, **39,** 1388.
27. C. Hurst, 1965, *Proc. Phys. Soc. (London)*, **86,** 193; D. Henderson, 1966, *ibid.*, **87,** 592; G. J. Throop and R. J. Bearman, 1966, *ibid.*, **88,** 539; R. O. Watts and D. Henderson, 1969, *Mol. Phys.*, **16,** 217; D. Henderson and R. O. Watts, 1970, *ibid.*, **18,** 429; G. Stell, 1969, *ibid.*, **16,** 209.
28. D. D. Carley and F. Lado, 1965, *Phys. Rev.*, **137,** A42; J. S. Rowlinson, 1965, *Mol. Phys.*, **9,** 217.
29. L. Verlet and J. J. Weis, 1972, *Phys. Rev.*, A**5,** 939.
30. E. W. Grundke and D. Henderson, 1972, *Mol. Phys.*, **24,** 269.
31. R. O. Watts, 1969, *J. Chem. Phys.*, **50,** 984; 1969, *Can. J. Phys.*, **47,** 2709.
32. P. G. Mikolaj and C. J. Pings, 1967, *J. Chem. Phys.*, **46,** 1401.
33. L. Verlet, 1967, *Phys. Rev.*, **159,** 98; 1968, *ibid.*, **165,** 201.
34. R. O. Watts, D. Henderson, and J. A. Barker, 1972, *J. Chem. Phys.*, **57,** 5391.
35. J. L. Lebowitz and J. K. Percus, 1966, *Phys. Rev.*, **144,** 251.
36. J. L. Yarnell, M. J. Katz, R. G. Wenzel, and S. H. Koenig, 1973, *Phys. Rev.*, A**7,** 2130.
37. J. A. Barker, R. A. Fisher, and R. O. Watts, 1971, *Mol. Phys.*, **21,** 657.
38. R. A. Fisher, 1972, Ph.D. Thesis, Australian National University.
39. J. A. Barker and D. Henderson, 1972, *Ann. Rev. Phys. Chem.*, **23,** 439.
40. F. R. Dintzis and R. S. Stein, 1964, *J. Chem. Phys.*, **40,** 1459; R. C. C. Leite, R. S. Moore, and S. P. S. Porto, 1964, *ibid.*, **40,** 3741; G. Dezelic, 1966, *ibid.*, **45,** 186.
41. R. P. Feynman and M. Cohen, 1956, *Phys. Rev.*, **102,** 1189; R. P. Feynman, 1954, *ibid.*, **94,** 262; J. Wilks, 1967, *The Properties of Liquid and Solid Helium*, Clarendon, Oxford.
42. B. A. Dasannacharya and K. P. Rao, 1965, *Phys. Rev.*, **137,** A417; N. Kroó, G. Borgonovi, K. Sköld, and K. E. Larsson, 1965, *Proceedings of the IAEA Symposium*

on Inelastic Scattering Neutrons Solids Liquids, Vienna, Vol. 2, p. 101, International Atomic Energy Agency, Vienna; K. E. Larsson, U. Dahlborg, and S. Holmryd, 1960, *Archiv. Fysik.*, **17,** 369; J. A. Janik, J. M. Janik, J. Mellor, and H. Palevsky, 1965, *J. Phys. Chem. Solids*, **25,** 1091; S. Hautecler and H. Stiller, 1962, *Z. Phys.*, **166,** 393; B. A. Dasannacharya and G. Venkataraman, 1967, *Phys. Rev.*, **156,** 196.

43. A. Rahman, 1964, *Phys. Rev.*, **136,** *A*405.
44. A. Rahman, 1966, *J. Chem. Phys.*, **45,** 2585.
45. D. Levesque and L. Verlet, 1970, *Phys. Rev.*, *A***2,** 2514.
46. M. Abramowitz and I. A. Stegun, 1965, *Handbook of Mathematical Functions*, Dover, New York.
47. M. V. Bobetic and J. A. Barker, 1970, *Phys. Rev.*, *B***2,** 4169.
48. R. A. Fisher and R. O. Watts, 1972, *Aust. J. Phys.*, **25,** 529.
49. R. W. Zwanzig, 1960, in *Lectures in Theoretical Physics*, W. E. Britten, B. W. Downs, and J. Downs, Eds., Vol. 3, p. 106, Interscience, New York; H. Mori, 1965, *Progr. Theor. Phys. (Kyoto)*, **33,** 423; 1965, *ibid.*, **34,** 399; A. Z. Akcasu and E. Daniels, 1970, *Phys. Rev.*, *A***2,** 962.
50. C. H. Chung and S. Yip, 1969, *Phys. Rev.*, **182,** 323.
51. G. H. Vineyard, 1958, *Phys. Rev.*, **110,** 999.
52. M. I. Barker and T. Gaskell, 1972, *J. Phys., C (London)*, **5,** 353; F. Lado, 1972, *Phys. Rev.*, *A***5,** 2238.
53. B. J. Alder and T. E. Wainwright, 1967, *Phys. Rev. Lett.*, **18,** 988; 1968, *J. Phys. Soc. Japan, Suppl.*, **26,** 267; 1970, *Phys. Rev.*, *A***1,** 18; T. E. Wainwright, B. J. Alder, and D. M. Gass, 1971, *ibid.*, *A***4,** 233.
54. See, for example, the review articles by Alder and by Deutsch in *Transport Phenomena*, 1973, J. Kestin, Ed., American Institute of Physics; M. H. Ernst, E. H. Hauge, and J. M. S. van Leeuwen, 1970, *Phys. Rev. Lett.* **25,** 1254; J. R. Dorfman and E. G. D. Cohen, *ibid.*, **25,** 1257; R. Zwanzig and M. Bixon, 1970, *Phys. Rev.*, *A***2,** 2005.
55. H. J. M. Hanley and R. O. Watts, 1975, *Mol. Phys.*, **29,** 1907; *Physica*, **79***A*, 351.
56. Y. W. Kim and J. E. Matta, 1973, *Phys. Rev. Lett.*, **31,** 208.
57. B. J. Alder, D. M. Gass, and T. E. Wainwright, 1970, *J. Chem. Phys.*, **53,** 3813.
58. D. Enskog, 1922, *Arkiv. Mat. Astron. Fys.*, **16,** 16.
59. D. Levesque, L. Verlet, and J. Kürkijarvi, 1973, *Phys. Rev.*, *A***7**, 1690.

Six

Bulk Properties of Model Fluids

Before discussing the intermolecular potentials for real liquids, we consider in this chapter the thermodynamic and transport properties of fluids generated by model potentials. These fluids are hypothetical systems whose molecules interact through specific mathematical forms. For definiteness we focus primarily on the much studied hard sphere and Lennard-Jones 12-6 potentials (LJ) [1], although on occasions we refer to results for the more recent m-6-8 potential [2]. The LJ and m-6-8 models are two- and four-parameter analytic functions, respectively, and represent a compromise between very simple models, such as the hard sphere and square well potentials, and the more realistic multiparameter potentials which we discuss in subsequent chapters. It is probably fair to say that the LJ potential has received more attention than any other semirealistic model, primarily because of its relative simplicity and the fact that it represents a good effective pair potential for a wide variety of substances in the gaseous, liquid, and solid states. As well as giving a good account of experimental data, the LJ fluid has many qualitative features found in real systems. Examples are a critical temperature, above which there is only one fluid phase and below which two fluid phases can coexist, and a solid-liquid phase transition. These properties are enjoyed by the m-6-8 fluid as well, and with its additional flexibility this potential is probably a better effective pair potential to use as a model for a real fluid. Many of the bulk properties discussed here have also been obtained for the square well [3] and other model potentials [4]. Most of these models behave qualitatively in a way similar to the LJ fluid and we do not consider them further.

The hard sphere fluid has been studied extensively, partly because it is relatively easy to obtain results for its bulk and structural properties using approximate theories, and partly because it is the reference system in a number of successful perturbation theories [5–19]. Also, at high temperatures the bulk properties of real fluids behave rather like those of a hard sphere fluid. For these reasons we include a discussion of hard sphere fluid bulk properties in this chapter. It is important to note, however, that although the hard sphere system shows a fluid-solid transition at high densities it does not have the critical temperature characteristic of real substances below which the fluid phase exists both as a gas and as a liquid.

6.1 The Law of Corresponding States

As noted in Section 2.3 the LJ potential can be written as

$$\phi(r) = 4\varepsilon\left[\left(\frac{\sigma}{r}\right)^{12} - \left(\frac{\sigma}{r}\right)^{6}\right] \tag{6.1}$$

where ε is the maximum depth of the potential and σ is the position of the zero. An alternative formulation of this model that is sometimes given is

$$\phi(r) = \varepsilon\left[\left(\frac{r_m}{r}\right)^{12} - 2\left(\frac{r_m}{r}\right)^{6}\right]$$

where $r_m/\sigma = 2^{1/6}$ is the position of the minimum in the potential. Although we do not use the potential in this form, it is important to note that the three parameters (ε, σ, r_m) give a useful characterization of any atom-atom potential.

Equation 6.1 suggests that it could be worthwhile to consider the potential in a dimensionless form

$$\phi^*(r^*) = 4[r^{*-12} - r^{*-6}] \tag{6.2}$$

where $\phi^*(r) = \phi(r)/\varepsilon$ and $r^* = r/\sigma$. Suppose we do this, then a plot of $\phi^*(r^*)$ against r^* will be independent of any values of ε and σ we may care to associate with a particular substance. That is, if the interaction potential between two atoms (or molecules) is accurately described by the LJ model, then that interaction can be written in terms of eq. 6.2. Suppose there exists a whole class of substances that is described by the LJ potential, each member of the class having different values for ε and σ. Then if the several interactions are written in the form of eq. 6.2 they will coincide, and the LJ model will be a universal interaction potential for that class. This reduction of the potential to a universal form is a special case of the Law of Corresponding States [20].

This law can be extended in two ways, first to any other two-parameter model of a system and second to the thermodynamic properties. We consider the extension to thermodynamic properties by examining the second virial coefficient for the LJ potential. From eq. 3.77 we have

$$B_2 = -\tfrac{1}{2}\int f(r)\, d\mathbf{r}$$

$$= -2\pi \int_0^\infty r^2 \left[\exp \frac{-\phi_{\text{LJ}}(r)}{kT} - 1 \right] dr \tag{6.3}$$

Now, suppose we define a reduced temperature, $T^* = kT/\varepsilon$, noting that kT has dimensions of energy, and a reduced volume, $V^* = V/\sigma^3$. Then as B_2 has dimensions of volume we can put $B_2^* = B_2/\sigma^3$ and by making these substitutions together with the change of variable $r^* = r/\sigma$ write eq. 6.3 as

$$B_2^* \sigma^3 = -2\pi\sigma^3 \int_0^\infty r^{*2} \left[\exp \frac{-\phi_{\text{LJ}}^*(r^*)}{T^*} - 1 \right] dr^*$$

or

$$B_2^* = -2\pi \int_0^\infty r^{*2} f_{\text{LJ}}(r^*)\, dr^* \tag{6.4}$$

From this equation we see that should a given class of substances be described by the LJ potential, then if their second virial coefficients are divided by σ^3, and the temperature at which B_2 is measured is multiplied by k/ε, the measurements should coincide. Similarly, other thermodynamic properties may be written in reduced units. A good second example is the equation for the internal energy, eq. 3.45

$$\frac{U}{NkT} = \frac{3}{2} + \frac{2\pi\rho}{kT} \int_0^\infty r^2 \phi(r) g(r)\, dr$$

where $g(r)$ is the radial distribution function. Writing $U^* = U/\varepsilon$ and $\rho^* = \rho\sigma^3 (= N/V^*)$, and other reduced units as before, this equation can be written as

$$\frac{U^*}{NT^*} = \frac{3}{2} + \frac{2\pi\rho^*}{T^*} \int_0^\infty r^{*2} \phi_{\text{LJ}}^*(r^*) g(r^*)\, dr^* \tag{6.5}$$

It is not difficult to extend the concept of reduced units to other thermodynamic properties, and in particular the pressure can be written as $p = p^* \varepsilon/\sigma^3$.

To see that the Law of Corresponding States is not specific to the Lennard-Jones potential we could repeat the above analysis for any model potential that can be written in the form $\phi(r) = \varepsilon\phi^*(r/\sigma)$. However, it is more useful to extend the concept further by considering the well known van der Waals theory. According to this theory, the equation of state of a fluid can be written in the form

$$\left(p+\frac{N^2a}{V^2}\right)(V-Nb)=NkT \tag{6.6}$$

where a and b are parameters representing the effect of attractive forces and the finite volume of the molecules, respectively (see Section 3.9). Note that although this model says very little about the pair potential it is a two parameter model of the system. It is well known that this model has a critical point given by [21]

$$p_c=\frac{a}{27b^2}, \qquad V_c=3Nb, \qquad T_c=\frac{8a}{27kb} \tag{6.7}$$

Thus we can define reduced variables $p^*=p/p_c$, $V^*=V/V_c$ and $T^*=T/T_c$, so that the critical point, in reduced units, is $p_c^*=1$, $V_c^*=1$, $T_c^*=1$. Once again, if a class of substances is described by the van der Waals theory, then when the thermodynamic properties are written in terms of reduced quantities they are described by the universal equation of state

$$\left(p^*+\frac{3}{V^{*2}}\right)\left(V^*-\frac{1}{3}\right)=\frac{8}{3}T^* \tag{6.8}$$

It is possible to obtain the two parameters (a, b) that give a good description of simple systems, such as the noble gases, by using their critical temperatures and pressures (or volumes) together with eq. 6.7.

The Law of Corresponding States is particularly simple for the thermodynamic properties of systems that can be described by a two-parameter model. Unfortunately, it is not so easy to use the concept for transport properties, where the mass of the molecules is important, or for systems where the thermodynamic properties are better described by a three-parameter model [22]. Nevertheless, it is a good method for using the experimental properties of real systems to test theories of liquids and thus we make use of the concept during this chapter.

6.2 Virial Expansions and the Equation of State

Coefficients in the density expansions for the equation of state and for the radial distribution function (see Chapter 3) have been studied extensively. The usefulness of a particular approximation to the statistical mechanics of a system may be judged by comparing its coefficients with the exact ones. Similarly, it is possible to examine the usefulness of a particular model potential by comparing its exact virial coefficients with corresponding experimental measurements.

In Section 3.7 we showed that the equation of state could be written

$$\frac{pV}{NkT} = 1 + B_2\rho + B_3\rho^2 + \cdots \tag{6.9}$$

where B_n, the nth virial coefficient, can be expressed in terms of the intermolecular forces. Unfortunately, it becomes very difficult to calculate virial coefficients for terms higher than ρ^5 in the equation of state, even if it is assumed that only pairwise additive interactions are important. Except for the special case of the Gaussian model, where terms up to the twelfth virial coefficient have been obtained analytically [23], and for hard spheres, where coefficients up to the seventh have been reported [24], no coefficients higher than the fifth have been published.

Exact results for the first seven virial coefficients for hard spheres have been reported by Ree and Hoover [24] and others [25]. The coefficients B_2, B_3, and B_4 can be obtained analytically and the higher order coefficients numerically. It is straightforward to calculate B_2 for hard spheres. Using eq. 3.77 we obtain

$$B_2 = -2\pi\int_0^\infty r^2 f(r)\, dr$$

and for hard spheres we have

$$\phi(r) = \begin{cases} \infty & r \leq \sigma \\ 0 & \sigma < r \end{cases}; \qquad f(r) = \begin{cases} -1 & r \leq \sigma \\ 0 & \sigma < r \end{cases} \tag{6.10}$$

so that $B_2 = 2\pi\sigma^3/3$. Note that B_2 is independent of temperature for hard spheres, due to the fact that the Boltzmann factor $\exp(-\phi(r)/kT)$ is either zero or one. Table 6.1 gives results for the first six nontrivial ($B_1 = 1$) virial coefficients for the hard sphere potential.

The most important approximate methods for calculating the equation of state of hard spheres are probably the integral equation approximations discussed in Section 3.8. Most of these methods give the first three coefficients exactly and then give approximate values for higher terms in the expansion. Table 6.1 also gives results for the Percus-Yevick (PY) and hypernetted chain (HNC) approximations to B_4 and B_5 [26]. Notice that two results are given for each approximation, the "compressibility" approximation (c) and the "virial" approximation (v). We showed in Section 5.3 that the PY equation gave different results for the equation of state of hard spheres if the corresponding approximation to the radial

Table 6.1 Virial coefficients for the hard sphere fluid, $b = 2\pi\sigma^3/3$

Method	B_2/b	B_3/b^2	B_4/b^3	B_5/b^4	B_6/b^5	B_7/b^6
Exact	1.000	0.625	0.2869	0.1103	0.0386	0.0138
PY(c)	1.000	0.625	0.2969	0.1211	0.0449	0.0156
PY(v)	1.000	0.625	0.2500	0.0859	0.0273	0.0083
HNC(c)	1.000	0.625	0.2092	0.0493	—	—
HNC(v)	1.000	0.625	0.4453	0.1447	—	—
BG2(c)	1.000	0.625	0.2869	0.1112	—	—
BG2(v)	1.000	0.625	0.2869	0.1090	—	—
PY2(c)	1.000	0.625	0.2869	0.1074	—	—
PY2(v)	1.000	0.625	0.2869	0.1240	—	—
HNC2(c)	1.000	0.625	0.2869	0.1230	—	—
HNC2(v)	1.000	0.625	0.2869	0.0657	—	—

distribution function was used with the "compressibility" equation, eq. 3.63, and with the "virial" equation, eq. 3.54. This difference also occurs for the HNC approximation, and is reflected by the different virial coefficients in Table 6.1. Table 6.1 also gives results for the Born-Green-Yvon hierarchy [27] truncated by approximating $g^{(4)}$ in terms of $g^{(3)}$—the BG2 approximation—and for the so-called PY2 and HNC2 approximations [28]. From Table 6.1 we see that the PY approximation is generally better than the HNC approximation and that the BG2 approximation is in very good agreement with the exact result for B_5.

The differences between the compressibility and virial equation results for the PY and HNC approximations have been used to generate new approximate theories in which the two equations give the same equations of state [29]. A simple way of doing this is to introduce a density dependent mixing parameter into the PY and HNC approximations. We noted in Section 5.3 that one such self-consistent approximation could be written in the form

$$g(r) = [1 - \alpha(\rho)]g_{\mathrm{PY}}(r) + \alpha(\rho)g_{\mathrm{HNC}}(r) \tag{6.11}$$

where $\alpha = 0$ for the PY approximation and $\alpha = 1$ for the HNC approximation. The self-consistent approximation for $g(r)$ can be obtained by determining the density dependent parameter α from the constraint that results for the equation of state calculated from the virial and compressibility equations be identical. Such approximations have proven to be very useful for purely repulsive model pair potentials but are less useful for more realistic models, where there is a gas-liquid phase transition at low temperatures [30].

The second virial coefficient, B_2, for inverse power potentials and for the LJ potential can be calculated analytically in terms of a series expansion. Details are given in Hirschfelder et al. [31] along with tables of second virial coefficients as a function of T^* for the LJ potential. The third and higher coefficients [32, 33] for the LJ potential must be evaluated numerically and such calculations become increasingly difficult as more and more cluster diagrams must be evaluated (see Section 3.7). Figure 6.1 shows the temperature dependence of the first few virial coefficients, as predicted by the LJ potential [33]. Note that all coefficients up to B_5 become large and negative at low temperatures, which brings into question the convergence of the virial series in this temperature regime. Fortunately, at these very low temperatures the vapor pressures of the liquid and solid are extremely small and so the behavior of the gas up to the coexistence curve is probably well reproduced.

As with the hard sphere system, identical results for both the equation of state and the virial coefficients would be obtained from the "virial" and "compressibility" approaches (eqs. 3.54 and 3.63) if no approximation to

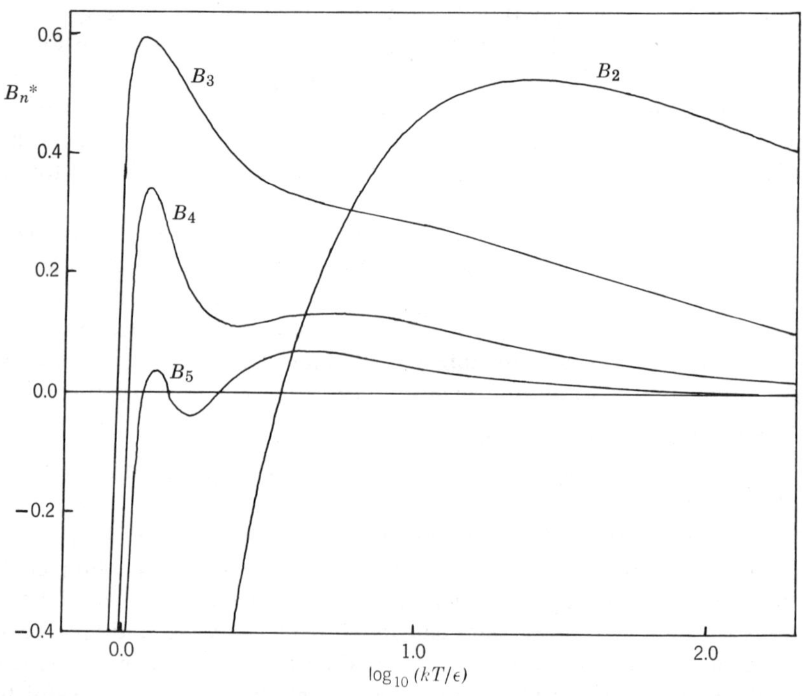

Figure 6.1 The second, third, fourth, and fifth virial coefficients for the LJ potential. Results are given for the scaled quantities $B_n^* = B_n(3/2\pi\sigma^3)^{n-1}$. (Adapted with permission from *Physics of Simple Liquids*, Chapter 3, p. 59, North Holland, Amsterdam.)

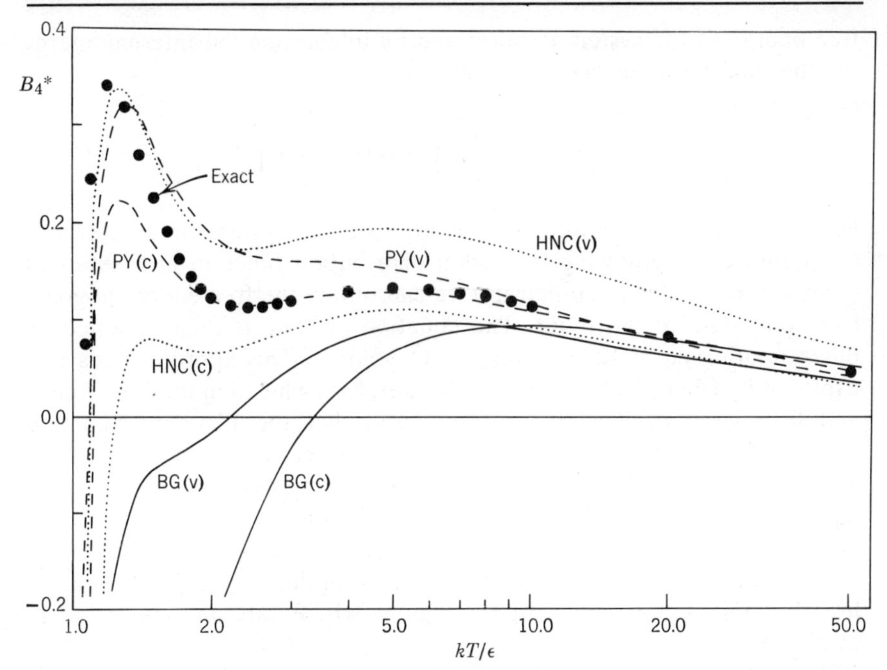

Figure 6.2 Fourth virial coefficient for the LJ potential computed exactly and using several approximations together with the virial (v) and compressibility (c) equations. The scaled variable B_4^* has the same meaning as in Fig. 6.1. (Adapted with permission from *Mol. Phys.*, **22,** 625, 1971.)

$g(r)$ were employed. Thus the higher virial coefficients can be used to examine the relative merits of various approximations for $g(r)$, and this is exemplified in Fig. 6.2 for the fourth virial coefficient [34]. Of the three integral equation approximations represented (BG, PY, HNC), the PY is the most satisfactory overall, although the HNC combined with the virial equation is better at intermediate temperatures. The BG approximation is clearly inferior at the intermediate temperatures but of comparable accuracy at very low and very high temperatures. It is found that the PY approximation at low temperatures gives virial coefficients that are too low when obtained from the virial equation and too high when obtained from the compressibility equation [34].

In Section 3.6 we derived three equations relating the radial distribution function to the thermodynamic properties, the virial and compressibility equations, and the energy equation, eq. 3.45. It has been demonstrated that the most successful integral equation method for obtaining the equation of state is to use the PY approximation with the energy equation [35–37]. In this somewhat indirect procedure, the Helmholtz free energy of the system is calculated by integrating the internal energy for the fluid between two temperatures:

$$\frac{A - A_0}{NkT} = \int_{1/T_0}^{1/T} \frac{2\pi\rho}{k} \left[\int_0^\infty r^2 \phi(r) g(r)\, dr \right] d\frac{1}{T} \tag{6.12}$$

Here A_0 is the Helmholtz free energy of the system at a reference temperature T_0. Knowing the radial distribution function at a series of temperatures and a given density, we can obtain the free energy in going from one state to the other. The equation of state is then obtained by differentiating eq. 6.12 with respect to volume. This approach was first explored by Chen, Henderson, and Barker, [35] who computed the fourth and fifth virial coefficients. It was found that the PY approximation combined with the energy equation is much closer to the exact values [34, 35] than results obtained using either the compressibility or the virial equation. As we see, the superior fit of the energy equation approach persists at lower temperatures and high densities [36, 37]. At very high temperatures the virial coefficients computed in this way tend to be rather high but this is not important for regions where experimental data are available.

As well as using the virial coefficients to test approximate theories we can use them to examine the usefulness of a pair potential as a model of a real system. We can do this for the Lennard-Jones potential by comparing predicted and experimental second virial coefficients. A necessary, but by no means sufficient, condition for an exact pair potential to be an accurate model of a real fluid is that it give the correct second virial coefficient. Remembering our discussion of the Law of Corresponding States earlier in this chapter, we can also use experimental data for second virial coefficients in appropriate reduced units to test the law. The LJ potential has been fitted to the second virial coefficient and other experimental data of a number of gases, and an extensive table of parameters is given by Hirschfelder et al. [31]. We use the LJ parameters for the inert gases He, Ne, Ar, Kr, and Xe, given in Table 2.1, to transform experimental data for the inert gas second virial coefficients [38] into reduced units. The results of this process are shown in Fig. 6.3 for data taken over a wide temperature range. We see that the LJ potential gives a good representation of the experimental second virial coefficients, although it tends to be poor at low temperatures. The one exception to this agreement is helium, but, as we discuss in Chapter 7, this breakdown is more a consequence of the use of the classical expression for the virial coefficient than a complete failure of the LJ potential. Obviously, in reduced units the inert gases have very similar second virial coefficients and it would appear that the

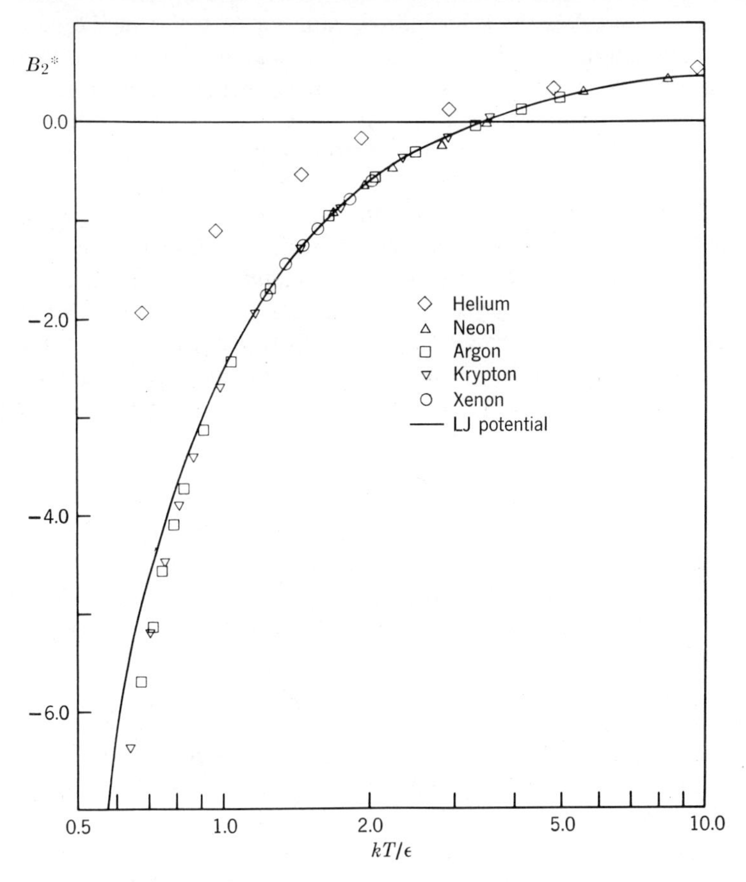

Figure 6.3 Second virial coefficients for the inert gases compared, in reduced units as in Fig. 6.1, with predictions of the LJ potential.

Law of Corresponding States is valid for these substances. We examine this point in more detail in a later chapter.

6.3 Thermodynamic Properties at High Densities

Once we move away from the very low density regime it is necessary to use more complex calculations than the virial coefficient expansion. At first sight it might appear that a useful approximation at high densities could be obtained by summing the first few terms in the virial expansion. Although this can be done, there are two problems associated with this approach. First, little information is available on the radius of convergence of the virial expansion [39] and there is no guarantee that the approach will work even if an infinite number of terms in the series are available. Second, at low temperatures a real fluid shows a gas-liquid phase transition and it is known that, at least in the region of the critical point, various thermodynamic quantities are nonanalytic functions of the state variables. As any simple truncation of the virial expansion gives a finite order polynomial such an approach cannot reproduce the critical point behavior. For most purposes the second problem is not important, and in fact most perturbation theories have a similar defect. If it is important to examine the critical point behavior some progress can be made by using Padé approximations to the truncated virial expansion [40]. However, the first problem, that of the radius of convergence, is more serious, particularly if we are interested in examining the liquid region. A number of studies of the truncated virial expansion at higher densities have been carried out, the most extensive being that of Barker and Henderson for the square well potential [41]. They showed that a five-term virial expansion gave reasonable results at temperatures down to the critical temperature and to densities somewhat higher than the critical density. However, the truncated expansion does not give very good results in the liquid region. It is likely that a similar conclusion would be reached for the LJ potential and so we do not consider the truncated virial expansion in any detail.

A number of methods are available for calculating the thermodynamic properties of fluids at high densities. First, and most important, are the computer simulation methods introduced in Section 3.10. The most important property of these methods is that for a given interaction potential it is possible to obtain essentially exact calculations of the bulk properties of the corresponding model fluid. This gives us some criteria for discussing the suitability of a particular approximation at high densities and low temperatures, where it is not possible to obtain analytic results. A second important property is that the methods give an excellent procedure for testing the accuracy of a given interaction potential as a model of a real system. As far as approximate methods are concerned, we examine two approaches, one based on the integral equations discussed in the previous section and the other based on perturbation theory (see Section 3.9). As perturbation theory is usually based on using the hard sphere fluid as a reference system, we begin by considering the properties of this system in more detail.

Hard sphere systems show a phase change from a fluid-like state at low densities to a solid or ordered phase at high densities with the transition density being $\rho^* = 0.943$. The equation of state of both phases has been determined accurately using Monte Carlo and molecular dynamics [42] methods; results for the fluid phase are given shortly.

Approximate results have been obtained from the HNC [43] and BG2 [27] equations as well as from the PY approximation. We noted in the preceding chapter (Section 5.3) that the PY equation could be solved analytically for the hard sphere model [44] and we recall the expressions for the equation of state:

$$\frac{pV}{NkT}=\frac{(1+2\eta+3\eta^2)}{(1-\eta)^2} \quad \text{(virial equation)} \tag{6.13}$$

$$\frac{pV}{NkT}=\frac{(1+\eta+\eta^2)}{(1-\eta)^3} \quad \text{(compressibility equation)} \tag{6.14}$$

where $\eta=\pi\rho^*/6$. It is interesting to note that at high densities the PY approximation gives negative values for the radial distribution function, and as this is nonphysical behavior it does, in some sense, recognize the onset of a solid phase. As far as the equation of state is concerned, the HNC approximation does not appear to be very useful [43] and, although the BG2 equation gives excellent results up to very high densities [27], it is very expensive to solve. Consequently, we restrict our detailed discussion to the PY and computer simulation results.

Table 6.2 gives the equation of state of hard spheres up to the limiting fluid density from machine simulation methods and from eqs. 6.13 and 6.14. The important points to notice are first that $pV/NkT \geqslant 1$ for all densities, a result expected for a purely repulsive potential, and second

Table 6.2 Equation of state of the hard sphere fluid as a function of density

$\rho\sigma^3$	Exact	Virial series	(3,3) Padé	Carnahan and Starling	PY(c)	PY(v)
0.0	1.00	1.00	1.00	1.00	1.00	1.00
0.1	1.24	1.24	1.24	1.24	1.24	1.24
0.2	1.55	1.55	1.55	1.56	1.56	1.55
0.3	1.97	1.97	1.97	1.96	1.97	1.95
0.4	2.52	2.52	2.52	2.52	2.54	2.48
0.5	3.27	3.24	3.27	3.26	3.31	3.17
0.6	4.29	4.21	4.29	4.28	4.38	4.09
0.7	5.71	5.48	5.71	5.71	5.90	5.32
0.8	7.73	7.16	7.73	7.75	8.12	7.00
0.9	10.7	9.34	10.65	10.8	11.5	9.33
1.0	15.0	12.15	14.99	15.3	16.6	12.6

that the PY compressibility and virial equation results bracket the simulation results. Also given is the equation of state predicted by the six-term virial expansion

$$\frac{pV}{NkT}=1+B_2\rho+B_3\rho^2+B_4\rho^3+B_5\rho^4+B_6\rho^5 \tag{6.15}$$

and from a (3, 3) Padé approximation to this polynomial Ree and Hoover [24] give

$$\frac{pV}{NkT}-1=\frac{b\rho(1+0.063499b\rho+0.017327b^2\rho^2)}{1-0.561501b\rho+0.081316b^2\rho^2} \tag{6.16}$$

where $b=2\pi\sigma^3/3$. It can be seen from the table that the virial series does not give very good results at high densities but that the Padé approximant is in good agreement with the simulation results. This result shows the power of Padé approximations as extrapolatory formulae and it is worth noting that they are often useful for extending the virial series (see, for example, Watts and Henderson [45]).

Carnahan and Starling [46] examined a number of analytic approximations to the equation of state of hard spheres, including the PY results (eqs. 6.13 and 6.14). On the basis of their investigation they suggested that a useful empirical equation of state for hard spheres could be obtained using the result

$$\frac{pV}{NkT}=\tfrac{1}{3}\,(\text{eq. }6.13)+\tfrac{2}{3}\,(\text{eq. }6.14)$$

or

$$\frac{pV}{NkT}=\frac{1+\eta+\eta^2-\eta^3}{(1-\eta)^3} \tag{6.17}$$

This equation is also represented in Table 6.2 and there is excellent agreement with the simulation results. Carnahan and Starling have used this equation to compute the enthalpy, entropy, Gibbs free energy, and fugacity of the hard sphere fluid [47] and their results can be used as the reference system properties in perturbation theories of more complex fluids.

The thermodynamic properties of the LJ fluid at high densities have been calculated in many ways, not all of which can be considered here [48]. In recent years, the more successful methods have been those based on either integral equation methods [30] or perturbation theory [5–19, 49]. Extensive machine simulation results also exist, from both Monte Carlo [50] and molecular dynamics methods [51]. We have seen in the foregoing section that once the radial distribution function is known the thermodynamic properties of the system can be calculated from the compressibility, virial, and energy equations. Our study of the hard sphere equation of state showed that at high densities the HNC and PY approximations were not good and that the machine simulation results were bracketed by the virial and compressibility equation predictions. A similar observation holds for these integral equations when used with the LJ potential, and apart from some interesting behavior in the region of the critical point [52, 53], neither the PY nor the HNC approximation is adequate when used with either the compressibility or virial equations in the liquid and dense fluid regions [30]. However, we noted in the previous section that the virial coefficients predicted by the PY approximation, when used with the energy equation, were very good down to low temperatures, and this suggests that a similar result may hold in the liquid region.

A thorough study of the equation of state for the LJ model was reported by Barker, Henderson, and Watts [36] using the PY energy equation route. They calculated the Helmholtz free energy of the system using eq. 6.12 and then obtained other thermodynamic properties by numerical differentiation. When obtaining their results they used $T^* = 2.74$ as the reference temperature where the PY results from both the pressure and the compressibility equations are in good agreement with the machine results [50]. The accuracy obtained close to the triple point of the LJ fluid, at $T^* = 0.72$, using the energy equation is evident from Fig. 6.4 where computer simulation results and experimental results for argon are also included. Agreement with exact results is quite striking and gives strong evidence for the accuracy of this approach. Also shown in this figure are the corresponding PY results from the compressibility and virial equations, to demonstrate that these methods are not good in the liquid region. Barker et al. [36] also obtained good agreement with argon data for the entropy and internal energy along the liquid-vapor coexistence curve. The agreement with exact results and with argon experiments in fact is very good at all temperatures and densities except near the critical point. This is well illustrated in Fig. 6.5 which shows the density for coexisting phases at various temperatures for the LJ fluid. The rather poor agreement with argon data at the critical point may either indicate the breakdown of the integral equation and machine simulation methods in this region or show that the LJ model is a poor effective pair potential at these densities. It is worth discussing the first of these possibilities in more detail.

In the critical region there are very large fluctuations in the density of the system, extending over considerable distances. These fluctuations manifest themselves experimentally in a number of ways, including the phenomenon of critical opalescence and the infinite value taken by the isothermal compressibility at the critical point. Computer simulation methods are based on a model of the fluid that is periodic over a short distance, usually less than about 10–15 molecular diameters (see Section 3.10). Consequently, such methods cannot show the very long wavelength disturbances found near the critical point. It may be that the differences shown in Fig. 6.5 reflect this problem, although as the LJ potential is not an accurate pair potential for argon the discrepancies may be due to other causes. A significant amount of information is available describing the properties of the PY and HNC approximations in the neighborhood of a phase change [52, 53]. However, before discussing this information we examine the usefulness of perturbation theories in the dense fluid region.

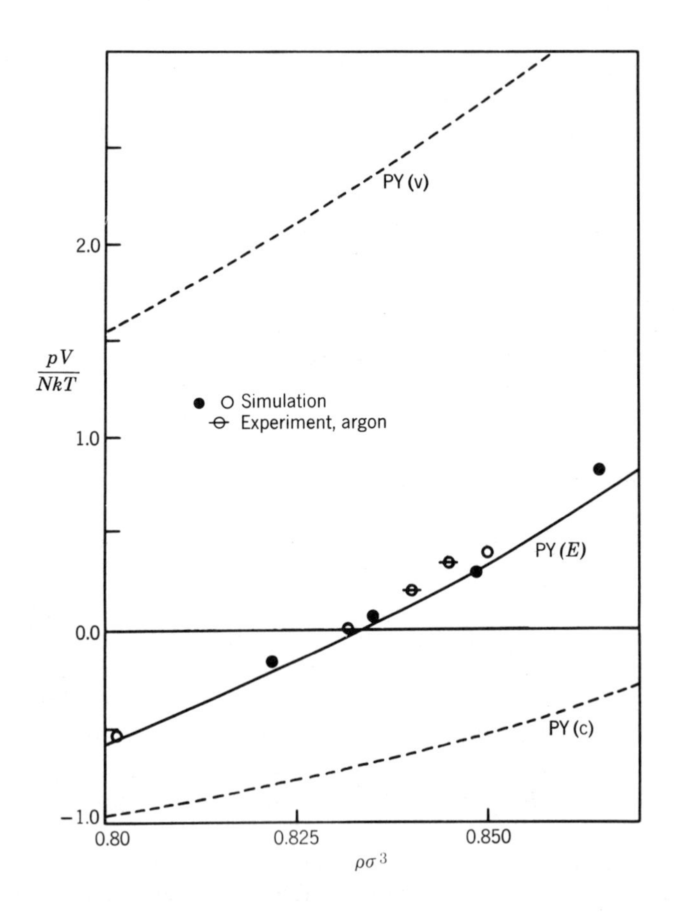

Figure 6.4 Equation of state for the LJ fluid at $kT/\varepsilon = 0.72$. (Adapted with permission from *IBM J. Res. Dev.*, **14,** 668, 1970.)

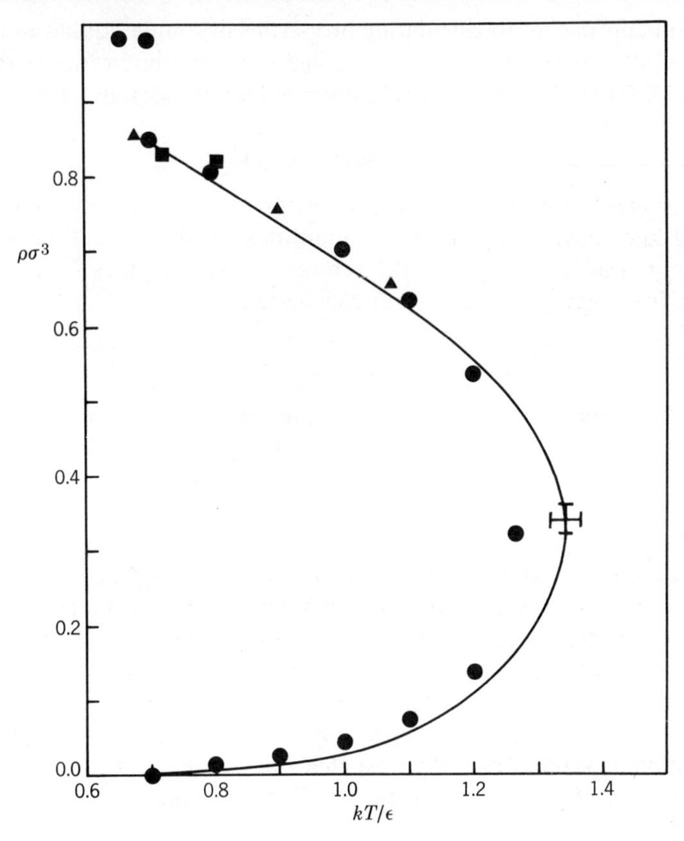

Figure 6.5 Densities for the coexisting phases of the LJ fluid calculated using the PY approximation with the energy equation. The various points represent simulation results and experimental argon values and the cross indicates the critical point estimated using simulation results. (Adapted with permission from *IBM J. Res. Dev.*, **14,** 668, 1970.)

6.4 Perturbation and Variational Theories

The rationale of perturbation and variational theories is the observation that at very high temperatures the equation of state of a fluid is determined primarily by the repulsive part of the potential, and only at lower temperatures does the effect of the attractive part become appreciable. Thus, if we perturb the partition function about the strong repulsive contribution we should be able to predict thermodynamic properties at lower temperatures, provided we have taken enough terms in the expansion. Recent perturbation theories have proven to be among the most successful approaches to calculating properties of simple liquids and liquid mixtures [49]. As we discussed in section 3.9, the method rewrites the pair potential $\phi(r)$ as the sum of a reference potential $u_0(r)$ and a perturbing term $u_1(r)$:

$$\phi(r) = u_0(r) + u_1(r) \tag{6.18}$$

If thermodynamic and structural properties of the reference system are assumed known, the corresponding quantities for the original system can be computed as corrections to the reference system values. The result is that the free energy of the system can be written

$$A = A_0 + A_1 + \cdots \tag{6.19}$$

where A_0 is the free energy of the reference system and $A_1, \ldots$ are correction terms. This expansion is discussed in detail in Section 3.9.

As the LJ potential is a good effective pair potential for argon at high densities, many perturbation calculations have been done for this model fluid. Since the properties of the hard sphere fluid are now well documented from extensive machine calculations, this fluid is the natural choice of reference system. A major difficulty with this reference system is the appropriate choice for the hard sphere diameter [8, 10, 54, 55]. A basic criterion for any perturbation or variational procedure is to have the reference system reasonably "close" to the original system to ensure that the subsequent expansion converges rapidly. However, by using the hard sphere fluid as a reference, the best that one can hope for is to calculate the properties of a system with the modified intermolecular potential:

$$\phi_m(r) = \begin{cases} \infty, & r < d \\ \phi(r), & r \geqslant d \end{cases} \tag{6.20}$$

Unless d is small, this potential will not generate the same properties as $\phi(r)$ itself. On the other hand if d is too small the perturbation term will be large and the series will converge rather slowly.

There are alternatives to using the hard sphere model as reference potential. One possible choice of the reference and perturbing potentials for the LJ fluid was made by McQuarrie and Katz [56]:

$$u_0(r) = 4\varepsilon\left(\frac{\sigma}{r}\right)^{12}$$

$$u_1(r) = -4\varepsilon\left(\frac{\sigma}{r}\right)^{6}$$

They calculated the properties of the "soft sphere" reference system using an expansion of the free energy in inverse powers of n, about the hard sphere system ($n = \infty$), described by Rowlinson [57]. Their results for the equation of state of the LJ potential were quite good at high temperatures but were inadequate at low temperatures.

As we discussed in Section 3.9 two particularly successful perturbation theories have been reported by Barker and Henderson (BH) [8] and Weeks, Chandler, and Andersen (WCA) [9, 10]. Barker and Henderson used a hard sphere fluid as reference system and expanded the free energy of the real system in terms of a steepness parameter, α, and a depth parameter, γ. Their result to first order is (eq. 3.119)

$$A = A_0 + 2\pi N\rho \int_{\sigma}^{\infty} r^2 \phi(r) g_{HS}(r)\, dr \tag{6.21}$$

with the hard sphere diameter, d, determined by (eq. 3.120)

$$d = -\int_0^{\sigma} \left[\exp \frac{-\phi(r)}{kT} - 1 \right] dr \tag{6.22}$$

where $\phi(r)$ is the potential function before perturbation (e.g., the LJ potential). Similarly, the WCA theory can also be written in terms of a hard sphere reference system, although the method for doing this is more complicated (see Section 3.9). The result is, to first order

$$A = A_0 + 2\pi N\rho \int_0^{\infty} r^2 u_1(r) g_0(r)\, dr \tag{6.23}$$

where $g_0(r)$ is given by

$$g_0(r) = \exp \frac{-u_0(r)}{kT}\, y_{HS}(r) \tag{6.24}$$

and $u_0(r)$ and the perturbing potential, are defined by eq. 3.121. The function $y_{HS}(r)$ is defined by eqs. 3.93 and 6.10, for hard spheres, and the hard sphere diameter, d, is chosen so that

$$\int_d^{r_m} r^2 y_{HS}(r)\, dr = \int_0^{r_m} r^2 \exp \frac{-u_0(r)}{kT} y_{HS}(r)\, dr \tag{6.25}$$

where r_m is the minimum in the potential $\phi(r)$.

The BH theory has the advantage of using the well studied hard sphere potential as the reference system. It has the disadvantage that it is only accurate at low temperatures if the second order correction to the free energy is included [7]. This term can be related to fluctuations in the pair distribution function [7, 8] and involves contributions from three-body and four-body correlations. Fortunately, Barker and Henderson have published convenient tables for calculating second order terms in the free energy [7] and so this difficulty is reduced. Results from the BH

Table 6.3 Equation of state of the LJ fluid. The exact results are taken from both MC and MD calculations

kT/ε	$\rho\sigma^3$	Exact	BH1	BH2	WCA	ORPA	Variational
2.74	0.65	2.22	2.24	2.22	2.18	2.20	2.54
	0.85	4.38	4.48	4.44	4.30	4.31	4.98
	0.95	6.15	6.41	6.40	6.10	6.10	6.97
1.35	0.20	0.50	0.55	0.52	0.53	0.51	0.56
	0.50	0.30	0.31	0.27	0.18	0.27	0.43
	0.85	3.37	3.54	3.36	3.28	3.30	4.24
	0.95	6.32	6.21	6.32	5.90	5.91	7.16
1.00	0.65	−0.25	−0.21	−0.36	−0.50	−0.38	0.04
	0.85	2.25	2.48	2.25	2.20	2.23	3.32
0.72	0.85	0.33±0.1	0.70	0.25	0.26	0.32	1.59

theory for the equation of state of the LJ fluid are tabulated in Table 6.3. Those listed as BH1 and BH2 are the Barker-Henderson first and second order results. Agreement with simulation results is excellent for the BH2 approximation, and in fact is superior to the PY energy equation approach [36]. For the LJ potential at $T^* = 1.35$ and $T^* = 0.72$, Levesque and Verlet [58] have shown that the BH theory converges very rapidly and yields accurate values at liquid temperatures down to the triple point. At higher temperatures their results were less satisfactory but Barker and Henderson [59] noted that this defect can be overcome by letting the upper limit in the equation defining the hard sphere diameter, eq. 6.22, be some parameter $\mu < \sigma$. This has the effect of making the hard sphere diameter d depend on density as well as on temperature at the higher temperatures.

The WCA theory has the perturbing potential $u_1(r)$ as a constant, $-\varepsilon$, for $r < r_m$ and equal to the full potential outside this distance (see eq. 3.121). This turns out to be an excellent division of the potential, because the perturbation $u_1(r)$ varies slowly, and consequently, the magnitude of higher order terms in the perturbation series is reduced. In their first study of the perturbation theory, Weeks, Chandler, and Andersen used PY values for the hard sphere system properties, but as Barker and Henderson pointed out, this introduced some errors [60]. Subsequently Weeks, Chandler, and Andersen repeated their calculations [10], using values of $g_{HS}(r)$ from the empirical fit to simulation results given by Verlet and Weis [61] (see Section 5.3). The WCA results for the thermodynamic properties and radial distribution function were in excellent agreement with machine simulations for the LJ potential. Results are shown in Table 6.3 for the equation of state at several temperatures and densities, along with results from other theories. More recently, Andersen et al. [62] have proposed an improvement on their theory of the thermodynamic properties and $g(r)$ which they call the optimized random phase approximation (ORPA). This approximation enhances the convergence of their perturbation series by an appropriate choice of the potential $u_1(r)$ in the region $r < d$. The method is closely related to the Mean Spherical Model of Lebowitz and Percus [63]. Results from this theory are also given in Table 6.3 and this approximation is excellent.

Mansoori and Canfield [18, 19] have proposed a variational method for calculating the free energy and equation of state of a fluid. Their observation, also made by Rasaiah and Stell [54] and Isihara [64], was that first order perturbation theory provides an upper bound on the Helmholtz free energy. If the reference fluid is the hard sphere system, the upper bound gives

$$A \leqslant A_0 + 2\pi N\rho \int_d^\infty u(r) g_0(r) r^2 \, dr \tag{6.26}$$

where A_0 and $g_0(r)$ are the hard sphere free energy and radial distribution function, respectively. The appropriate choice for the hard sphere diameter, d, is the one that minimizes the right hand side of eq. 6.26. Mansoori and Canfield reported calculations for the LJ fluid based on the PY expressions for the hard sphere system. Some of these variational results for the equation of state, obtained using the Verlet and Weiss [61] fit to the simulation values for $g_0(r)$, are also shown in Table 6.3. Unfortunately, replacing the PY $g_0(r)$ by the more accurate Verlet and Weiss model worsens the agreement with machine results. Results for the variational theory are best at the lower temperatures but tend to overestimate values of pV/NkT at the higher temperatures.

At this point, it is worth summarizing the relative merits of the several theories mentioned in this section. The BH and WCA perturbation theories are of comparable accuracy and are superior to the various integral equation approximations discussed at the end of the previous section. When comparing results from the two perturbation theories for the equation of state in Table 6.3, we note that the WCA theory is essentially a first order theory whereas the BH expansion must be taken to second order to yield comparable accuracy. This reflects the fact that the WCA theory is more rapidly convergent, primarily because of a different (and more complicated) choice of reference potential than the BH theory. The variational procedure suffers from difficulties associated with making the optimum choice of hard sphere diameter, d [65]. For the LJ fluid, the WCA choice of d is usually larger than either the BH or variational choices. At high densities this can sometimes present problems because the hard sphere system that must be considered is in the solid region. The other theories can, however, be easily applied in the high density regime.

6.5 Critical Point and Triple Point

We have mentioned earlier that the hard sphere model shows a phase transition from a fluid state of density $\rho^* = 0.943$ to a solid state at a density $\rho^* = 1.041$. For the LJ model the corresponding phase changes are more complex. First, there is a critical point in the fluid phase described by a critical temperature, T_c^*, a critical density, ρ_c^*, and a critical pressure, p_c^*. At temperatures higher than T_c^* the fluid can be continuously compressed until at a rather high density (~ 0.9) an ordered solid phase is more stable. Below T_c^* there is a region of densities outside of which the fluid exists either as a gas (low ρ^*) or as a liquid (high ρ^*). Once again, at sufficiently high densities there is a stable solid phase. All the approximate theories discussed here recognize the gas-liquid phase change in some way. The perturbation theories usually show a point of inflection along the critical isotherm, so that

$$\frac{\partial p^*}{\partial V^*} = 0; \qquad \frac{\partial^2 p^*}{\partial V^{*2}} = 0$$

at the critical point. Below this temperature any isotherm will show a minimum and a maximum between which the isothermal compressibility is negative. Such a prediction is consistent with the behavior shown by the van der Waals equation of state, and by equating the free energies on either side of the nonphysical, or two-phase, region it is possible to determine the thermodynamic properties of the coexisting vapor and liquid phases.

The integral equation approximations also recognize the existence of a two-phase region, but this time the behavior is more complex. Initially it is easier to discuss the behavior of such approximations in terms of the radial distribution function as this function is the solution to the basic integral equation. Once $g(r)$ is obtained it is possible to obtain thermodynamic properties using any one of the compressibility, virial, and energy equations, but this increases the complexity of the discussion.

For the LJ potential, the PY equation has solutions for $g(r)$ at all densities above $T^* \simeq 1.32$ [52, 53], at least up to $\rho^* \sim 1.3$. Below this temperature, there is a range of densities where no solutions exist. Watts [53] has investigated this region in some detail using Baxter's formulation of the PY equation [66]. He has shown that two solutions to the equation

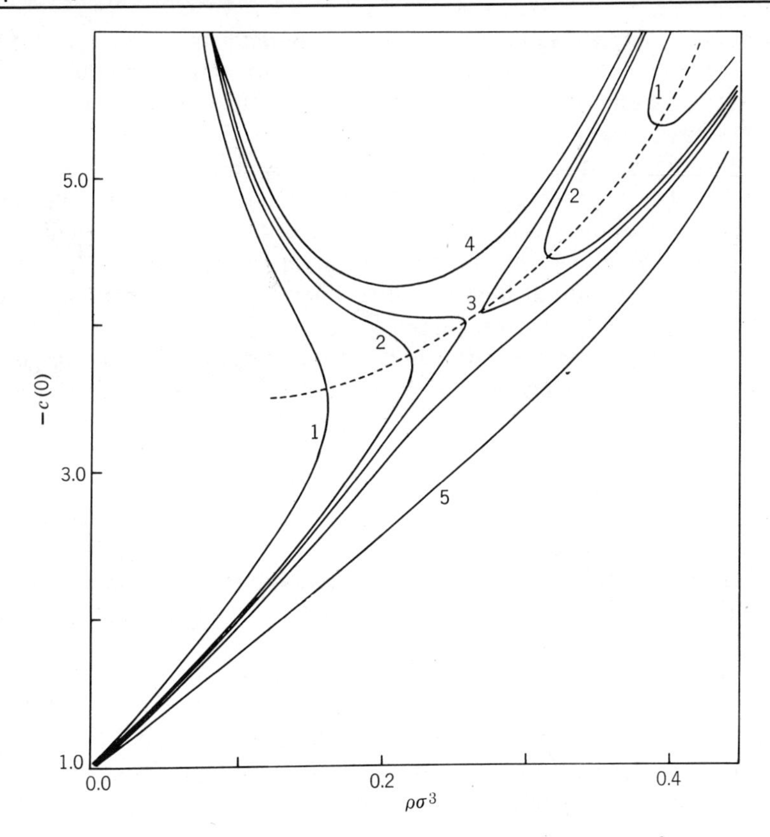

Figure 6.6 Direct correlation function at $r = 0$ as a function kT/ε and $\rho\sigma^3$ for the LJ fluid and PY approximation 1: $kT/\varepsilon = 1.2$; 2: $kT/\varepsilon = 1.263$; 3: $kT/\varepsilon = 1.275$; 4: $kT/\varepsilon = 1.3$; 5: $kT/\varepsilon = 1.5$. (Adapted with permission from *J. Chem. Phys.*, **48,** 50, 1968.)

are obtained, only one of which is physically meaningful. Figure 6.6 illustrates this behavior by displaying the direct correlation function, $c(r)$, at $r = 0$ as a function of T^* and ρ^*. Given that $c(0) = -1$ at zero density, one can use thermodynamic arguments to show that only solutions below the dotted line in this figure are physically meaningful. The critical point is a saddle point in the diagram. Below the critical temperature no solutions exist for a range of intermediate densities. Thus the PY theory gives a critical point and a gas liquid coexistence region characteristic of that found in real fluids. Using the compressibility equation (eq. 3.63) it is found that the isothermal compressibility becomes infinite at the critical point [53]. Similar results have been found for the HNC approximation, where the isothermal compressibility becomes at least quite large [52]. Results computed from the virial equation are not so satisfactory near the critical point, and the equation of state computed using this method does not show a point of inflection [53]. In an attempt to overcome this problem, Levesque [52] has extrapolated his results into the region where no solutions to the PY approximation exist in order to extract values for the critical temperature and density. The energy equation predicts infinite compressibility when used with the PY approximation [36]. Results from the energy equation are interesting for the critical point obtained using this method lies above that obtained from the compressibility equation. Thus there is a region at high densities where the energy equation gives negative isothermal compressibilities [36].

In Table 6.4 we compare the critical constants for the LJ fluid obtained using various approaches, including integral equation theories, perturbation theories, and computer simulation methods. Included in the table are results for an improved PY approximation, PY2, obtained through a functional differentiation approach [28]. This extension, and the corresponding HNC improvement, HNC2 [28], is more complicated and more difficult to solve numerically. It is apparent from the table that the PY approximation with the energy equation gives the most satisfactory results from an integral equation. Perturbation theory estimates are also in good agreement with machine results.

In order to study the fluid-solid transition some model for the solid state must be used, usually either a computer simulation method [67] or cell theory [68]. A detailed study of the liquid-gas and solid-fluid transitions for the LJ model has been made by Hansen and Verlet [69]. Such calculations are not within the scope of this book.

Table 6.4 Critical point values of thermodynamic properties for the LJ fluid. The range of values for the exact results reflects the uncertainty in simulation methods in the critical region

Method	kT_c/ε	$\rho_c\sigma^3$	$p_c\sigma^3/\varepsilon$	$p_c/\rho_c kT_c$
Exact	1.32–1.36	0.32–0.36	0.13–0.17	0.30–0.36
PY(E)	1.34	0.34	0.14	0.31
PY(c)	1.32	0.28	0.13	0.36
PY(v)	1.25	0.29	0.11	0.30
HNC(c)	1.41	0.28	—	—
HNC(v)	1.25	0.26	—	0.35
PY2(c)	1.33	0.33	0.15	0.34
PY2(v)	1.36	0.36	0.15	0.31
Variational	1.36	0.33	0.17	0.37
BH2	1.38	0.33	0.16	0.35

6.6 Transport Coefficients

Most theoretical studies of the transport coefficients of dense fluids have been restricted to the self-diffusion coefficient of the hard sphere and LJ systems. Thus in his early paper on the molecular dynamics of the LJ system Rahman [51] reported a calculation of this quantity. More recently, results have been reported by Levesque and Verlet [70], Bruin [71], and also by Fisher and Watts [72], who based their study on the Barker-Bobetic potential [73]. There have been several attempts to compute the shear and bulk viscosities and the thermal conductivity of the dense fluid using a number of methods [74–76]. In general, the standard approach to computing these quantities, using the Kubo formulae given in Section 4.2, requires a great deal of computing time and other methods are more effective [77].

The self-diffusion coefficient can be calculated either by computing the mean square displacement of a particle as a function of time (eq. 4.58)

$$D = \lim_{t\to\infty} \frac{\langle [\mathbf{r}_1(t) - \mathbf{r}_1(0)]^2 \rangle}{6t} \tag{6.27}$$

or by integrating the velocity autocorrelation function (eq. 4.51)

$$D = \int_0^\infty C(t)\, dt = \tfrac{1}{3}\int_0^\infty \langle \mathbf{v}_1(0) \cdot \mathbf{v}_1(t) \rangle\, dt \tag{6.28}$$

Providing a sufficiently accurate molecular dynamics calculation is available, both methods give the same answer. The standard approach to calculating D from eq. 6.27 is to plot the mean square displacement as a function of time and estimate D from the slope of the linear part of the graph, found for longer times [51, 70–72]. If the length of the molecular dynamics run is not sufficiently long the estimate of D obtained using this method may be rather high, particularly in the liquid region [72, 78]. In general the agreement with experimental results for the self-diffusion coefficients of the inert gases is reasonably good when the LJ potential is used. Levesque and Verlet [70] fitted their results for D to an interpolatory formula

$$D = (0.3438 T^* \rho^{*-2} + 0.1188 - 0.1499 \rho^*) \times 10^{-9}\ \mathrm{m^2 s^{-1}} \tag{6.29}$$

and this can be used to obtain self-diffusion coefficients of the LJ potential in the liquid region.

In Section 5.5 we considered the long time behavior of $C(t)$ and discussed Alder and Wainwright's findings [79] that $C(t) \sim t^{-3/2}$ for long times. This behavior must be taken into account when computing D from eq. 6.28. We can see this most easily by considering the case of a two-dimensional system, where it has been suggested that at long times $C(t) \sim t^{-1}$ [79]. If this is so, the integral in eq. 6.28 does not converge and there is some confusion about the meaning of the two-dimensional self-diffusion coefficient [77, 80]. For the three-dimensional case, the integral is slowly convergent and so the long time tail can make a significant contribution to the value of D. The size of this contribution has been considered theoretically, using both hydrodynamic arguments and kinetic theory [81], and as can be seen by integrating eq. 5.50, it takes the form [81, 82]

$$D_L = \frac{4kT}{3m\rho} t_0^{-1/2} \left(4\pi \left[\frac{\eta}{m\rho} + D \right] \right)^{-3/2} \tag{6.30}$$

where η is the shear viscosity and D is the full self-diffusion coefficient, both obtained (in principle) from the molecular dynamics calculation. The time t_0 is the time beyond which the long time tail becomes important and has to be estimated heuristically [82, 83]. Hanley and Watts [83] have made a detailed study of the size of the long range correction, D_L, for the m-6-8 potential [2]. They showed that the tail correction is particularly important at densities typical of those found for a fluid in the region of its critical point and becomes less important at higher densities. This phenomenon is not closely related to the existence of the critical point, as might be expected, and is also found for hard sphere fluids at similar densities [84]. In fact, the predominance of the long time tail at intermediate densities can be associated with the absence of the negative region in $C(t)$ observed at liquid densities [77, 79, 84].

We mentioned earlier in this section that the other transport coefficients were not as accurately known as the self-diffusion coefficient. The reason for this is readily understood if we examine the Kubo expressions for D and the shear viscosity, η, given in Section 4.2. We saw there that both coefficients are time integrals of flux-flux correlation functions, where for D the flux is the velocity of a single particle and for η the flux takes the form

$$J_\eta(xy) = \left(\frac{1}{VkT} \right)^{1/2} \sum_{i=1}^{N} [m\dot{x}_i\dot{y}_i + x_iF_i(y)] \tag{6.31}$$

We can estimate D by computing the velocity autocorrelation function for all N particles in the molecular dynamics calculation and then improve the statistics by averaging over the particles. Such an approach is not possible for η as its flux depends on the coordinates of all the particles in the system. The corresponding fluxes for bulk viscosity, ϕ, and thermal conductivity, λ, also depend on the coordinates of all the particles in the system and so they, too, are difficult to compute. It appears that reasonably accurate estimates of η, ϕ, and λ can only be obtained using linear response theory if a molecular dynamics calculation for a large number of particles is extended over a considerable period of time. Thus Levesque et al. [74] used a system of 864 particles and followed their motions for an interval of 10^{-9} s in order to obtain reasonably accurate values of these three transport coefficients. Similarly, extensive computations were reported by Alder et al. for the hard sphere fluid [84]. By contrast, Fisher and Watts [72] used a system of 108 particles and followed their time development for 5×10^{-11} s to obtain comparable accuracy for the self diffusion coefficient.

Two alternative methods have been suggested for computing the viscosities and thermal conductivity of dense fluids, both involving the application of external fields to the molecular dynamics calculations. Gosling, McDonald, and Singer [75] used a steady state method in which they applied a sinusoidal shearing force along one direction in the unit cell used in molecular dynamics calculations. The flow of momentum in such a system, from the hydrodynamic viewpoint, is determined by the Navier-Stokes equation [85], and for the work of Gosling et al. this can be written

$$\rho\frac{\partial v_x}{\partial t}=\eta\frac{\partial^2 v_x}{\partial z^2}+\frac{\rho}{m}F_x(z) \tag{6.32}$$

where v_x is the velocity in the x direction and where the shear force is

$$F_x(z)=F_0\sin\frac{2\pi z}{L} \tag{6.33}$$

with L the length of the side of the unit cell across which the shear force is imposed. This force was applied to the molecules during the molecular dynamics calculation and computation continued until a steady state drift velocity profile was established across the cell, namely

$$v_x(z)=v_0\sin\frac{2\pi z}{L} \tag{6.34}$$

This profile can also be calculated by integrating the Navier-Stokes equation, eq. 6.32, and on equating the two expressions we find

$$\eta=\frac{\rho L^2F_0}{4\pi^2mv_0} \tag{6.35}$$

Thus the shear viscosity can be computed if the amplitude of the steady state velocity profile is obtained. Gosling et al. [75] found that this device, which bypasses the linear response theory outlined in Chapter 4, enabled values of η to be obtained from a molecular dynamics calculation extending over about 5×10^{-11} s.

The second method of computing viscosities and thermal conductivities was developed by Ashurst and Hoover [76, 77] and has some features in common with the method described by Gosling et al. [75]. In one set of calculations [76], these authors used a nonequilibrium molecular dynamics simulation to model the phenomenon of Couette flow, which occurs when a fluid is confined between two parallel walls in steady relative motion. In this situation momentum gradients are established across the fluid and these may be used to compute viscosity coefficients. The shear viscosity can be calculated from the measured velocity gradient across the cell together with the wall shear force, or equivalently the momentum flux across the wall. Ashurst and Hoover also reported calculations for the thermal conductivity [77] by using a similar method with a temperature gradient replacing the momentum gradient. They were able to compute the thermal conductivity from the measured temperature gradients and heat fluxes. As with the method proposed by Gosling et al. [75], this method is comparable in accuracy with the Kubo approach used by Levesque et al. [74] and uses significantly less computer time.

References

1. J. E. Lennard-Jones, 1924, *Proc. Roy. Soc. (London), Ser. A*, **106,** 463.
2. M. Klein and H. J. M. Hanley, 1970, *J. Chem. Phys.*, **53,** 4722; H. J. M. Hanley and M. Klein, 1972, *J. Phys. Chem.*, **76,** 1743.
3. A. Rotenberg, 1965, *J. Chem. Phys.*, **43,** 1198; F. Lado and W. W. Wood, 1968, *ibid.*, **49,** 4244; B. J. Alder, D. A. Young, and M. A. Mark, 1972, *ibid.*, **56,** 3013; D. Levesque, 1966, *Physica*, **32,** 1985; S. Katsura and K. Nishihara, 1969, *J. Chem. Phys.*, **50,** 3579; W. R. Smith, D. Henderson, and J. A. Barker, 1971, *ibid.*, **55,** 4027; W. G. Rudd and H. L. Frisch, 1971, *ibid.*, **54,** 3479.
4. W. G. Hoover, M. Ross, K. W. Johnson, D. Henderson, J. A. Barker, and B. C. Brown, 1970, *J. Chem. Phys.*, **52,** 4931; W. G. Hoover, S. G. Gray, and K. W. Johnson, 1971, *ibid.*, **55,** 1128; J. L. Lebowitz and J. K. Percus, 1966, *Phys. Rev.*, **144,** 251; E. Waisman and J. L. Lebowitz, 1970, *J. Chem Phys.*, **52,** 4307; C. T. Chen, 1971, *ibid.*, **54,** 1515; L. L. Lee and H. M. Hulburt, 1973, *ibid.*, **58,** 44, 61; A. E. Sherwood and J. M. Prausnitz, 1964, *ibid.*, **41,** 413, 429.
5. R. W. Zwanzig, 1954, *J. Chem. Phys.*, **22,** 1420.
6. E. B. Smith and B. J. Alder, 1959, *J. Chem. Phys.*, **30,** 1190.
7. J. A. Barker and D. Henderson, 1972, *Ann. Rev. Phys. Chem.*, **23,** 439.
8. J. A. Barker and D. Henderson, 1967, *J. Chem. Phys.*, **47,** 2856, 4714.
9. D. Chandler and J. D. Weeks, 1970, *Phys. Rev. Lett.*, **25,** 149.
10. J. D. Weeks, D. Chandler, and H. C. Andersen, 1971, *J. Chem. Phys.*, **54,** 5237; 1971, *ibid.*, **55,** 5422.
11. J. A. Barker and D. Henderson, 1971, *Phys. Rev.*, **A4,** 806.
12. K. E. Gubbins and C. G. Gray, 1972, *Mol. Phys.*, **23,** 187.

13. K. E. Gubbins, W. R. Smith, M. K. Tham, and E. W. Tiepel, 1971, *Mol. Phys.*, **22,** 1089.
14. H. C. Andersen, J. D. Weeks, and D. Chandler, 1971, *Phys. Rev.*, **A4,** 1597.
15. D. Henderson and J. A. Barker, 1968, *J. Chem. Phys.*, **49,** 3377.
16. W. R. Smith, 1971, *Mol. Phys.*, **22,** 105.
17. D. Henderson and P. J. Leonard, 1971, *Proc. Natl. Acad. Sci.*, **68,** 2354.
18. G. A. Mansoori and F. B. Canfield, 1969, *J. Chem. Phys.*, **51,** 4958, 4967; 1970, *ibid.*, **53,** 1618.
19. G. A. Mansoori and F. B. Canfield, 1970, *Ind. Eng. Chem. Fundam.*, **62,** 12.
20. E. A. Guggenheim, 1945, *J. Chem. Phys.*, **13,** 253.
21. L. D. Landau and E. Lifshitz, 1958, *Statistical Physics*, Pergamon, London.
22. M. K. Tham and K. E. Gubbins, 1969, *I. and E. C. Fundamentals*, **8,** 791.
23. G. E. Uhlenbeck and G. W. Ford, 1962, in *Studies in Statistical Mechanics*, J. de Boer and G. E. Uhlenbeck, Eds., Vol. 1, North Holland, Amsterdam.
24. F. H. Ree and W. G. Hoover, 1964, *J. Chem. Phys.*, **40,** 939; 1967, *ibid.*, **46,** 4181.
25. J. S. Rowlinson, 1965, *Rep. Progr. Phys.*, **28,** 169; S. Katsura and Y. Abe, 1963, *J. Chem. Phys.*, **39,** 2068; J. E. Kilpatrick and S. Katsura, 1966, *ibid.*, **45,** 1866; F. H. Ree, R. N. Keeler, and S. L. McCarthy, 1966, *ibid.*, **44,** 3407.
26. J. S. Rowlinson, 1964, *Contemp. Phys.*, **5,** 359.
27. Y. T. Lee, F. H. Ree, and T. Ree, 1968, *J. Chem. Phys.*, **48,** 3506.
28. L. Verlet, 1965, *Physica*, **31,** 959; 1964, *ibid.*, **30,** 95; 1966, *ibid.*, **32,** 304; H. S. Green, 1965, *Phys. Fluids*, **8,** 1; A. R. Allnatt, 1966, *Physica*, **32,** 133; M. S. Wertheim, 1967, *J. Math, Phys.*, **8,** 927; R. J. Baxter, 1968, *Ann. Phys.*, **46,** 509.
29. G. S. Rushbrooke and P. Hutchinson, 1961, *Physica*, **27,** 647; D. D. Carley and F. Lado, 1965, *Phys. Rev.*, **137, A**42; J. S. Rowlinson, 1965, *Mol. Phys.*, **9,** 217; C. Hurst, 1965, *Proc. Phys. Soc. (London)*, **86,** 193; G. J. Throop and R. J. Bearman, 1966, *ibid.*, **88,** 539; R. O. Watts and D. Henderson, 1969, *Mol. Phys.*, **16,** 217; T. Morita, 1969, *Progr. Theor. Phys. (Kyoto)*, **41,** 339.
30. R. O. Watts, 1973, Chapter 1 in *Specialist Periodical Reports of the Chemical Society: Statistical Mechanics*, K. Singer, Ed., Vol. 1, Chemical Society London.
31. J. O. Hirschfelder, C. F. Curtiss, and R. B. Bird, 1964, *Molecular Theory of Gases and Liquids*, Wiley, New York.
32. R. Bergeon, 1952, *C. R.*, **234,** 1039; J. S. Rowlinson, 1963, *Mol. Phys.*, **6,** 75, 429; S. F. Boys and I. Shavitt, 1960, *Proc. Roy. Soc. (London), Ser. A*, **254,** 487; J. A. Barker and J. J. Monaghan, 1962, *J. Chem. Phys.*, **36,** 2564; D. Henderson and L. Oden, 1966, *Mol. Phys.*, **10,** 405; J. A. Barker, P. J. Leonard, and A. Pompe, 1966, *J. Chem. Phys.*, **44,** 4206.
33. J. S. Rowlinson, 1968, Chapter 3 in *Physics of Simple Liquids*, H. N. V. Temperley, J. S. Rowlinson, G. S. Rushbrooke, Eds., North Holland, Amsterdam.
34. D. Henderson, S. Kim, and L. Oden, 1967, *Discuss. Faraday Soc.*, **43,** 26; S. Kim, D. Henderson, and L. Oden, 1969, *Trans. Faraday Soc.*, **65,** 2308; D. D. Fitts and W. R. Smith, 1971, *Mol. Phys.*, **22,** 625; S. Kim, D. Henderson, and L. Oden, 1966, *J. Chem. Phys.*, **45,** 4030.
35. M. Chen. D. Henderson, and J. A. Barker, 1969, *Can. J. Phys.*, **47,** 2009; D. Henderson and M. Chen, 1970, *ibid.*, **48,** 634.
36. J. A. Barker, D. Henderson, and R. O. Watts, 1970, *Phys. Lett.*, **31A,** 48; D. Henderson, J. A. Barker, and R. O. Watts. 1970, *IBM J. Res. Dev.*, **14,** 668.
37. R. J. Bearman, F. Theeuwes, M. Y. Bearman, F. Mandel, and G. J. Throop, 1970, *J. Chem. Phys.*, **52,** 5486; F. Theeuwes and R. J. Bearman, 1970, *J. Chem. Phys.*, **53,** 3114.

38. J. H. Dymond and E. B. Smith, 1969, *The Virial Coefficients of Gases: A Critical Compilation*, Clarendon, Oxford.
39. F. H. Ree, 1967, *Phys. Rev.*, **155,** 84; J. L. Lebowitz and O. Penrose, 1964, *J. Math. Phys.*, **53,** 3114.
40. M. E. Fisher, 1967, *Rep. Progr. Phys.*, **30,** 615; L. P. Kadanoff, W. Götze, D. Hamblen, R. Hecht, E. A. S. Lewis, V. V. Palciauskas, M. Rayl, J. Swift, D. Aspnes, and J. Kane, 1967, *Rev. Mod. Phys.*, **39,** 395.
41. J. A. Barker and D. Henderson, 1967, *Can. J. Phys.*, **45,** 3959.
42. W. W. Wood and J. D. Jacobson, 1957, *J. Chem. Phys.*, **27,** 1207; M. N. Rosenbluth and A. W. Rosenbluth, 1954, *ibid.*, **22,** 881; J. P. Hansen and L. Verlet, 1969, *Phys. Rev.*, **184,** 151; B. J. Alder and T. E. Wainwright, 1962, *Phys. Rev.*, **127,** 359; B. J. Alder and T. E. Wainwright, 1960, *J. Chem. Phys.*, **33,** 1439.
43. M. Klein, 1963, *J. Chem. Phys.*, **39,** 1388; L. Verlet and D. Levesque, 1962, *Physica*, **28,** 1124.
44. M. S. Wertheim, 1963, *Phys. Rev. Lett.*, **10,** 321; 1964, *J. Math. Phys.*, **5,** 643; E. Thiele, 1963, *J. Chem. Phys.*, **39,** 474.
45. R. O. Watts and D. Henderson, 1969, *J. Chem. Phys.*, **50,** 1651.
46. N. F. Carnahan and K. E. Starling, 1969, *J. Chem. Phys.*, **51,** 635.
47. N. F. Carnahan and K. E. Starling, 1970, *J. Chem. Phys.*, **53,** 600.
48. J. A. Barker, 1963, *Lattice Theories of the Liquid State*, Pergamon, Oxford.
49. W. R. Smith, 1973, Chapter 2 in *Specialist Periodical Reports*: *Statistical Mechanics*, K. Singer, Ed., Vol. 1, Chemical Society, London.
50. W. W. Wood and F. R. Parker, 1957, *J. Chem. Phys.*, **27,** 720; I. R. McDonald and K. Singer, 1967, *Discuss. Faraday Soc.*, **43,** 40; L. Verlet and D. Levesque, 1967, *Physica*, **36,** 254.
51. A. Rahman, 1964, *Phys. Rev.*, **136,** *A*405; L. Verlet, 1967, *Phys. Rev.*, **159,** 98; 1967, *ibid.*, **165,** 201.
52. D. Henderson and R. D. Murphy, 1972, *Phys. Rev.*, *A***6,** 1224; M. I. Guerrero, G. Saville, and J. S. Rowlinson, 1975, *Mol. Phys.*, **29,** 1941; D. Levesque, 1966; *Physica*, **32,** 1985; R. O. Watts, 1969, *J. Chem. Phys.*, **50,** 1358.
53. R. O. Watts, 1968, *J. Chem. Phys.*, **48,** 50.
54. J. Rasaiah and G. Stell, 1970, *Mol. Phys.*, **18,** 249.
55. E. Wilhelm, 1973, *J. Chem. Phys.*, **58,** 3558.
56. D. A. McQuarrie and J. L. Katz, 1966, *J. Chem. Phys.*, **44,** 2393.
57. J. S. Rowlinson, 1964, *Mol. Phys.*, **8,** 107.
58. D. Levesque and L. Verlet, 1969, *Phys. Rev.*, **182,** 307.
59. D. Henderson and J. A. Barker, 1970, *Phys. Rev.*, *A***1,** 1266.
60. J. A. Barker and D. Henderson, 1971, *Phys. Rev.*, *A***4,** 806.
61. L. Verlet and J. J. Weis, 1972, *Phys. Rev.*, *A***5,** 939.
62. H. C. Andersen, D. Chandler, and J. D. Weeks, 1972, *J. Chem. Phys.*, **56,** 3812.
63. J. L. Lebowitz and J. K. Percus, 1966, *Phys. Rev.*, **144,** 251.
64. A. Isihara, 1968, *J. Phys., A (London)*, **1,** 539.
65. D. Henderson and J. A. Barker, 1970, *J. Chem. Phys.*, **52,** 2315.
66. R. J. Baxter, 1967, *Phys. Rev.*, **154,** 170.
67. J. P. Hansen, 1970, *Phys. Rev.*, *A***2,** 221; W. G. Hoover, S. G. Gray, and K. W. Johnson, 1971, *J. Chem. Phys.*, **55,** 1128.
68. D. Henderson and J. A. Barker, 1968, *Mol. Phys.*, **14,** 587.
69. J. P. Hansen and L. Verlet, 1969, *Phys. Rev.*, **184,** 151; 1969, *ibid.*, **188,** 314.
70. D. Levesque and L. Verlet, 1970, *Phys. Rev.*, *A***2,** 2514.
71. C. Bruin, 1969, *Phys. Lett.*, **28***A*, 777.

72. R. A. Fisher and R. O. Watts, 1972, *Aust. J. Phys.*, **25**, 529.
73. M. V. Bobetic and J. A. Barker, 1970, *Phys. Rev.*, *B***2**, 4169.
74. D. Levesque, L. Verlet, and J. Kürkijarvi, 1973, *Phys. Rev.*, *A***7**, 1690.
75. E. Gosling, I. R. McDonald, and K. Singer, 1973, *Mol. Phys.*, **26**, 1475.
76. W. G. Hoover and W. T. Ashurst, 1973, *Phys. Rev. Lett.*, **31**, 206; W. T. Ashurst and W. G. Hoover, 1975, *Phys. Rev.*, A**11**, 658; 1975, *A. I. Ch. E. J.*, **21**, 410.
77. W. G. Hoover and W. T. Ashurst, 1975, in *Advances in Theoretical Physics*, H. Eyring and D. Henderson, Eds., Vol. 1, Academic, New York.
78. R. O. Watts, 1972, *Faraday Soc. Symp.*, **6**, 166.
79. B. J. Alder and T. E. Wainwright, 1967, *Phys. Rev. Lett.*, **18**, 988; 1968, *J. Phys. Soc. Japan Suppl.*, **26**, 267; 1970, *Phys. Rev.*, A**1**, 18; T. E. Wainwright, B. J. Alder, and D. M. Gass, 1971, *Phys., Rev.*, A**4**, 233.
80. T. Keyes and I. Oppenheim, 1973, *Phys. Rev.*, A**8**, 937; W. W. Wood, 1973, *Acta Phys. Austr., Suppl.*, **10**, 451.
81. See, for example, the review articles by Alder and by Deutsch in *Transport Phenomena*, 1973, J. Kestin, Ed., American Institute of Physics; M. H. Ernst, E. H. Hauge, and J. M. S. van Leeuwen, 1970, *Phys. Rev. Lett.*, **25**, 1254; J. R. Dorfman and E. G. D. Cohen, *ibid.*, **25**, 1257; R. Zwanzig and M. Bixon, 1970, *Phys. Rev.*, A**2**, 2005.
82. N. K. Ailawadi, 1974, in *Molecular Motions in Liquids*, J. Lascombe, Ed., p. 71, D. Reidel Publishers, Boston.
83. H. J. M. Hanley and R. O. Watts, 1975, *Mol. Phys.*, **29**, 1907; 1975, *Physica*, **79***A*, 351.
84. B. J. Alder, D. M. Gass, and T. E. Wainwright, 1970, *J. Chem. Phys.*, **53**, 3813.
85. L. M. Milne-Thomson, 1967, *Theoretical Hydrodynamics*, MacMillan, London.

Seven

Exact Pair Potentials and the Properties of the Noble Gases

As the particles in the noble, or inert, gases are spherically symmetric they provide an ideal testing ground both for theories of intermolecular potentials and for theoretical calculations of their bulk properties. Semiempirical interaction potentials can be computed using the most accurate available experimental data taken from the gas, liquid, and solid phases together with theoretical estimates of certain atomic properties. The study of argon is perhaps the best example of such a complete investigation of atom-atom interactions.

Many empirical and semiempirical models for the argon-argon pair potential have been proposed, the most successful being a model developed by Barker and co-workers [1–3] by fitting a very wide range of experimental data. This potential, which was mentioned briefly in Chapter 2, has the form

$$\phi(r)=\varepsilon\left\{\exp\left[\alpha(1-R)\right]\sum_{i=0}^{5} A_i(R-1)^i-\sum_{i=0}^{2}\frac{C_{2i+6}}{R^{2i+6}+\delta}\right\} \tag{7.1}$$

where $R = r/R_m$ and R_m is the position of the minimum in the potential. The coefficients A_i, ε, α, R_m, and δ were chosen to obtain the best fit to certain experimental data, and the coefficients C_j are the leading terms in the multipole expansion of the dispersion energy. Values of the coefficients in this equation have since been determined for krypton and xenon and a discussion of them is given in later sections of this chapter.

Since quantum effects are relatively small in argon, and there is a good deal of high quality experimental data available, classical statistical mechanics can be used to analyze the bulk data except at low temperatures. As a consequence, the potential energy function for argon is known to within an error of probably less than 1%. That is, a single potential function of the type shown in eq. 7.1 is in accord with theoretical long range interactions and reproduces virtually all experimental data in the gas, liquid, and solid states over a very wide range of temperatures and densities [1–4].

As can be inferred from the many types of data used to determine the argon-argon pair potential, different kinds of experiments provide information on different regions of the function describing this interaction. If sufficient data is used in conjunction with statistical mechanical methods, the complete potential energy function can be determined. Computing interaction potentials using a wide range of experimental data frequently focuses attention on inconsistencies between two experiments. For example, viscosity measurements on dilute gases at high temperature and the high energy total scattering cross-section measurements are both sensitive to the short range repulsive region of the potential. For argon, at one time such measurements were found to be inconsistent in the sense that a single potential function could not reproduce both sets of data [1,5]. Eventually, the accumulated evidence suggested that there were errors in the viscosity data of a few percent [1,6]. This observation prompted new, more accurate experiments [7] so that currently the viscosity data are probably more reliable than those obtained from high energy scattering cross-section measurements. In subsequent sections of this chapter we discuss the properties of the inert gases in the solid, liquid, and gaseous states and show how a combination of theory and experiment can lead to very accurate interatomic potentials. We show that the potential energy function for the noble gases is described adequately by two-body interactions together with a part of the three-body interaction, the Axilrod and Teller [8] triple-dipole dispersion term. In the last section we discuss once more the Law of Corresponding States, and show that with care the massive amount of information on the noble gases can be represented by a two-parameter potential function.

In Section 2.2 we stated that quantum mechanical methods can be used to compute the leading terms in the asymptotic expansion of the potential energy of the inert gases. These terms have the functional form

$$-\frac{C_6}{r^6}-\frac{C_8}{r^8}-\frac{C_{10}}{r^{10}}\cdots$$

and are present in eq. 7.1. The coefficient C_6 represents the induced dipole-induced dipole interaction, C_8 the induced dipole-induced quadrupole term, and so on. The coefficient C_6 can be expressed in terms of the *dipole oscillator strengths* [9], which in turn can be obtained experimentally from optical data. Approximate values of C_8 and C_{10} can be derived from information on higher multipole oscillator strengths. Table 7.1 gives values of C_6, C_8, and C_{10}, together with the coefficient in the Axilrod-Teller interaction, for the noble gases. These values are used in the many semiempirical potential functions discussed in the remainder of this chapter.

Table 7.1 Coefficients in the multipole expansion of the dispersion interactions between inert gas atoms. The values given in this table are based on those given by A. I. M. Rae, 1975, *Mol. Phys.*, **29,** 467

	C_6(K Å^6)	C_8(K Å^8)	C_{10}(K Å^{10})	ν(K Å^9)
He-He	1.01×10^4	2.72×10^4	9.87×10^4	1.52×10^3
Ne-Ne	4.45×10^4	1.42×10^5	5.29×10^5	1.23×10^4
Ar-Ar	4.45×10^5	2.27×10^6	1.61×10^7	5.32×10^5
Kr-Kr	8.86×10^5	4.98×10^6	3.59×10^7	1.50×10^6
Xe-Xe	2.01×10^6	6.04×10^6	—	5.76×10^6
He-Ne	2.12×10^4	5.05×10^4	1.99×10^5	—
He-Ar	6.83×10^4	1.31×10^5	1.28×10^6	—
He-Kr	9.46×10^4	1.99×10^5	2.10×10^6	—
He-Xe	1.41×10^5	3.33×10^5	—	—
Ne-Ar	1.42×10^5	5.06×10^5	2.50×10^6	—
Ne-Kr	1.95×10^5	7.48×10^5	4.09×10^6	—
Ar-Kr	6.56×10^5	3.25×10^6	2.16×10^7	—

7.1 Dilute Gas Properties

In the past, studies of the dilute gas pair interactions were based on bulk properties such as the virial coefficients, viscosity, thermal conductivity, thermal diffusion, and self-diffusion coefficients. Recently, an increasing amount of information on the potential function has been extracted from molecular beam scattering measurements at high and low energies and from spectroscopic data on the bound state energies of inert gas dimers.

In Section 3.7 we showed that the second virial coefficient for a spherically symmetric classical gas is related directly to the pair potential by the equation

$$B_2(T) = -2\pi\int_0^\infty \left[\exp - \frac{\phi(r)}{kT} - 1\right] r^2\, dr \tag{7.2}$$

Except for helium, and possibly for neon at very low temperatures, this formula can be used to calculate the second virial coefficient for all the inert gases with the addition of the first one or two correction terms in the *semiclassical expansion* of the partition function [9,10]. In Section 3.4 we used heuristic arguments to make the transition from quantum statistical mechanics to classical statistical mechanics. This transition can be made more rigorous by expanding the partition function in even powers of the thermal wavelength $\lambda = (h^2/2\pi mkT)^{1/2}$. Unfortunately, the details of this expansion lie outside the scope of this book and we are restricted to quoting the results only. The important point to remember is that at reasonably high atomic masses and reasonably high temperatures λ is small and so only the first few terms in the expansion are significant.

The first correction term in the semiclassical expansion to the second virial coefficients is given by

$$B_2^{(1)}(T) = \frac{\lambda^2}{12(kT)^2} \int_0^\infty r^2 \left(\frac{d\phi}{dr}\right)^2 \exp - \frac{\phi(r)}{kT}\, dr \tag{7.3}$$

and is often discussed in terms of the so-called reduced de Broglie wavelength $\Lambda^* = h/\sigma(m\varepsilon)^{1/2}$ where σ is a characteristic length, usually the position of the zero in the pair potential, and ε is the maximum attractive energy. Values of Λ^* for the inert gases are discussed at the end of this chapter, when we reexamine the Law of Corresponding States. Examining the expression for Λ^*, we see that the size of the quantum correction will increase as the mass of the atom decreases. For helium, the lightest of the noble gases, the series expansion is very slowly convergent even at quite high temperatures, and the computation of $B_2(T)$ requires a full quantum mechanical treatment at temperatures less than about 50 K.

It can be shown that the quantum mechanical expression for the second virial coefficients is [9,11]

$$B_2(T) = \pm 2^{-5/2}\lambda^3 - 2^{3/2}\lambda^3 \sum_l (2l+1) \sum_n \exp\left(-\frac{E_{nl}}{kT}\right)$$
$$- 2^{3/2}\lambda^3 \sum_l (2l+1) \int_0^\infty \left[\frac{1}{\pi}\frac{\partial \eta_l}{\partial q}\right] \exp\left(-\frac{\lambda^2 q^2}{2\pi}\right) dq \tag{7.4}$$

where the minus sign is used for He-4 atoms which have zero nuclear spin, and the plus sign is used for He-3 atoms which have nuclear spin $\frac{1}{2}$. Only even values of l are used for He-4 and only odd values of l for He-3. The quantities η_l, q, and E_{nl} in these equations are the phase shifts, wave number, and bound state energies discussed in Sections 2.5–2.7. For helium at temperatures above 50 K using the first few semiclassical corrections yields just as accurate a result as using eq. 7.4 [12], and is less time consuming. At higher temperatures, above 700 K, the classical expression given in eq. 7.1 is adequate. For neon the quantum effects are smaller [13], and for argon it is necessary to use the semiclassical calculation only at very low temperatures. In Table 7.2, we have listed

Table 7.2 Classical second virial coefficients and the first two correction terms in the semiclassical expansion. Results are in cm³ mole⁻¹

Helium				Neon			
T(K)	$B_2^{(0)}$	$B_2^{(1)}$	$B_2^{(2)}$	T(K)	$B_2^{(0)}$	$B_2^{(1)}$	$B_2^{(2)}$
27.3	−5.00	9.07	−3.17	44.2	−50.2	3.38	−0.24
40.9	2.43	4.92	−0.98	70.2	−18.6	1.19	−0.04
61.3	6.80	2.76	−0.33	123.2	−0.0	0.42	−0.01
81.8	8.71	1.88	−0.16	273.2	10.6	0.12	0
122.6	10.29	1.12	−0.06	573.2	13.6	0.04	0
				873.2	13.9	0.03	0

values of the first few terms in the semiclassical contributions to the second virial coefficient for the lighter inert gases as a function of temperature. The dependence of successive corrections on increasing powers of Λ^{*2}/T is evident, and the table illustrates the onset of classical behavior for each species.

It is apparent from eq. 7.2 that the classical relation between the virial coefficient and interaction potential predicts that $B_2(T)$ depends essentially on the area under the attractive part of the potential curve. Changes in the short range repulsion have a relatively small effect; for example, a 10% increase in the repulsive energy can be compensated for by a 2–3% increase in the potential well depth [14]. Consequently, it is not sufficient to use second virial coefficient data alone if an accurate pair potential is required [15]. At low temperatures, where quantum mechanical calculations are needed, the relationship between $\phi(r)$ and $B_2(T)$ is much more complicated. Detailed studies have shown that for helium at very low temperatures (2–4 K) the second virial coefficient is quite sensitive to the value assumed for the long range dipole-dipole interaction [16].

Gas phase data can also be used to examine the importance of many body interactions between molecules. Striking evidence that the pair potential alone is not sufficient to describe the inert gas interaction is given by calculations of the third virial coefficient of argon. As we showed in Section 3.7, the third virial coefficient depends on both the pair and triplet potentials

$$B_3(T) = -\tfrac{1}{3}\int\int [f_{12}f_{23}f_{13} + f_{123}(1+f_{12})(1+f_{13})(1+f_{23})]\, d\mathbf{r}_{12}\, d\mathbf{r}_{23} \quad (7.5)$$

where $f_{ij} = \exp[-\phi(r_{ij})/kT] - 1$ and $f_{123} = \exp[-\phi^{(3)}(\mathbf{r}_1, \mathbf{r}_2, \mathbf{r}_3)/kT] - 1$. Barker and Pompe [1] developed an accurate pair potential for argon that correlated most of the transport data for the dilute gas together with the second virial coefficient measurements and zero temperature solid state data. They found that values of $B_3(T)$ calculated by ignoring three-body effects could be substantially in error. This problem was corrected by including in their calculation the Axilrod-Teller three-body term [8]

$$\phi^{(3)} = \frac{\nu(1+3\cos\theta_1\cos\theta_2\cos\theta_3)}{r_{12}^3 r_{23}^3 r_{13}^3} \tag{7.6}$$

where θ_1, θ_2, θ_3, r_{12}, r_{23}, and r_{13} are the angles and sides of the triangle joining the centers of the three atoms. As mentioned earlier in this chapter, Table 7.1 gives values of ν for the noble gases. Figure 7.1 illustrates the influence of triplet interactions on the third virial coefficient for argon.

The gas phase transport properties can be calculated using the well established Chapman-Enskog solution of the Boltzmann equation [9]. As was discussed in Section 2.4, these properties are obtained classically in

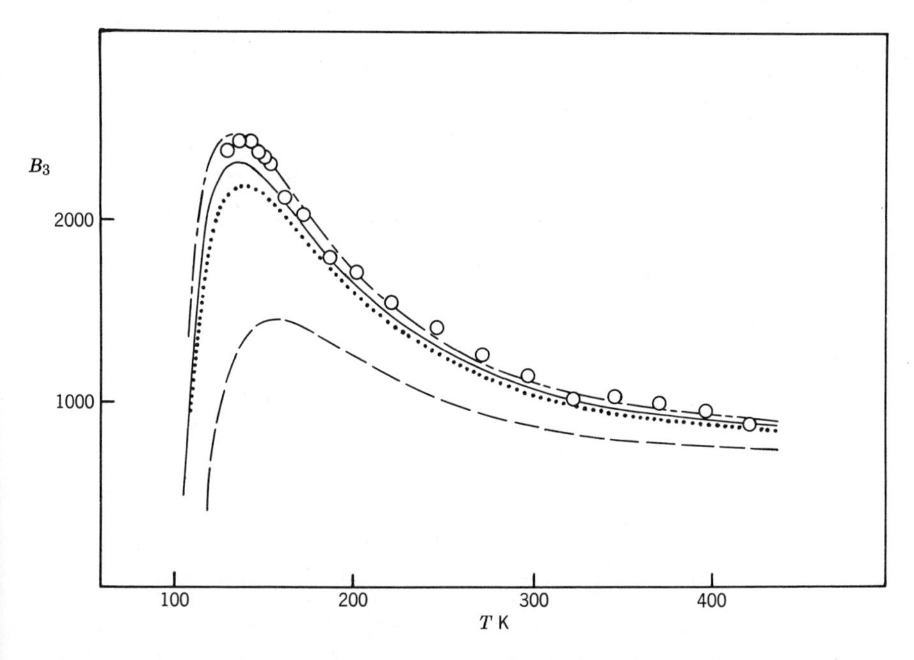

Figure 7.1 Third virial coefficient for argon in cm^3 mole^{-1} computed using Barker-Fisher-Watts potential. ○ experimental results; ---- two-body term only; ······ two-body+Axilrod-Teller term; —-—- two-body+Axilrod-Teller+third order dipole-quadrupole term; ——— two-body+Axilrod-Teller+third order dipole-quadrupole+fourth order tripole-dipole terms. (Adapted with permission from J. A. Barker, 1976, chapter 4 in *Rare Gas Solids*, M. L. Klein and J. A. Venables, Eds., Vol. 1, Academic, New York.)

terms of the collision integrals (see eqs. 2.47–2.49). Quantum mechanical calculations must be used for the transport coefficients of helium at temperatures lower than about 80 K [17]. Unfortunately, there is no useful semiclassical formulation of the transport coefficients comparable to that found for equilibrium properties [9,10], so that calculations are made using either the fully classical or the fully quantum mechanical expressions. The classical calculation appears to be sufficient for the heavier inert gases, and those quantum calculations that have been done for neon and argon are within 1% of their classical counterparts [13,18].

The viscosity is usually the most accurately measured dilute gas transport coefficient. Experimental values for the viscosity of argon, along with

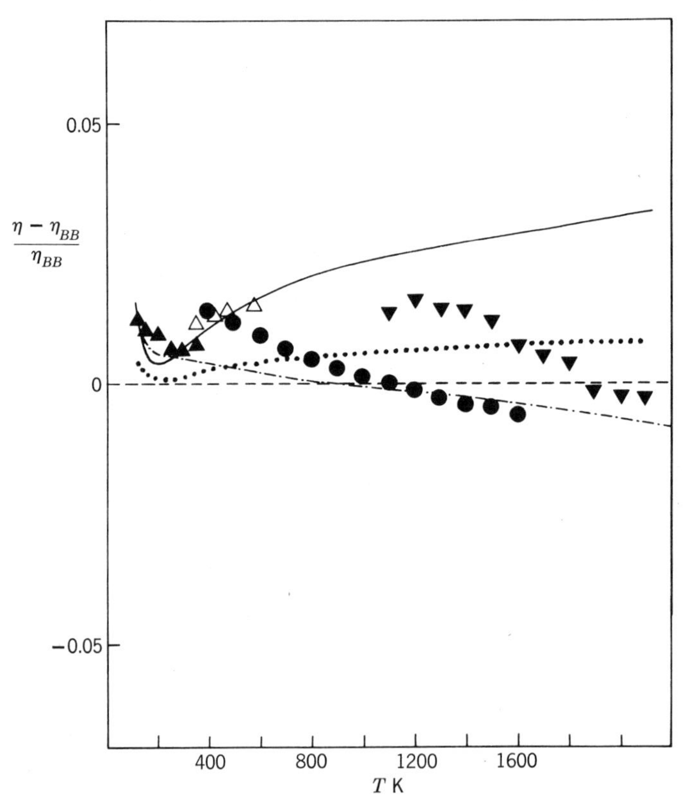

Figure 7.2 Comparison of calculated and experimental viscosities for argon; the quantity plotted is the relative difference from the values given for the Barker-Bobetic potential. —— Barker-Pompe potential; -·-·-·- Dymond-Alder potential; ······· Barker-Fisher-Watts potential. Several sets of experimental data are also shown. (Adapted with permission from J. A. Barker, 1976, chapter 4 in *Rare Gas Solids*, M. L. Klein and J. A. Venables, Eds., Vol. 1, Academic, New York.)

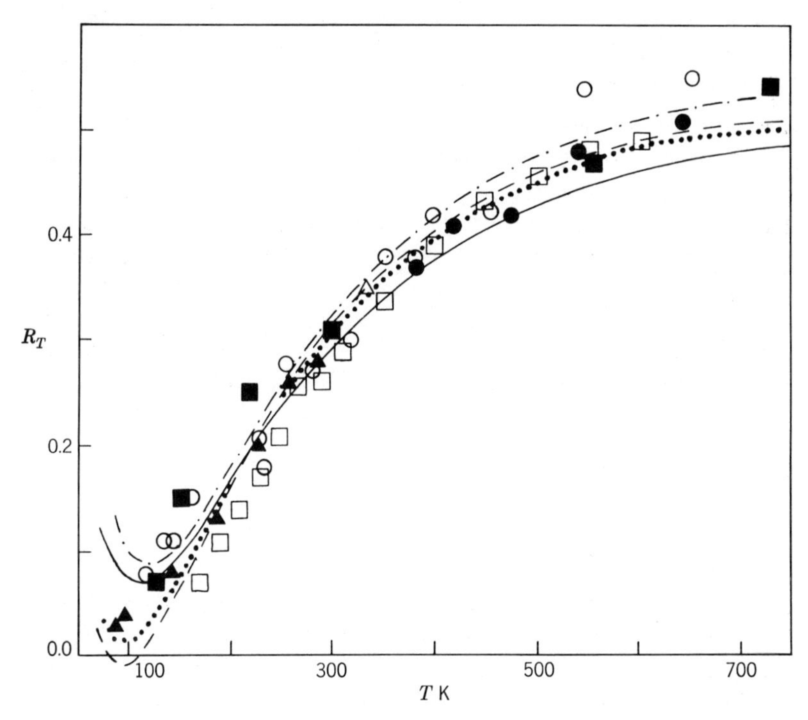

Figure 7.3 Dimensionless ratio of thermal diffusion coefficient to self-diffusion coefficient, $R_T = D_T/(N\,dD_s)$, where d is the mass density for argon. ——— Barker-Pompe potential; ---- Barker-Bobetic potential; –·–·– Dymond-Alder potential; ······ Barker-Fisher-Watts potential. Several sets of data are also shown. (Adapted with permission from J. A. Barker, 1976, chapter 4 in *Rare Gas Solids*, M. L. Klein and J. A. Venables, Eds., Vol. 1, Academic, New York.)

the predictions of several pair potentials, are shown in Fig. 7.2. It can be seen that the viscosity gives a reasonable method for examining the usefulness of pair potentials. The thermal conductivity depends on the same collision integrals as the viscosity and hence does not give an independent test of model intermolecular potential functions. In any case, this property has not been measured to the same accuracy as the gas viscosities. Experimental values of the thermal diffusion ratio for argon are shown in Fig. 7.3 together with results from several model pair interactions. It is evident that improved experimental measurements at very low temperatures would give a better criterion for preferring one potential over another. The thermal diffusion ratio is quite sensitive to the shape of the intermolecular potential function, particularly if a comparison is made over a large temperature range. Unfortunately, experiments to date have not been as accurate as desirable, although this situation will undoubtedly improve. Finally, the gas phase self-diffusion coefficient measurements are neither sufficiently accurate nor available over a sufficiently wide temperature range at the present time to provide definitive tests of interatomic potentials.

In general, the combined use of second virial coefficients and transport properties is not sufficient to give an unambiguous preference for one potential function over another. Hanley and Klein [19] discussed this point in detail when deriving parameters for the m-6-8 potential. They show that measurements of bulk gas phase data can be reproduced by a number of rather different potential functions over a reasonably large temperature range. The extraction of the third and higher virial coefficients from experimental data is quite difficult. Although these coefficients give information on the importance of many body interactions they are not usually known with sufficient accuracy to enable quantitative comparisons to be made between various model pair potentials. It is probably preferable to examine directly the bulk properties of the solid state and of liquid and dense gases using either machine calculations or other accurate theories.

Molecular beam scattering experiments of the type discussed in Sections 2.5 and 2.6 can explore the potential energy of the colliding atoms over a wide range of separations. Although high energy molecular beam total scattering cross-section experiments were reported over 30 years ago [20], there has been a marked increase in the number of experiments recently [21–27] largely due to the advent of supersonic nozzles and accurate collimating devices which generate intense beams of low energy atoms. This advance has allowed not only total cross-sections to be measured accurately as a function of incident energy or velocity but also enabled much more refined differential scattering cross-section experiments to be performed. In these latter experiments the intensity of the scattered particles is measured as a function of energy and scattering angle, thus giving considerably more information on the pair potential. The differential scattering cross-section for argon measured by Parson, Siska, and Lee [21] is shown in Fig. 7.4 as a function of scattering angle and compared with a number of models of the Ar-Ar pair potential. It can be seen that the scattering data give a good test of the accuracy of various model pair potentials.

The result of the many accurate measurements of scattering cross-sections is a wealth of data on both the pure noble gases, and on their mixtures, which can be fitted to a sufficiently flexible analytic potential. If the experimental data are coupled with theoretical calculations of the interaction potential a valuable cross-check on the approximations employed in the calculations is available. In addition to this use of scattering data, the recent advances in experimental methods have enabled a

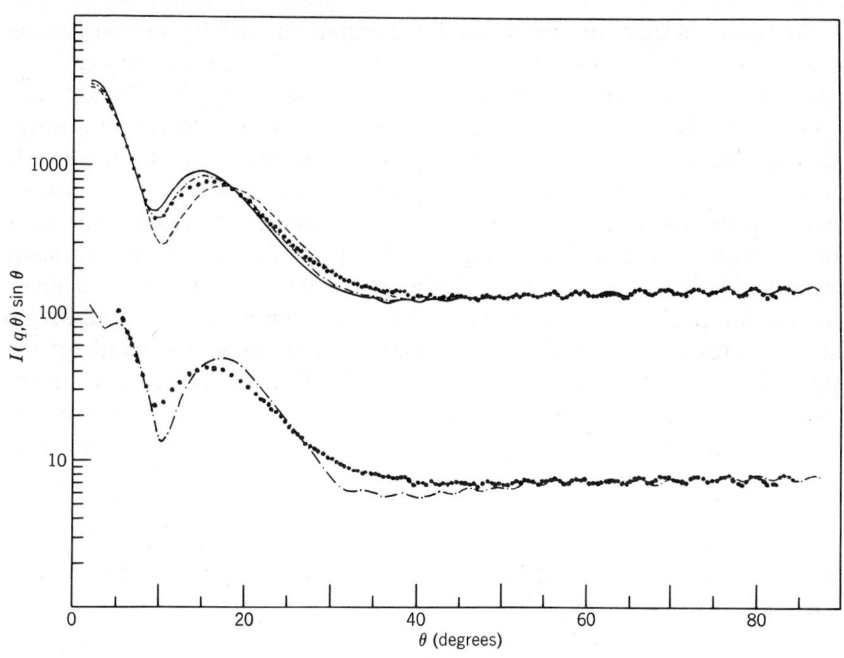

Figure 7.4 Molecular beam differential scattering cross-section (arbitrary units) for argon compared with theoretical predictions. ●●● experimental data [21]; ——— Barker-Bobetic potential; - - - - - - - Barker-Pompe potential; -●-●- Barker-Fisher-Watts potential in the upper curve and Dymond-Alder potential in the lower curve. (Adapted with permission from *Mol. Phys.*, **21,** 657, 1971.)

number of other phenomena to be examined. One example of such measurements is the recent observation of the Ramsauer-Townsend effect in He-He scattering [23,25]. This purely quantum mechanical effect occurs in scattering experiments where the relative collision energy is so low that only s wave phase shifts (i.e., $l = 0$) are important. As the s wave phase passes through zero the total cross-section almost vanishes.

Although scattering data alone are still not sufficient to determine the interaction potential uniquely, their use with other dilute gas data such as virial coefficients, quantum calculations of the dispersion forces, and solid state data can supply valuable information. Siska et al. [27] have found it advantageous to use piecewise continuous analytic functions to fit their measurements. Data sensitive to the short, intermediate, and long range regions of the potential were fitted almost independently and joined smoothly using cubic spline functions. The result was a realistic yet flexible potential curve—a so-called ESMSV potential (Exponential-Spline-Morse-Spline-van der Waals), which combines theoretical and experimental data to obtain a single curve.

Another method of obtaining information about the intermolecular potential is to measure the bound state energies, or vibrational level spacings, of the van der Waals dimer (see Section 2.7). Tanaka and Yoshino [28] have measured the band structure in the ultraviolet absorption spectra of the inert gas dimers. These experiments make it possible to determine the vibrational level spacings as side bands on the transition between the electronic ground state and the first excited electronic state of the inert gas dimer. The energy levels can be determined theoretically by calculating the eigenvalues of the radial wave equation for a given interaction potential. Alternatively, the experimental values of the bound state energy levels can be used directly to determine the width of the potential well as a function of energy relative to the energy at the minimum [29]. Both experimental and theoretical studies of the bound state energies of the inert gases have been reported. There appears to be no helium dimer to support a measurable vibrational level [30,31], but Tanaka and Yoshino have observed one level in the He-Ne dimer [32]. Two vibrational levels have been observed in Ne_2 [32] and six in Ar_2 [28], all of which have served to augment existing information on the shape of the well in the pair potential.

7.2 Liquid State Properties

When a fluid is compressed so that it condenses from the gaseous to the liquid phase it assumes a local short range order that is suitably measured by the radial distribution function $g(r)$. As we described in Chapter 5, this function can be measured experimentally using either X-ray or neutron diffraction studies. The radial distribution function can be computed using either Monte Carlo and molecular dynamics calculations [33] or from various approximations [34], and so it might be thought useful for determining intermolecular potential functions. Unfortunately, at liquid densities $g(r)$ is insensitive to the detailed form of the interaction potential and is determined primarily by the short range repulsion. As we observed in Chapter 5, calculations based on an assembly of hard spheres are capable of giving a good qualitative description of the main features of the radial distribution function. Consequently, measurements of $g(r)$ in the liquid state give a rather insensitive test of the total intermolecular potential. In the limit of very low densities, $g(r)$ approaches the form exp $[-\phi(r)/kT]$ so that a precise measurement of $g(r)$ could, in principle, lead to a determination of $\phi(r)$ [35]. Accurate low density measurements are difficult to perform and although it has been possible to obtain some information this type of data inversion is not competitive with the molecular beam and spectroscopic techniques. The situation is even more complicated for liquid He-3 and He-4, where quantum statistical mechanics must be used. Although experimental measurements have been reported [36], exploratory calculations of $g(r)$ are in their infancy [37–40]. Consequently, we will not say very much about the agreement between experimental and calculated radial distribution functions.

Perhaps the most severe test of model intermolecular potentials afforded by liquid state properties is the reproduction of experimental pressures at various temperatures [3,41]. Contributions to the pressure include a positive component from the short range repulsion balanced by a negative component due to the attractive portion of the potential. Hence a small shift in the core radius has a large effect on the calculated pressure. In addition, three-body contributions can be large and must be included in realistic calculations. Quantum corrections, although small for argon, krypton, and xenon, should be incorporated into the calculations for all the inert gases. Liquid helium has to be examined using a full quantum mechanical treatment; details of this procedure are left until Section 7.4.

The most accurate method available for calculating the bulk properties are the computer simulation methods [33]. Although perturbation theories appear to be accurate for potential functions of the inert gas type [41], there is no guarantee that they have converged. Integral equation approximations, and other approximate methods, are not sufficiently accurate to allow a quantitative study of a given model potential to be made. Despite the accuracy of simulation methods they have two major disadvantages. First, the pair interaction has to be truncated at a relatively short distance, usually about 3σ, and second, it is not possible to include many body interactions explicitly. These disadvantages could be reduced, in principle, by using larger and faster computers but, unfortunately, even the next generation of machines will probably be inadequate for this task. It follows that if quantitative studies are to be made some corrections must be included for these missing terms. A satisfactory method for doing this is to use a simple perturbation theory with the truncated two-body potential as the reference system. We can write the total potential energy of the system as

$$\Phi(\mathbf{r}_1, \ldots \mathbf{r}_N) = \Phi_0(\mathbf{r}_1, \ldots \mathbf{r}_N) + \Phi_T(\mathbf{r}_1, \ldots \mathbf{r}_N) + \Phi_M(\mathbf{r}_1, \ldots \mathbf{r}_N) \tag{7.7}$$

where if $\phi(r)$ is the pair potential and R is the distance at which $\phi(r)$ is truncated

$$\Phi_0 = \begin{cases} \sum_{i<j} \phi(r_{ij}) & r_{ij} \leqslant R \\ 0 & R < r_{ij} \end{cases}$$

$$\Phi_T = \begin{cases} 0 & r_{ij} \leqslant R \\ \sum_{i<j} \phi(r_{ij}) & R < r_{ij} \end{cases} \tag{7.8}$$

and Φ_M is the many body contribution, usually a sum over triplets of the three-body Axilrod-Teller potential, eq. 7.6. If Φ_0 is treated as the reference system, and Φ_T and Φ_M as small perturbing terms, it is possible to use the simple perturbation theory outlined at the beginning of Section 3.9 to obtain estimates of the correction terms. The first quantum correction in the semiclassical expansion of the partition function [9,10] can also be included in this perturbation scheme. Barker, Fisher, and Watts [3] used this method to first order to obtain the following results for the thermodynamic internal energy and the equation of state:

$$\begin{aligned} U = {} & \tfrac{3}{2}NkT + \langle\Phi_0\rangle + \langle\Phi_T\rangle + \langle\Phi_M\rangle \\ & - \frac{1}{kT}[\langle\Phi_0\Phi_T\rangle - \langle\Phi_0\rangle\langle\Phi_T\rangle + \langle\Phi_0\Phi_M\rangle - \langle\Phi_0\rangle\langle\Phi_M\rangle] \\ & + 2\langle\Phi_Q\rangle - \frac{1}{kT}[\langle\Phi_Q\Phi_0\rangle - \langle\Phi_Q\rangle\langle\Phi_0\rangle] \end{aligned} \tag{7.9}$$

and

$$\begin{aligned} pV = {} & NkT + \langle P_0\rangle + \langle P_T\rangle + \langle P_M\rangle + \langle P_Q\rangle \\ & - \frac{1}{kT}[\langle P_0P_T\rangle - \langle P_0\rangle\langle P_T\rangle + \langle P_0P_M\rangle \\ & - \langle P_0\rangle\langle P_M\rangle + \langle P_0P_Q\rangle - \langle P_0\rangle\langle P_Q\rangle] \end{aligned} \tag{7.10}$$

In these equations the averages $\langle\chi\rangle$ are computed using positions generated from a Monte Carlo or molecular dynamics calculation with the truncated potential Φ_0. The terms, P_0, P_T, and P_M are related to the derivatives of the potential with respect to distance, so that from eqs. 3.54 and 7.8

$$P_0 = -\frac{1}{3}\sum_{i<j} r_{ij}\frac{\partial\phi}{\partial r_{ij}}, \qquad r_{ij} \leqslant R$$

$$P_T = -\frac{1}{3}\sum_{i<j} r_{ij}\frac{\partial\phi}{\partial r_{ij}}, \qquad r_{ij} > R$$

The many body term, P_M, can only be written down once the particular contributions to be included are specified; for a three-body interaction we have

$$P_3 = -\frac{1}{3} \sum_{i<j<k} \left(r_{ij} \frac{\partial}{\partial r_{ij}} + r_{ik} \frac{\partial}{\partial r_{ik}} + r_{jk} \frac{\partial}{\partial r_{jk}} \right) \phi^{(3)}(\mathbf{r}_i, \mathbf{r}_j, \mathbf{r}_k)$$

Finally, the quantum corrections can be calculated from the leading terms in the semiclassical expansion [9,10] mentioned in the previous section

$$\Phi_Q = \frac{h^2}{48\pi^2 mkT} \sum_{i<j} \boldsymbol{\nabla}_i^2 \phi(r_{ij})$$

and

$$P_Q = -\frac{h^2}{48\pi^2 mkT} \frac{1}{3} \sum_{i<j} \boldsymbol{\nabla}_i^2 \phi(r_{ij})$$

with $\boldsymbol{\nabla}_i^2$ the Laplacian for the ith particle.

It is important to note that a quantitative description of the bulk properties of a simple liquid cannot be obtained using an accurate pair potential unless all these contributions are included. This is well illustrated for argon in Table 7.3, where contributions to the pressure from various parts of the interaction potential are given using the Barker, Fisher, and Watts (BFW) [3] pair potential together with the Axilrod-Teller three-body term. It can be seen that the total pressure is the sum of large, canceling terms. Internal energies are much less sensitive than the pressure and the contributions from long range and three-body terms are smaller. The dependence of this property on the potential form is more like that of the second virial coefficient—a weighted integral over the attractive potential well. Consequently, calculations of U provide a less sensitive test of model interaction potentials.

Table 7.3 Contributions to the pressure, in atmospheres, of liquid argon from various parts of the interaction potential

T(K)	V(cm^3 mole^{-1})	p_0	p_T	p_3	p_Q	Total	Experimental
100	27.04	454.6	−214.7	364.2	42.2	646	652
100	29.66	14.1	−162.1	238.8	25.3	116	105
140	30.65	495.6	−146.7	214.3	16.7	580	583
140	41.79	23.7	−57.4	49.0	2.7	18	37
151	70.73	46.3	−11.8	13.2	1.2	49	49

7.3 Solid State Properties

Table 7.3 has demonstrated that when condensed phase data are used to determine the pair potential it is essential that many body interactions be included. Satisfactory results have been obtained for the inert gases by assuming that the Axilrod-Teller three-body contribution is the only important many body interaction, although recent calculations have shown that there are small but not negligible contributions from the third order dipole-quadrupole and fourth order triple-dipole interactions. These terms are of opposite sign but almost equal magnitude, and so cancel for argon and krypton although they may be more important in xenon [42–44].

These remarks also hold for the crystalline inert gases. Thermodynamic properties of the solids depend rather strongly on details of the intermolecular forces. In particular, solid state data, such as the static lattice energy, bulk modulus, and phonon dispersion curves [2, 45], provide a more precise determination of the position of the potential minimum R_m than does gas phase data. The lattice spacing predicted for the crystal depends strongly on R_m and is nearly equal to it for the heavier inert gases [46]. As is the case for liquids, comparing measured and experimental values of the internal energies and pressures for the solid provides a sensitive test of any potential. Contributions to these quantities from various parts of the interaction, calculated along the melting curve at temperatures between 40 K and 80 K, are given in Tables 7.4 and 7.5 for argon, again using the BFW potential for the pair interaction [4]. Measurements of the adiabatic elastic constants of the inert gas solids give further data with which to test the accuracy of intermolecular potential models. Fisher and Watts [4] calculated these quantities for solid argon using the BFW potential and at the time obtained poor agreement with measured values. This was disappointing because of the otherwise complete success of this particular potential and it was suggested that the experimental data were in error [4]. However, the elastic constants were subsequently remeasured by Stoicheff and co-workers [47] and agree very well with the computed values.

The problem of determining the crystal structure of the inert gas solids is still not completely resolved, although it is apparent that many body forces are the important terms. Experimentally, the gases are observed to crystallize into the face centered cubic structure (except for helium), whereas current calculations of the binding energies predict that the hexagonal close packed structure is slightly favored. The difference between the static lattice energies of the two structures is very small and hence a very accurate calculation is required to determine their relative stabilities [44].

Table 7.4 Contributions to the internal energy of solid argon from various parts of the interaction potential. Energies are given in degrees Kelvin

T(K)	V(cm^3 mole^{-1})	U_0	U_T	U_3	U_Q	Total	Experimental
40.0	23.06	−974.8	−48.4	68.8	24.0	−930.4	−944.0
50.0	23.29	−953.9	−47.8	66.5	18.3	−916.9	−920.4
60.0	23.61	−933.9	−46.0	64.1	14.8	−901.0	−904.3
70.0	23.98	−909.1	−44.2	61.7	12.3	−879.3	−884.7
80.0	24.43	−884.5	−43.5	57.9	9.7	−860.4	−863.0

Table 7.5 Contributions to the pressure, in atmospheres, of solid argon from various parts of the interaction potential

T(K)	V(cm^3 mole^{-1})	p_0	p_T	p_3	p_Q	Total
40.0	23.06	−609.0	−355.0	693.0	239.0	33.0
50.0	23.29	−458.0	−351.0	656.0	168.0	15.0
60.0	23.61	−410.0	−330.0	621.0	131.0	12.0
70.0	23.98	−344.0	−312.0	590.0	103.0	38.0
80.0	24.43	−313.0	−315.0	536.0	71.0	21.0

7.4 Bulk Properties of Helium

A wealth of experimental data exists on the bulk properties of helium in all three phases, but unfortunately theoretical calculations of these quantities usually require a full quantum mechanical analysis. Consequently, the He-He interaction is less accurately known than the other inert gas pair potentials. Many model potentials have been proposed for the helium interaction, and most of these are obtained empirically using dilute gas data such as second virial coefficients and viscosities [16, 30, 48]. More recently, interaction potentials have been obtained from scattering experiments [49] and from quantum mechanical treatments [50, 51]. When information from experimental data is used to determine $\phi(r)$ it is often supplemented by the known theoretical long and short range behavior [52, 53]. The resulting semiempirical potentials have the advantages of simplicity and fair accuracy but would be superseded by full fledged *ab initio* quantum mechanical calculations. In a step toward this goal, Snook and Watts [54] showed that good agreement with experiment in both the dilute gas and liquid phases could be obtained by combining *ab initio* calculations of the intermediate range of the potential with accurate long and short range forms.

Parameters for the Lennard-Jones 12-6 potential appropriate to helium were reported in 1939 by de Boer and Michels [55] who determined parameters $\varepsilon/k = 10.22$ K and $\sigma = 2.556$ Å. Even though this model is known to have the incorrect long range form it has proved to be as good a two-parameter representation of the helium potential as those developed later [56]. Beck developed a four-parameter analytic potential function using second virial coefficient data from 50 K to 1500 K together with *a priori* calculations of the high energy potential [48]. This potential was shown subsequently to be in good agreement with solid state data [57]. Bruch and McGee [16, 30], who modified a piecewise continuous pair potential of Bernstein and Morse [58], reported semiempirical potentials that fitted dilute gas data to within 6% over the temperature range 2–2000 K.

The Bruch-McGee potentials were constructed by using a parameterized representation of the bowl of the potential to join smoothly the theoretically determined short and long range forms. Two of their potentials were given by the formulae

$$\phi_{\text{BM1}}(r) = \begin{cases} \phi_{\text{SR}}(r) & r < 0.644\ \text{Å} \\ -\varepsilon\left[1 + c\left(1 - \dfrac{1}{x}\right)\right] \exp\left[-c(x-1)\right] & 0.644\ \text{Å} \leqslant r < 3.511\ \text{Å} \\ \phi_{\text{LR}}(r) & 3.511\ \text{Å} \leqslant r \end{cases} \tag{7.11}$$

where $\varepsilon = 12.5$ K, $x = r/2.98$ Å, and $c = 8.00877$ and

$$\phi_{\text{BM2}}(r) = \begin{cases} \phi_{\text{SR}}(r) & r < 0.8919\ \text{Å} \\ \varepsilon[\exp\left[2c(1-x)\right] - 2\exp\left[c(1-x)\right]] & 0.8919\ \text{Å} \leqslant r < 3.682\ \text{Å} \\ \phi_{\text{LR}}(r) & 3.682\ \text{Å} \leqslant r \end{cases} \tag{7.12}$$

where $\varepsilon = 10.75$ K, $x = r/3.0238$ Å, and $c = 6.127768$. For the short range behavior Bruch and McGee used a fit to the results of Gilbert and Wahl [53]

$$\phi_{\text{SR}}(r) = 1.987 \times 10^6 \exp\left[-3.4573r - 0.26206r^2\right] \text{K} \tag{7.13}$$

The long range form was given by the first two terms in the multipole expansion [52]

$$\phi_{\text{LR}}(r) = -\frac{C_6}{r^6} - \frac{C_8}{r^8} \tag{7.14}$$

where $C_6 = 1.021 \times 10^4$ K Å^6 and $C_8 = 2.77 \times 10^4$ K Å^8. The two Bruch-McGee potentials and the Lennard-Jones potential of de Boer and Michels are shown in Fig. 7.5. It can be seen that both ϕ_{BM1} and ϕ_{BM2} are "harder" and deeper than the Lennard-Jones potential. As both the Bruch-McGee potentials fit gas phase data equally well, and as one is the deepest and the other the shallowest of their family, it is likely that together these two models give a measure of the uncertainty in the helium pair potential.

The Bruch-McGee, the Beck, the Lennard-Jones, and some of the *a priori* potentials have been used in calculations of condensed phase data. Variational calculations for solid helium [57] clearly demonstrated the inadequacy of the usual Lennard-Jones potential and showed that the Beck potential gave better overall agreement with experimental data. Similarly, it was shown in Monte Carlo calculations of the density dependence of the ground state energy of liquid helium [39] that the

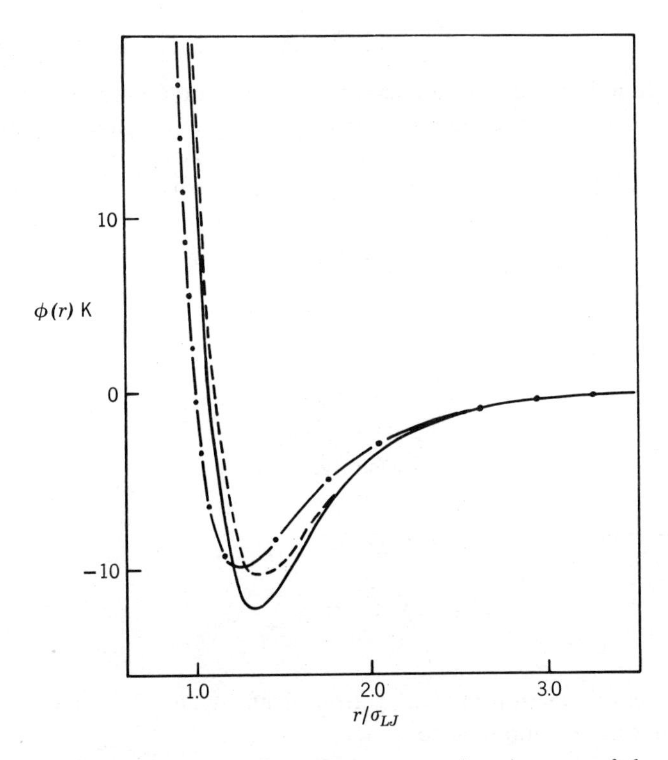

Figure 7.5 Pair potentials for helium. Distances are given in terms of the zero of the Lennard-Jones potential, σ_{LJ}. –·–·– LJ potential; ——— BM1 potential; ---- BM2 potential. (Adapted with permission from *J. Low Temp. Phys.*, **2,** 507, 1970.)

Lennard-Jones potential was less satisfactory than the potential ϕ_{BM1} of Bruch and McGee. As noted by Bruch and Certain [59], progress in the refinement of the helium potential will probably come from high resolution scattering experiments and from additional theoretical understanding of the origin of the potential minimum.

We stated earlier that low temperature studies of the properties of bulk helium have to be based on quantum mechanical calculations. In this book we have largely restricted ourselves to the use of classical statistical mechanics, introducing quantum effects using perturbation theory. Such an approach is not possible for liquid helium and thus some alternative must be used. It is possible to use the computer simulation methods and the approximate theories discussed in Chapter 3 to calculate some properties of liquid helium provided we make certain assumptions. We assume that the ground state wave function of the liquid can be written as a product of pair wave functions

$$\Psi(\mathbf{r}_1, \ldots \mathbf{r}_N) = \prod_{i<j} f(\mathbf{r}_{ij}) \tag{7.15}$$

The pair function may be an assumed form, containing one or more variational parameters chosen to minimize the ground state energy [60], or it may be a form based on a solution to the two-particle Schrödinger equation [37]. Suppose we write the pair function $f(r_{ij})$ in the form

$$f(r_{ij}) = \exp\left[u(r_{ij})\right] \tag{7.16}$$

then the ground state energy of the fluid is given by

$$E_0 = \frac{\displaystyle\int \Psi^* H_N \Psi \, d\mathbf{r}_1 \cdots d\mathbf{r}_N}{\displaystyle\int \Psi^* \Psi \, d\mathbf{r}_1 \cdots d\mathbf{r}_N} \tag{7.17}$$

where the Hamiltonian operator is

$$H_N = -\sum_i \frac{\hbar^2}{2m} \nabla_i^2 + \sum_{i<j} \phi(r_{ij}) \tag{7.18}$$

When the wave function takes the form given by eqs. 7.15 and 7.16 this expression can be simplified to

$$E_0 = 2\pi N\rho \int_0^\infty r^2 \tilde{\phi}(r) g(r) \, dr \tag{7.19}$$

where the radial distribution function is given by

$$g(r)=\frac{N(N-1)}{\rho^2}\frac{\int \Psi\Psi\, d\mathbf{r}_3\cdots d\mathbf{r}_N}{\int \Psi\Psi\, d\mathbf{r}_1\cdots d\mathbf{r}_N}$$

and

$$\tilde{\phi}(r)=\phi(r)+\frac{\hbar^2}{2m}\nabla^2 u(r)$$

Given the form for $\Psi(\mathbf{r}_1,\ldots \mathbf{r}_N)$ it is apparent that $g(r)$ is determined by a fictitious classical gas interacting through a pair potential $V(r)=2kT_{\mathrm{eff}}u(r)$ and consequently we can determine the distribution function using machine simulation methods or other procedures. It is important to note that the radial distribution function is determined directly by the form assumed for the wave function rather than by the pair potential. The pair potential has an indirect influence on $g(r)$ as it determines $\Psi(\mathbf{r}_1,\ldots \mathbf{r}_N)$ through the Schrödinger equation.

Two methods have been used to obtain the pair wave function $f(r)$. In the first method, the variational method, an analytical form is assumed that contains a variational parameter. Calculations of the ground state energy are then repeated for several values of the variational parameter until the ground state energy takes up a minimum value. Although the functional form assumed for $f(r)$ is not changed, for different pair interactions the variational parameter will depend on the details of the potential. Consequently, the predicted ground state properties of the liquid, including both $g(r)$ and E_0, will depend on $\phi(r)$. The second method used to obtain $f(r)$ is to use a solution to the two-particle Schrödinger equation for a particular interaction potential as the approximation. This approach ignores any many body contributions to the total wave function, but the form of $\Psi(\mathbf{r}_1,\ldots \mathbf{r}_N)$, and hence the ground state properties, depend on the interaction potential.

The most common assumption in the first method for the calculation of $g(r)$ and E_0 is

$$f(r)=\exp\frac{-a^5}{r^5} \tag{7.20}$$

where a is the variational parameter. This assumption has been used in both machine calculations [39,40] and in solutions to integral equations [39,61]. Typical results for $g(r)$, obtained using the PY, HNC, and Monte Carlo methods [34,39] together with the Lennard-Jones potential, are shown in Fig. 7.6. In addition, results for E_0 from the PY, HNC, and Monte Carlo methods obtained using several pair potentials are given in

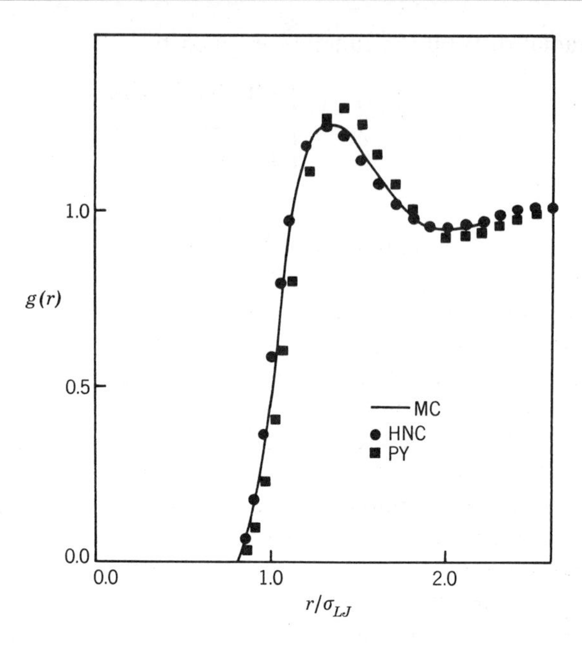

Figure 7.6 Comparison of PY, HNC, and MC calculations of the radial distribution function of He-4 for $\alpha = 2.60$ Å, $\rho\sigma_{\mathrm{LJ}}^3 = 0.365$. Distances are measured as in Fig. 7.5. (Adapted with permission from *J. Low Temp. Phys.*, **2,** 507, 1970.)

Table 7.6. It can be seen that the HNC results are in better agreement with results from the Monte Carlo calculations than are those from the PY approximation. This observation is rather unusual in that the PY approximation is usually superior to the HNC approximation [34]. The variational method has been used to examine the accuracy of various pair potentials for helium, and from Table 7.6 it is probable that one of the Bruch-McGee family is the best semiempirical model currently available [39].

Published work on the ground state properties of liquid helium using wave functions obtained from the Schrödinger equation is less extensive. McGee and Murphy [62] have constructed a product wave function for the liquid using the zero energy s wave (i.e., $l = 0$) solution to the two-body Schrödinger equation for the Lennard-Jones potential. Their results were similar to those obtained from the variational method. However, they argued that if a more realistic potential had been used, better agreement with experiment would be obtained. More recently Bruch, McGee, and Watts [38] have extended the approach used for the ground state properties of the liquid to higher temperatures. To do this the pair wave function was replaced by the two-body Slater sum. This

Table 7.6 Density dependence of the ground state energy of liquid helium-4. Energies are given in Kelvin (E_0/k) and densities in atoms Å^{-3}

Method	$\phi(r)$	ρ: .0180	.0195	.0210	.0219	.0231	.0240	.0252
MC	LJ	−5.86	−5.96	−5.83	−5.70	−5.47	−5.19	−4.91
	BM1	−6.97	−7.10	−7.07	−6.95	−6.69	−6.44	−6.15
	BM2	−5.67	−5.62	−5.32	−5.06	−4.65	−4.24	−3.77
HNC	LJ	−5.16	−5.02	−4.73	−4.47	−4.01	−3.57	−2.86
	BM1	−6.14	−6.02	−5.70	−5.41	−4.87	−4.35	−3.51
	BM2	−4.72	−4.38	−3.85	−3.40	−2.66	−1.97	−0.90
PY	LJ	—	—	−7.53	−7.72	−7.93	−8.06	−8.20
	BM1	—	—	−9.08	−9.31	−9.58	−9.74	−9.90
	BM2	—	—	−7.41	−7.52	−7.61	−7.64	−7.61
Experimental		—	—	—	−7.21	−7.15	−7.05	−6.81

quantity is the quantum statistical mechanical equivalent of the Boltzmann factor, $\exp[-\phi(r)/kT]$, and so can be used in pseudoclassical calculations. Bruch et al. calculated the Slater sums using a method developed by Klemm and Storer [63], who had reported some preliminary calculations for the PY approximation for He-4. Bruch et al. [38] used the PY and HNC approximations to calculate the critical point parameters of He-3 and He-4, obtaining reasonable agreement with experiment.

7.5 The Argon-Argon Interaction

Since quantum effects are relatively small in argon, this gas has been widely used as the standard by which theories of classical statistical mechanics are judged. In addition, as there has been a very wide range of experimental data reported for this substance, argon has also been used to analyze the pair interaction between atoms. Much of the work on pair interactions has been carried out using the traditional Lennard-Jones 12-6 potential, and parameters for this model have been derived from the gas, liquid, and solid phases [9, 46, 64]. The values $\varepsilon/k = 119.8$ K and $\sigma = 3.405$ Å are the most widely used values of the parameters for this model of the argon pair potential. However, the work of Guggenheim and McGlashan [65] in particular showed clearly that the intermolecular forces in argon are considerably more complicated than that described by a simple Lennard-Jones model. Over the years there have been many studies of the shape of the argon pair potential and it is interesting to compare some of these. Table 7.7 gives values of the well depth, ε/k,

Table 7.7 Parameters important in determining the shape of the argon pair potential. All the models represented in the table were fitted to experimental data. The curvature, γ^*, is dimensionless (see text). This table was adapted from a more extensive table given by Parson, Siska, and Lee [21]

$\phi(r)$	ε/k(K)	σ(Å)	R_m(Å)	γ^*	C_6(K Å^6)
LJ 12-6	119.8	3.405	3.822	72.0	7.468×10^5
LJ 16-6	148.9	3.315	3.721	96.0	6.324×10^5
Morse	132.6	3.554	4.040	66.3	—
Exp-6	152.1	3.302	3.644	99.0	5.342×10^5
12-6-8	153.0	3.292	3.669	84.0	1.493×10^5
Kingston	146.5	3.321	3.728	72.0	4.538×10^5
Munn-Smith	153.0	3.31	3.64	—	4.829×10^5
Dymond-Alder	138.2	3.28	3.813	—	4.855×10^5
Barker-Pompe	147.7	3.341	3.756	79.5	4.437×10^5
Barker-Bobetic	140.2	3.367	3.763	81.9	4.458×10^5
BFW	142.1	3.360	3.761	—	4.452×10^5
MSV III	140.8	3.345	3.760	78.9	4.695×10^5
Kihara	147.2	3.314	3.675	88.9	3.847×10^5

position of the zero of the potential, σ, position of the minimum, R_m, the second derivative of the potential at its minimum, $\gamma^* = \sigma^2\phi''(R_m)/\varepsilon$, and the value of the coefficient C_6 in the term in r^{-6}. The potential reported by Barker, Fisher, and Watts [3] gives the best fit to all the available argon data in all three phases, and in addition it has the correct value of C_6 [66]. Even so, the other potentials represented in the table were also fitted to various experimental properties, and it is interesting to note that no one set of experimental data can give the complete pair potential.

Dymond and Alder [14] made an extensive study of the argon interaction and reported a pair potential that fitted the available gas phase data, including virial coefficients and transport coefficients. This potential was presented in tabular form, and no attempt was made to represent their model as an analytic function. Consequently, it was not very useful for calculating properties depending on the derivatives of the potential such as the equation of state of the liquid or the elastic constants of the solid. The Dymond-Alder potential was characterized by a broader bowl than that of most argon potentials and did not go smoothly to the correct value of the potential at longer distances. Subsequently, it has been shown to be in disagreement with spectroscopic data and differential scattering cross-sections [3, 21, 67].

The importance of including three-body interactions in calculations of the bulk properties of argon was clearly demonstrated by Barker and

Pompe [1]. They were unable to obtain a single pair potential that represented the gas phase data and the static lattice energy and spacing of the solid unless the three-body Axilrod-Teller potential [8] was included. Specifically, they found that none of their pair potentials would predict the third virial coefficient for argon unless this three-body term was included. Subsequently, most calculations of the argon interaction potential have recognized that three-body terms must be included [2, 18, 45, 68, 70]. In particular, Barker and Bobetic [2] found that when this term was included it was possible to derive a pair potential that gave excellent agreement with the thermodynamic and transport properties of the gas and with the very low temperature properties of the solid. Following these findings, the Barker-Pompe and Barker-Bobetic models were used in liquid state calculations by Barker, Fisher, and Watts [3, 69]. These authors showed that although both models gave good agreement with the internal energy and self-diffusion coefficient of the liquid, agreement with the equation of state was poor. Consequently, they proposed a new model as a linear combination of the Barker-Pompe and Barker-Bobetic models which follows:

$$\phi_{BFW}(r) = 0.75\phi_{BB}(r) + 0.25\phi_{BP}(r) \qquad (7.21)$$

Both ϕ_{BB} and ϕ_{BP} are given by eq. 7.1, and parameters for the three models are given in Table 7.8. The BFW potential for argon is in excellent agreement with the scattering data of Parson, Siska, and Lee [21] and of Scoles et al. [22], as was shown in Fig. 7.4. In addition, the potential gives excellent agreement with the bound state energies of the van der Waals dimers as is shown by the comparison with the spectroscopic data of Tanaka and Yoshino in Fig. 7.7.

Since the BFW potential was reported, it has been used in other studies, including work on the solid state. Fisher and Watts [4] reported internal energies, pressures, and elastic constants along the solid-vapor coexistence curve for temperatures in the range 40–80 K. They found that the theoretical results for the internal energy were within experimental error and that, although the pressures had rather large computational errors, they were also in good agreement with experiment. The elastic constants differed by about 15% from measured ones, and at the time this discrepancy was not resolved. Fisher and Watts [4] suggested that the experimental results were in error and subsequent work [47] has confirmed this attribution. Barker and Klein [71] extended the solid state studies to higher temperatures and found excellent agreement between theory and experiment for pressures of both phases computed on the solid-fluid coexistence curve.

Table 7.8 Parameters to be used with eq. 7.1 for the Barker-Pompe (BP), Barker-Bobetic (BB), and Barker-Fisher-Watts (BFW) pair potentials for argon, and for krypton and xenon. The parameters given for the BFW model were shown by P. D. Neufeld and R. A. Aziz, (*J. Chem. Phys.*, **58,** 1877, 1973), to give a model that was numerically indistinguishable from that given in eq. 7.21

	BP	BB	BFW	Kr-Kr	Xe-Xe
ε/k(K)	147.7	140.2	142.1	201.9	282.4
R_m(Å)	3.756	3.763	3.761	4.007	4.362
σ(Å)	3.341	3.367	3.361	3.573	3.890
A_0	0.235	0.292	0.278	0.235	0.240
A_1	−4.774	−4.415	−4.504	−4.787	−4.817
A_2	−10.219	−7.702	−8.331	−9.2	−10.9
A_3	−5.291	−31.93	−25.27	−8.0	−25.0
A_4	0.0	−136.0	−102.0	−30.0	−50.7
A_5	0.0	−151.0	−113.3	−205.8	−200.0
C_6	1.070	1.120	1.107	1.063	1.054
C_8	0.164	0.172	0.170	0.170	0.166
C_{10}	0.013	0.014	0.013	0.014	0.032
α	12.5	12.5	12.5	12.5	12.5
δ	0.01	0.01	0.01	0.01	0.01

To summarize, in addition to having the correct asymptotic argon-argon interaction, the BFW potential is in agreement with a very wide range of experimental data. Gas phase properties include virial coefficients, transport properties, differential and total scattering cross-sections, and measurements of the van der Waals dimer bound state energies. In the liquid state there is good agreement with thermodynamic and structural properties as well as with current estimates of the self-diffusion coefficient. Finally, the potential gives excellent agreement with a considerable amount of solid state data, including thermodynamic properties, elastic properties, and phonon dispersion curves. Probably the most important conclusion to be reached from this study of the properties of argon is that it is possible to derive a single interaction potential from bulk data that is consistent with all properties of the substance.

7.6 Other Inert Gas Potentials

The neon-neon potential has been analyzed in detail [72–76] during recent years and the accuracy of various models has improved as more

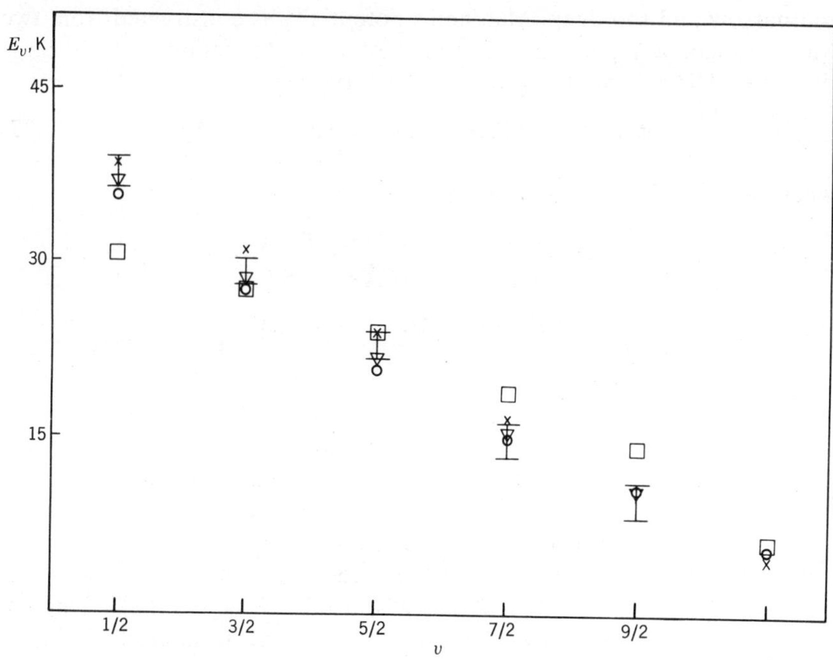

Figure 7.7 Vibrational level spacings of argon. I experimental data of Tanaka and Yoshino [28]; ▽ Barker-Fisher-Watts potential; ○ Barker-Bobetic potential; × Barker-Pompe potential; □ Dymond-Alder potential. (Adapted with permission from *Mol. Phys.*, **21**, 657, 1971.)

precise experiments have been performed [27, 32]. As with argon, earlier studies used the Lennard-Jones 12-6 potential, and the parameters $\varepsilon/k = 36.76$ K and $\sigma = 2.786$ Å [77] give reasonably good agreement with experimental data for the solid and liquid phases [78]. However, the coefficient of the r^{-6} term predicted by this potential is much larger than Leonard's calculations of C_6 [66], and this, together with other evidence, makes it clear that the Lennard-Jones potential is inadequate. Fortunately, quantum mechanical effects are smaller for neon than for helium and they can be included by using the semiclassical expansion of the partition function discussed in Section 7.1 [9, 10]. Table 7.2 gives values of the quantum correction to the second virial coefficient for neon. A recent study of the neon pair potential reported by Maitland [73] is probably the most comprehensive, and it combines experimental data of the Ne_2 bound state energy levels taken from spectroscopic measurements together with measurements of second virial coefficients and viscosities over the temperature range 80–2000 K. His conclusions were that most previous interaction functions have potential wells which are both too narrow and too deep. Maitland's potential was constructed from two Morse functions joined smoothly to the known dispersion terms at large distances. His potential is given by the formulae

$$\phi(r)=\varepsilon\{\exp[2\beta(1-x)]-2\exp[\beta(1-x)]\} \tag{7.22}$$

where $\varepsilon=45.8$ K, $x=r/3.07$ Å and

$$\beta=\begin{cases}6.87 & 0<r\leqslant 3.07\ \text{Å}\\ 5.00 & 3.07\ \text{Å}<r\leqslant 5.0\ \text{Å}\end{cases}$$

$$\phi(r)=-\frac{4.54\times10^4}{r^6}-\frac{11.1\times10^4}{r^8}-\frac{37.9\times10^4}{r^{10}}\ \text{K} \qquad 5.0\ \text{Å}<r$$

Earlier Siska et al. [27] reported a potential based on a piecewise continuous function (Morse-Spline-van der Waals: MSV) that fitted their elastic scattering differential cross-sections, augmenting this data with experimental second virial coefficients and theoretical long range dispersion coefficients [66]. Their potential has a well depth $\varepsilon/k=45.8$ K that is about 30% larger than that of the Lennard-Jones potential derived from the bulk properties of the dilute gas, while σ is only 2% smaller. Although Barker [13] showed that the potential of Siska et al. [27] gave good agreement with viscosity data, Klein [79] noted that its agreement with solid state data was poor, even after the triple-dipole dispersion interaction was included. Following this, Goldman and Klein [75] reported two other potentials, derived using solid state data, and obtained good agreement with the elastic constants near 0 K as well as fitting second virial coefficients and differential scattering cross-sections. Farrar et al. [76] determined ESMSV potentials (see Section 7.1), fitting the same data as Goldman and Klein. In a subsequent analysis of the spectroscopic data Le Roy et al. [74] supported the conclusion that the Goldman-Klein potentials are the most reliable neon pair potentials currently available. Both potential models of Goldman and Klein were of the MSV type but the first assumes pairwise additivity of the intermolecular potential whereas the second assumes that the nonpairwise additivity can be approximated by the Axilrod-Teller three-body interaction [8]. The second potential gives better agreement with the low temperature heat capacity measurements on solid neon and has the form

$$\phi(r)=\begin{cases}\varepsilon\{\exp[2\beta(1-x)]-2\exp[\beta(1-x)]\} & r<3.744\ \text{Å}\\ \varepsilon(a_1+(x-x_1)\{a_2+(x-x_2)[a_3+(x-x_1)a_4]\}) & 3.744\ \text{Å}\leqslant r<4.033\ \text{Å}\\ -\dfrac{4.55\times10^4\ \text{KÅ}^6}{r^6}-\dfrac{11.2\times10^4\ \text{KÅ}^8}{r^8} & 4.033\ \text{Å}\leqslant r\end{cases} \tag{7.23}$$

where $\varepsilon = 42.0$ K, $x = r/3.102$ Å, $\beta = 6.27$, and spline parameters $a_1 = -0.471$, $a_2 = 1.954$, $a_3 = -5.657$, and $a_4 = -1.565$.

At the time of writing, the only extensive computer simulation studies of liquid neon reported have used the Lennard-Jones potential [78] so it is not known how good these models are in the liquid state.

The success in obtaining an argon pair potential that reproduces a wide range of experimental data has been followed by several attempts to determine the ground state potential for krypton-krypton and xenon-xenon interactions [27, 44, 49, 80–84]. There is strong evidence to show that when the pair potential functions of the inert gases are scaled appropriately, they are virtually identical [80]. Using a well depth based on arguments from the Law of Corresponding States (see Section 6.1) together with a solid state estimate of $R_m = 3.99$ Å, Gough et al. [80] were able to obtain an accurate pair potential for krypton from gas phase properties using data inversion techniques. Their potential was shown later [81] to be in excellent agreement with the vibrational spectrum of Tanaka et al. [85].

Bobetic et al. [82] and, subsequently, Barker et al [44], using methods similar to those used previously for argon, derived pair potentials based on the properties of solid and gaseous krypton. Data used in those studies included scattering cross-sections, vibrational levels, second virial coefficients, and viscosity coefficients, as well as thermodynamic properties of the low temperature solid. Potentials of comparable accuracy were obtained by Buck et al. and by Farrar et al. [83] using scattering data and the vibrational levels reported by Tanaka et al. [85].

The krypton potential of Barker et al. [44] has the analytic form used for argon, eq. 7.1, together with small additional terms added in the region beyond the minimum

$$\begin{aligned}
\phi(r) &= \varepsilon[u_0(R) + u_1(R)] \\
u_0(R) &= \exp[\alpha(1-R)]\sum_{i=0}^{5} A_i(R-1)^i - \sum_{i=0}^{2} \frac{C_{6+2i}}{(R^{6+2i}+\delta)} \\
u_1(R) &= [P(R-1)^4 + Q(R-1)^5]\exp[\alpha(1-R)] \qquad R \geqslant 1 \\
&= 0 \qquad R < 1
\end{aligned} \tag{7.24}$$

where $P = -9.0$, $Q = 68.67$, $R = r/R_m$, and R_m is the position of the minimum in the potential. Values of the other parameters are listed in Table 7.8. As is the case for the argon potential, this model must be supplemented with the Axilrod-Teller interaction [8] if calculations are carried out at higher densities.

One would expect the pair potential for xenon to be very similar to the pair interactions constructed for argon and krypton. Gough et al. [80]

determined interaction potentials for xenon by inversion of viscosity measurements, using an estimate of the position of the minimum in the potential obtained from solid state data. They reported a number of pair potentials ranging in well depth from 265 K to 290 K, and attributed this spread to uncertainties in the viscosity. Barker et al. [44] have reported a potential which agrees with measured differential scattering cross-sections, second virial coefficients, vibrational energy levels of the Xe_2 dimer, and compression data for solid xenon. The potential has the form

$$\phi(r) = \varepsilon[u_0(R) + u_1(R) + u_2(R)] \tag{7.25}$$

where u_0 and u_1 are given by eq. 7.24 with $P = 59.3$ and $Q = 71.1$; the other parameters are listed in Table 7.8:

$$u_2(R) = [2.08(R-1)^2 - 6.24(R-1)^3] \exp[-50(R-1)^2] \tag{7.26}$$

Apart from problems with experimental data, the residual discrepancies between theory and experiment may be due to inaccurate estimates of the many body interactions. Future experiments, such as low temperature virial coefficient measurements and more accurate solid state properties, will undoubtedly reduce the uncertainties in the xenon-xenon interaction.

7.7 Unlike Pairs of Inert Gas Atoms

To a certain extent the study of pairs of unlike noble gas atoms has followed that of the like atoms. Thus during the past few years there have been several attempts to obtain interaction potentials for mixtures from differential scattering cross-section measurements [24, 86, 87] and from liquid state data [88]. One problem that is usually examined in studies of simple mixtures is that of determining "combining rules" for estimating parameters of unlike interactions from those of like interactions. Usually, these rules attempt to describe the well depth, ε_{ij}, and the position of the potential zero, σ_{ij}, in terms of those for like pairs, ε_{ii}, ε_{jj}, and σ_{ii}, σ_{jj}, in the hope of finding a simple prescription for deducing interaction potentials for unlike atoms from those of the pure substances [89]. As an example, the position of the zero in unlike interactions has frequently been estimated from the arithmetic mean of the zeros in the like interactions, $\sigma_{ij} = \frac{1}{2}(\sigma_{ii} + \sigma_{jj})$, and the well depth estimated using the geometric mean $\varepsilon_{ij} = (\varepsilon_{ii}\varepsilon_{jj})^{1/2}$. If it is then assumed that the unlike interaction has the same functional form as the like interactions, $\phi_{ij}(r)$ can readily be obtained.

Several variations on these combining rules have been proposed and tested [89]. Recent papers on unlike inert gas pairs have generally found poor agreement with the simple combining rules, particularly where there is a large difference in the masses of the two atoms, such as for He-Xe. Perhaps the main reason for the lack of success of the simple combining rules is that they take no account of possible differences in shape of the potentials.

A recent account of the helium series of unlike pairs, He-Ne, He-Ar, He-Kr, and He-Xe was given by Chen et al. [24] who also summarize much of the earlier work on such systems. These authors reported ESMSV-type potentials based on simultaneous fits to the second virial coefficients and to measured differential scattering cross-sections.

Potentials for the neon series, Ne-Ar, Ne-Kr, and Ne-Xe have also been reported [86]. The results of Ng et al. [86] are probably more accurate than those of Chen et al. [24], since they fit diffusion coefficients as well as differential scattering data and second virial coefficients. The interaction potentials between unlike pairs all have wider wells than that of the He-He interaction. A striking feature of the He-Ar, He-Kr, and He-Xe gas potentials is that they appear to have nearly equal well depths ε. This feature has also been observed in the neon-rare gas system [86] as well as in rare gas-alkali atom combinations [90]. In these latter studies it was reported that for systems with the same rare gas atom and different alkali atom, the well depths are nearly equal, whereas with the same alkali atom and different rare gas atoms, the results differ considerably. This observation implies that the less polarizable atom plays the more important role in determining the well depth for asymmetric systems. This result is in disagreement with the commonly used geometric combining rule for well depths, which predicts a significantly different result for, say, Ne-Xe than that for Ne-Ar.

Less experimental data are available for the heavier inert gas systems, Ar-Kr, Ar-Xe, and Kr-Xe. Hogevorst [91] made accurate measurements of the diffusion coefficient of inert gas mixtures and determined from them parameters for the Lennard-Jones 12-6 model and for the exp-6 model. Parson et al. [87] derived potentials from their differential scattering cross-section measurements for Ar-Kr and Ar-Xe but these were not fitted to second virial coefficients or to the theoretical long range interaction. In a more elaborate fitting procedure, Lee et al. [88] derived multiparameter potentials for Ar-Kr and Kr-Xe using equations determined for the like atom pairs. They calculated the excess thermodynamic properties of the liquid mixtures using the Barker-Henderson perturbation theory [41], obtaining good agreement with experiment at all concentrations.

7.8 Law of Corresponding States Revisited

In Chapter 6 we discussed the Law of Corresponding States in some detail showing that if the pair potential for a number of different substances could be represented as a two-parameter form then by suitable scaling the thermodynamic properties of those substances could be made coincident. It was suggested that the inert gases form a class of substances that obey the Law of Corresponding States and there is strong experimental evidence to suggest that this is so [92]. This chapter has recommended pair interactions for all the inert gases, and it may have struck the reader that each gas has been described by a different functional form. Consequently, it is of interest to see if the several pair potentials can be scaled to become coincident.

Figure 7.8 is a graph of all the "best" pair potentials given in this

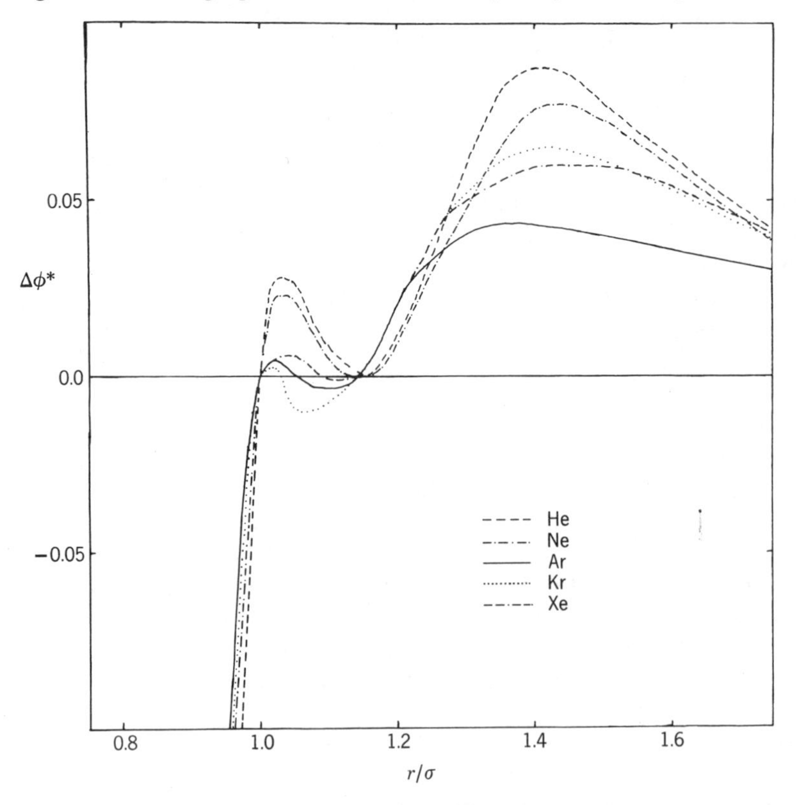

Figure 7.8 The quantity $\Delta\phi^* = \phi^* - \phi^*_{LJ}$ for the inert gases. Note that the distance variables in all potentials are scaled by the corresponding value of σ, so that $\phi^*(1) = 0$ for all models.

chapter together with the Lennard-Jones 12-6 model. We have chosen as suitable scaling parameters the maximum attractive energy, ε, and the position of the zero, σ, and all the curves are of the form

$$\Delta\phi^*(r^*) = \frac{\phi(r/\sigma)}{\varepsilon} - \frac{\phi_{LJ}(r/\sigma_{LJ})}{\varepsilon_{LJ}} \tag{7.27}$$

It can be seen that the accurate pair potentials, when written in this way, are in very good agreement and that the Lennard-Jones model is significantly different. Despite the apparent agreement, there are small differences between the various models, as can be seen from Table 7.9. In this table values of ε/k and σ are given together with reduced values of the dipole-dipole dispersion coefficient, $C_6^* = C_6/(\varepsilon\sigma^6)$, the position of the minimum $R_m^* = R_m/\sigma$, and the second and third derivatives of $\phi(r)$ at R_m, $\phi''^* = \phi''(R_m)\sigma^2/\varepsilon$ and $\phi'''^* = \phi'''(R_m)\sigma^3/\varepsilon$. Also given are reduced values of the Axilrod-Teller coefficient $\nu^* = \nu/(\varepsilon\sigma^9)$ and of the reduced de Broglie wavelength $\Lambda^* = h/\sigma(m\varepsilon)^{1/2}$. The most significant differences appear to be in the values of the dispersion coefficient C_6, and values of the other pair potential attributes are in remarkably good agreement. More important are the different values obtained for ν^* and Λ^*.

In Sections 7.2 and 7.3 we showed that it was necessary to introduce three-body interactions and quantum correction terms into studies of the thermodynamic properties of liquid and solid argon. Values of Λ^* in Table 7.9 show that the quantum corrections given in Tables 7.3–7.5 do

Table 7.9 Properties of the inert gas pair potentials in reduced units (see text)

Gas	ε/k(K)	σ(Å)	R_m^*	C_6^*	ϕ''^*	ϕ'''^*	ν^*	Λ^*
He	10.4	2.636	1.126	2.90	58.6	—	0.024	2.58
Ne	42.0	2.764	1.122	2.38	63.2	−1067.0	0.031	0.544
Ar	142.1	3.361	1.119	2.17	64.9	−1214.0	0.068	0.172
Kr	201.9	3.573	1.121	2.11	65.0	−1190.0	0.078	0.094
Xe	282.4	3.890	1.121	2.09	62.5	−1119.0	0.100	0.058

not follow the Law of Corresponding States. Consequently, if any attempt is made to use liquid or solid state results for argon, say, to predict condensed phase properties of the other noble gases it is important that the quantum correction be scaled correctly. Fortunately, this correction is small for Ar, Kr, and Xe, and is not excessively large for neon, and so the law can be used quite successfully to correlate the properties of these gases. The differences between values of ν^* for the heavier inert gases are rather small and consequently contributions from three-body interactions for one gas will scale reasonably well to give estimates of this term for the other gases.

References

1. J. A. Barker and A. Pompe, 1968, *Aust. J. Chem.*, **21,** 1683.
2. M. V. Bobetic and J. A. Barker, 1970, *Phys. Rev.*, *B***2,** 4169.
3. J. A. Barker, R. A. Fisher, and R. O. Watts, 1971, *Mol. Phys.*, **21,** 657.
4. R. A. Fisher and R. O. Watts, 1972, *Mol. Phys.*, **23,** 1051.
5. J. A. Barker, W. Fock, and F. Smith, 1964, *Phys. Fluids*, **7,** 897.
6. E. A. Mason and W. E. Rice, 1954, *J. Chem. Phys.*, **22,** 522, 843; H. J. M. Hanley and G. E. Childs, 1968, *Science*, **159,** 1114.
7. F. A. Guevara, B. B. McInteer, and W. E. Wageman, 1969, *Phys. Fluids*, **12,** 2493; R. A. Dawe and E. B. Smith, 1970, *J. Chem. Phys.*, **52,** 693.
8. B. M. Axilrod and E. Teller, 1943, *J. Chem. Phys.*, **11,** 299.
9. J. O. Hirschfelder, C. F. Curtiss, and R. B. Bird, 1964, *Molecular Theory of Gases and Liquids*, Wiley, New York.
10. J. G. Kirkwood, 1933, *Phys. Rev.*, **44,** 31; E. P. Wigner, 1932, *ibid.*, **40,** 749; H. S. Green, 1951, *J. Chem. Phys.*, **19,** 955; L. D. Landau and E. M. Lifshitz, 1958, *Statistical Physics*, Addison-Wesley, Reading, Mass.
11. J. E. Kilpatrick, W. E. Keller, E. F. Hammel, and N. Metropolis, 1954, *Phys. Rev.*, **94,** 1103.
12. R. Haberlandt, 1964, *Phys. Lett.*, **8,** 172.
13. J. A. Barker, 1972, *Chem. Phys. Lett.*, **14,** 242.
14. J. H. Dymond and B. J. Alder, 1969, *J. Chem. Phys.*, **51,** 308; 1968, *Chem. Phys. Lett.*, **2,** 541.
15. H. J. M. Hanley and M. Klein, 1968, *Trans. Faraday Soc.*, **64,** 2927.
16. L. W. Bruch and I. J. McGee, 1970, *J. Chem. Phys.*, **52,** 5884.
17. E. A. Mason, 1954, *J. Chem. Phys.*, **22,** 169; W. E. Keller, 1957, *Phys. Rev.*, **105,** 41; L. Monchick, E. A. Mason, R. J. Munn, and F. J. Smith, 1965, *Phys. Rev.*, **139,** *A*1016; R. J. Munn, F. J. Smith, E. A. Mason, and L. Monchick, 1965, *J. Chem. Phys.*, **42,** 537; S. Imam-Rahajoe, C. F. Curtiss, and R. B. Bernstein, 1965, *ibid.*, **42,** 530.
18. J. A. Barker, M. V. Bobetic, and A. Pompe, 1971, *Mol. Phys.*, **20,** 347.
19. H. J. M. Hanley and M. Klein, 1972, *J. Phys. Chem.*, **76,** 1743.
20. I. Amdur and H. Pearlman, 1940, *J. Chem. Phys.*; **8,** 7; 1941, *ibid.*, **9,** 503; I. Amdur, 1943, *ibid.*, **11,** 157.
21. J. M. Parson, P. E. Siska, and Y. T. Lee, 1972, *J. Chem. Phys.*, **56,** 1511.
22. M. Cavallini, G. Gallinaro, L. Meneghetti, G. Scoles, and U. Valbusa, 1970, *Chem. Phys. Lett.*, **7,** 303; M. Cavallini, M. G. Dondi, G. Scoles, and U. Valbusa, 1971, *Proc. Int. Symp. Mol. Beams*, Cannes, France; M. G. Dondi, G. Scoles, F. Torello, and H. Pauly, 1969, *J. Chem. Phys.*, **51,** 392.
23. R. Feltgen, H. Pauly, F. Torello, and H. Vehmeyer, 1973, *Phys. Rev. Lett.*, **30,** 820.
24. C. H. Chen, P. E. Siska, and Y. T. Lee, 1973, *J. Chem. Phys.*, **59,** 601.
25. W. Aufm Kampe, D. E. Oates, W. Schrader, and H. G. Bennewitz, 1973, *Chem. Phys. Lett.*, **18,** 323.

26. H. G. Bennewitz, H. Busse, H. D. Dohman, D. E. Oates, and W. Schrader, 1972, *Z. Phys.*, **253,** 435.
27. P. E. Siska, J. M. Parson, T. P. Schafer, and Y. T. Lee, 1971, *J. Chem. Phys.*, **55,** 5762.
28. Y. Tanaka and K. Yoshino, 1970, *J. Chem. Phys.*, **53,** 2012.
29. W. G. Richards and R. F. Barrow, 1964, *Proc. Phys. Soc. (London)*, **83,** 1045; 1964, *Trans. Faraday Soc.*, **60,** 797; G. C. Maitland and E. B. Smith, 1970, *Mol. Phys.*, **22,** 861.
30. L. W. Bruch and I. J. McGee, 1967, *J. Chem. Phys.*, **46,** 2959.
31. Y. Tanaka and K. Yoshino, 1969, *J. Chem. Phys.*, **50,** 3087.
32. Y. Tanaka and K. Yoshino, 1972, *J. Chem. Phys.*, **57,** 2964.
33. I. R. McDonald and K. Singer, 1970, *Quart. Rev.*, **24,** 238; N. Metropolis, A. W. Rosenbluth, M. N. Rosenbluth, E. Teller, and A. H. Teller, 1953, *J. Chem. Phys.*, **21,** 1087; B. J. Alder and T. E. Wainwright, 1957, *ibid.*, **27,** 1208.
34. R. O. Watts, 1973, Chapter 1 in *Specialist Periodical Reports of the Chemical Society: Statistical Mechanics*, K. Singer, Ed., Vol. 1. Chemical Society, London.
35. P. A. Mikolaj and C. J. Pings, 1967, *J. Chem. Phys.*, **46,** 1401; C. D. Andriesse and E. Legrand, 1972, *Physica*, **57,** 191.
36. E. K. Achter and L. Meyer, 1969, *Phys. Rev.*, **188,** 291.
37. R. D. Murphy and I. J. McGee, 1973, *Phys. Lett.*, **45***A*, 323.
38. L. W. Bruch, I. J. McGee, and R. O. Watts, 1974, *Phys. Lett.*, **50***A*, 315.
39. R. D. Murphy and R. O. Watts, 1970, *J. Low Temp. Phys.*, **2,** 507.
40. W. E. Massey and C. W. Woo, 1967, *Phys. Rev.*, **164,** 256; W. L. McMillan, 1965, *ibid.*, **138,** *A*442; D. Schiff and L. Verlet, 1967, *ibid.*, **160,** 208.
41. J. A. Barker and D. Henderson, 1972, *Ann. Rev. Phys. Chem.*, **23,** 439.
42. C. H. J. Johnson and T. H. Spurling, 1971, *Aust. J. Chem.*, **24,** 2205; J. A. Barker, C. H. J. Johnson, and T. H. Spurling, 1972, *ibid.*, **25,** 1811.
43. M. B. Doran and I. J. Zucker, 1971, *J. Phys., C (London)*, **4,** 307.
44. J. A. Barker, R. O. Watts, J. K. Lee, T. P. Schafer, and Y. T. Lee, 1974, *J. Chem. Phys.*, **61,** 3081.
45. J. A. Barker, M. L. Klein, and M. V. Bobetic, 1970, *Phys. Rev.*, *B***2,** 4176.
46. G. K. Horton, 1968, *Am. J. Phys.*, **36,** 93.
47. S. Gewurtz, H. Kiefte, D. Landheer, R. A. McLaren, and B. P. Stoicheff, 1972, *Phys. Rev. Lett.*, **29,** 1454.
48. D. Beck, 1968, *Mol. Phys.*, **14,** 311.
49. J. M. Farrar and Y. T. Lee, 1972, *J. Chem. Phys.*, **56,** 5801.
50. P. Bertoncini and A. C. Wahl, 1970, *Phys. Rev. Lett.*, **25,** 991; D. R. McLaughlin and H. F. Schafer, 1971, *Chem. Phys. Lett.*, **12,** 244.
51. J. N. Murrell and G. Shaw, 1968, *Mol. Phys.*, **15,** 328; T. H. Spurling and I. K. Snook, 1975, *J. Chem. Soc., Faraday Trans. II*, **71,** 852.
52. W. D. Davidson, 1966, *Proc. Phys. Soc. (London)*, **87,** 133; Y. M. Chan and A. Dalgarno, 1965, *ibid.*, **86,** 777; R. J. Bell, 1965, *ibid.*, **86,** 17.
53. T. L. Gilbert and A. C. Wahl, 1967, *J. Chem. Phys.*, **47,** 3425; P. E. Phillipson, 1962, *Phys. Rev.*, **125,** 1981.
54. I. K. Snook and R. O. Watts, 1972, *Aust. J. Phys.*, **25,** 735.
55. J. de Boer and A. Michels, 1939, *Physica*, **5,** 945.
56. R. Haberlandt, 1965, *Phys. Lett.*, **14,** 197; J. L. Yntema and W. G. Schneider, 1950, *J. Chem. Phys.*, **18,** 646; W. E. Massey, 1966, *Phys. Rev.*, **151,** 153; E. A. Mason and W. E. Rice, 1954, *J. Chem. Phys.*, **22,** 522.
57. J. P. Hansen and E. L. Pollock, 1972, *Phys. Rev.*, *A***5,** 2561.

58. R. B. Bernstein and F. A. Morse, 1964, *J. Chem. Phys.*, **40,** 917.
59. L. W. Bruch and P. R. Certain, 1972, "Intermolecular Forces," in *MTP International Review of Science, Theoretical Chemistry,* W. Byers Brown, Ed., Vol. 1, p. 113, Butterworths, London.
60. R. Jastrow, 1955, *Phys. Rev.*, **98,** 1479.
61. M. H. Lee, 1969, *Physica,* **43,** 132; D. K. Lee, 1970, *Phys. Rev.*, *A***2,** 278; W. P. Francis, G. V. Chester, and L. Reatto, 1970, *ibid.*, *A***1,** 86.
62. I. J. McGee and R. D. Murphy, 1972, *J. Phys., C, (London)*, **5,** L311.
63. A. D. Klemm and R. G. Storer, 1973, *Aust. J. Phys.*, **26,** 43; A. D. Klemm, 1971, Ph. D. thesis, University of Adelaide, Australia, unpublished.
64. G. K. Horton and J. W. Leech, 1963, *Proc. Phys. Soc. (London)*, **82,** 816.
65. E. A. Guggenheim and M. L. McGlashan, 1960, *Proc. Roy. Soc. (London), Ser. A,* **255,** 456; M. L. McGlashan, 1965, *Discuss. Faraday Soc.*, **40,** 59.
66. P. J. Leonard, 1968, Masters thesis, University of Melbourne; G. Starkschall and R. G. Gordon, 1972, *J. Chem. Phys.*, **57,** 3213; 1972, *ibid.*, **56,** 2801; 1971, *ibid.*, **54,** 663.
67. L. W. Bruch and I. J. McGee, 1970, *J. Chem. Phys.*, **53,** 4711.
68. J. A. Barker, D. Henderson, and W. R. Smith, 1969, *Mol. Phys.*, **17,** 579; W. Götze and H. Schmidt, 1966, *Z. Phys.*, **192,** 409; I. J. Zucker and G. C. Chell, 1968, *J. Phys., C, (London)*, **1,** 1505; G. C. Chell and I. J. Zucker, 1968, *ibid.*, **1,** 35; A. Hüller, W. Götze, and H. Schmidt, 1970, *Z. Phys.*, **231,** 173.
69. R. A. Fisher and R. O. Watts, 1972, *Aust. J. Phys.*, **25,** 529.
70. T. G. Gibbons, M. L. Klein, and R. D. Murphy, 1973, *Chem. Phys. Lett.*, **18,** 325.
71. J. A. Barker and M. L. Klein, 1973, *Phys. Rev.*, *B***7,** 4707.
72. D. D. Konowalow and D. S. Zakheim, 1972, *J. Chem. Phys.*, **57,** 4375: J. H. Dymond, 1971, *ibid.*, **54,** 3675.
73. G. C. Maitland, 1973, *Mol. Phys.*, **26,** 513.
74. R. J. LeRoy, M. L. Klein, and I. J. McGee, 1974, *Mol. Phys.*, **28,** 587.
75. V. V. Goldman and M. L. Klein, 1973, *J. Low Temp. Phys.*, **12,** 101.
76. J. M. Farrar, Y. T. Lee, V. V. Goldman, and M. L. Klein, 1973, *Chem. Phys. Lett.*, **19,** 359.
77. J. S. Brown, 1966, *Proc. Phys. Soc. (London)*, **89,** 987.
78. J. P. Hansen, 1968, *Phys. Rev.*, **172,** 919; J. P. Hansen and L. Verlet, 1969, *ibid.*, **184,** 151; J. P. Hansen and J. J. Weis, 1969, *ibid.*, **188,** 314.
79. M. L. Klein, 1973, *Chem. Phys. Lett.*, **18,** 203.
80. D. Gough, E. B. Smith, and G. C. Maitland, 1973, *Mol. Phys.*, **25,** 1433.
81. D. Gough, E. B. Smith, and G. C. Maitland, 1974, *Mol. Phys.*, **27,** 867.
82. M. V. Bobetic, J. A. Barker, and M. L. Klein, 1972, *Phys. Rev.*, *B***5,** 3185.
83. U. Buck, M. G. Dondi, U. Valbusa, M. L. Klein, and G. Scoles, 1973, *Phys. Rev.*, *A***8,** 2409; J. M. Farrar, T. P. Schafer, and Y. T. Lee, 1973, *AIP Conf. Proc.*, **11,** 279.
84. K. Docken and T. P. Schafer, 1973, *J. Mol. Spectrosc.*, **46,** 454.
85. Y. Tanaka, K. Yoshino, and D. E. Freeman, 1973, *J. Chem. Phys.*, **59,** 5160.
86. C. Y. Ng, Y. T. Lee, and J. A. Barker, 1974, *J. Chem. Phys.*, **61,** 1996; J. M. Parson, T. P. Schafer, F. P. Tully, P. E. Siska, Y. C. Wong, and Y. T. Lee, 1970, *J. Chem. Phys.*, **53,** 2123; 1973, *ibid.*, **58,** 4044.
87. J. M. Parson, T. P. Schafer, F. P. Tully, Y. C. Wong, and Y. T. Lee, 1970, *J. Chem. Phys.*, **53,** 3755.
88. J. K. Lee, D. Henderson, and J. A. Barker, 1975, *Mol. Phys.*, **29,** 429.
89. D. Henderson and P. J. Leonard, 1971, Chapter 7 in *Physical Chemistry—An Advanced Treatise,* H. Eyring, D. Henderson, and W. Jost, Eds., Vol. 4, Academic,

New York; H. M. Lin and R. L. Robinson, Jr., 1971, *J. Chem. Phys.*, **54,** 52; E. A. Mason, M. Islam, and S. Weissman, 1964, *Phys. Fluids*, **7,** 1011; R. C. Reid and T. W. Leland, Jr., 1965, *A. I. Chem. E. J.*, **11,** 228.

90. P. Barwig, U. Buck, E. Hundhausen, and H. Pauly, 1966, *Z. Phys.*, **196,** 343; D. Beck and H. J. Loesch, 1966, *ibid.*, **195,** 444; M. Holstein and H. Pauly, 1967, *ibid.*, **201,** 10; E. W. Rothe, P. K. Rol, and R. B. Bernstein, 1963, *Phys. Rev.*, **130,** 2333.
91. W. Hogevorst, 1971, *Physica*, **51,** 59, 77, 90.
92. W. B. Streett and M. S. Constantino, 1974, *Physica*, **75,** 283; J. M. H. Levelt, 1960, *ibid.*, **26,** 361; W. B. Streett and L. A. K. Staveley, 1967, *J. Chem. Phys.*, **47,** 2449.

Eight

Statistical Mechanics of Nonspherical Molecules

In general, the interaction potential between two molecules depends on their relative orientations as well as on their relative positions. For example, if we note that the benzene molecule is a flat regular hexagon, then we can expect different interaction potentials between the three configurations shown in Fig. 8.1 even though the distances between the centers of mass of the two molecules is unaltered. This complexity in the interaction potential is matched by an increased complexity in the statistical mechanics of molecular systems. Once a single particle contains more than one atom, a number of effects that are nonexistent in atomic systems become important. In particular, we must consider the importance of internal degrees of freedom, such as rotational and vibrational states. In this chapter we consider the changes to the statistical mechanics needed to accommodate rotational degrees of freedom, but in general we do not consider the influence of vibrational states on bulk properties. Thus our molecules are approximated as rigid entities so that we need consider only intermolecular interactions. If we extended the study to vibrating molecules we would have to include intramolecular atom-atom interactions and so make the problem even more difficult.

Once we have decided to consider molecular systems the field of study becomes immense. It is probably better to begin with simple molecules, such as oxygen or nitrogen, before going on to treat very large molecules such as polymers that are important in, say, biochemistry. In the previous chapters we saw that recent advances in the theory of simple fluids have culminated in a set of extremely accurate interaction potentials for the inert gases. This chapter gives an account of recent attempts to extend the methods used in earlier chapters to include shape dependent interactions. As these extensions have so far covered only simple molecules, such as

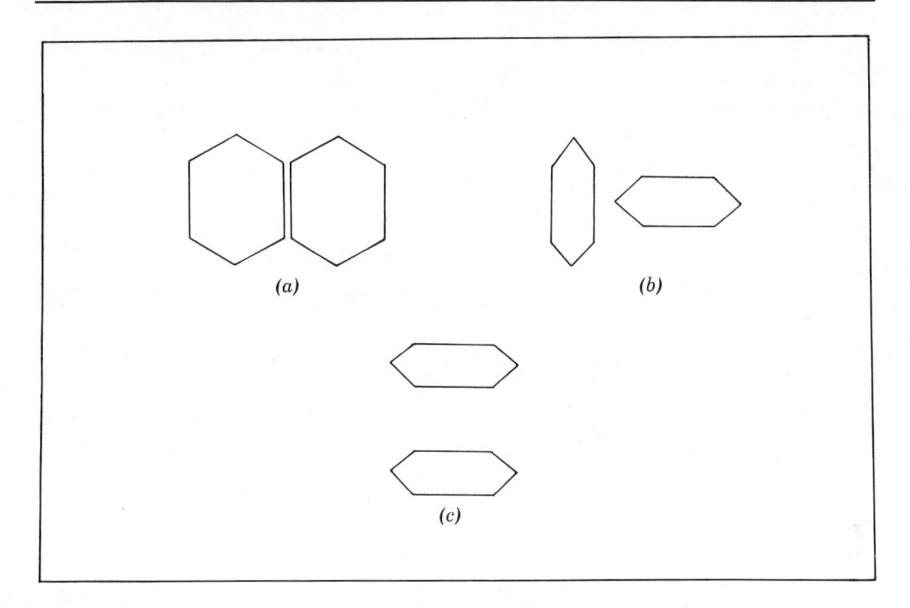

Figure 8.1 Three configurations for benzene molecules in which the center-of-mass distances are equal. Configuration (*b*) is that found in the benzene crystal.

O_2, N_2, CO, H_2O, CH_4, and C_6H_6, we restrict ourselves to this range of substances. Furthermore, as water and ionic solutions are so important, we devote the following chapter to an account of recent advances in the theory of these systems. Finally, the theory of polymers, and the development of accurate interaction potentials for these substances, is so far removed from other work in this book that no attempt is made to give an account of this field. The interested reader is referred to other publications for an account of polymer theory [1].

The effect of orientational dependent interactions on the properties of a fluid depends to some extent on the density and temperature being considered. In the gas phase, where molecules are widely separated, there is essentially free rotation. Thus the virial coefficients and transport properties of nitrogen gas, say, or even for benzene or cyclohexane vapor, can be reproduced to a fair degree of precision using a spherically symmetric Lennard-Jones potential [2]. At higher densities and in the liquid phase the asymmetry of the molecules generates significant correlations between the molecular translational and rotational motions. It is not possible to describe these effects in terms of spherically symmetric interactions [3]. Consequently, we must consider in more detail the way in which nonspherical effects are included in the interaction potential.

As we discussed in Section 2.3, attempts to introduce nonspherical terms into the potential have taken two routes. The first is to consider molecules that have essentially spherical symmetry at short distances and introduce the nonspherical behavior using permanent electrostatic interactions between point multipoles. As an example, a common model for examining the properties of simple polar fluids is the hard sphere potential with embedded point dipoles

$$\phi(r, \theta_1, \theta_2, \phi_1-\phi_2)=\phi_{HS}(r)-\frac{\mu^2}{r^3}D(\theta_1, \theta_2, \phi_1-\phi_2) \tag{8.1}$$

where $\phi_{HS}(r)$ is the usual hard sphere potential and

$$D(\theta_1, \theta_2, \phi_1-\phi_2)=2\cos\theta_1\cos\theta_2-\sin\theta_1\sin\theta_2\cos(\phi_1-\phi_2) \tag{8.2}$$

The angles θ_1, θ_2 are the angles between the dipole vectors $\boldsymbol{\mu}_1$, $\boldsymbol{\mu}_2$ and the line joining the centers of the molecules, and $(\phi_1-\phi_2)$ is the relative rotation of these vectors about the line (see Fig. 2.2). This model has been used to examine the thermodynamic properties of aqueous systems as well as the dielectric properties of polar fluids [4].

Another simple potential used for polar molecules is the Stockmayer potential [5], consisting of a spherically symmetric Lennard-Jones 12-6 potential together with dipole-dipole terms

$$\phi(r, \theta_1, \theta_2, \phi_1-\phi_2)=4\varepsilon\left[\left(\frac{\sigma}{r}\right)^{12}-\left(\frac{\sigma}{r}\right)^{6}\right]-\frac{\mu^2}{r^3}D(\theta_1, \theta_2, \phi_1-\phi_2) \tag{8.3}$$

This model has been used to account for interactions between dipolar substances such as steam [6], carbon monoxide [7], and hydrogen chloride [8]. Many molecules, such as nitrogen, carbon dioxide, and benzene, have quite large permanent quadrupole moments whereas others (e.g., carbon tetrachloride, methane) have significant octupole moments. Hence an obvious generalization of the Stockmayer potential is to include quadrupole-quadrupole interactions and other electrostatic terms. Thus for carbon dioxide or nitrogen we could assume

$$\begin{aligned}\phi(r, \theta_1, \theta_2, \phi_1-\phi_2)=\phi_{LJ}(r)+\frac{3Q^2}{4r^5}[1&-5\cos^2\theta_1\\ &-5\cos^2\theta_2-15\cos^2\theta_1\cos^2\theta_2\\ &+2(\sin\theta_1\sin\theta_2\cos(\phi_1-\phi_2)-4\cos\theta_1\cos\theta_2)^2]\end{aligned} \tag{8.4}$$

where Q is the size of the permanent quadrupole moment and the angles are defined in terms of the symmetry axes of the molecules.

Potentials with central forces and added multipole terms cannot be expected to represent adequately the absence of spherical symmetry associated with the highly repulsive core of a molecule. In such circumstances a second method is used in which the asymmetry is included specifically in the core term. This approach has been used both for model fluids, in which the "molecules" are composed of fused hard spheres, or dumb-bells [9], and for realistic models of fluids, such as water [10] and benzene [11], where atom-atom or group-group terms have been included specifically. For example, a two-atom molecule such as N_2 has been modeled by two Lennard-Jones 12-6 interactions, of diameter σ and strength ε, with the centers of interaction a distance L from each other [12]. This model is a three-parameter potential and is usually specified by the quantities (ε, σ, $d = L/\sigma$). Several workers have used this model to represent nitrogen-nitrogen interactions, but there does not seem to be a consensus on the best values of ε, σ, and d to use.

In this chapter we examine the modifications to both equilibrium and nonequilibrium theory needed to account for the properties of nonspherical molecular fluids. The next section examines the problems associated with computing the virial coefficients for nonspherical interaction potentials. Subsequent sections examine the static structure and thermodynamic properties of dense fluids. Finally, problems associated with the transport properties of nonspherical molecules in the fluid phase are discussed.

8.1 Second Virial Coefficient of Nonspherical Molecules

In Section 3.7 we derived expressions for the first few coefficients in the density expansion of the equation of state of a dilute gas. The first nontrivial term in this expansion, the second virial coefficient B_2, was given as the integral

$$B_2 = -2\pi \int_0^\infty r^2 \left[\exp \frac{-\phi(r)}{kT} - 1 \right] dr$$

Unfortunately, it is not possible to generalize this expression to nonspherical molecules merely by replacing the integral over distance by integrals over both distance and orientation. When we derived the expressions for $B_2, B_3, \ldots$, we expanded both the grand canonical partition function and the number density in powers of the parameter $z = \lambda^{-3} \exp(\mu/kT)$ where μ is the chemical potential of the gas and λ is the thermal wavelength $\lambda = (h^2/2\pi mkT)^{1/2}$. The temperature terms in λ arose from the integration over momenta in eq. 3.31, and it is this term that changes for nonspherical molecules.

To proceed, we need some method of specifying the orientation of a nonspherical rigid molecule in a laboratory frame of reference. The usual choice of coordinates for specifying orientations are the Euler angles [13], $\boldsymbol{\psi} = (\psi_1, \psi_2, \psi_3)$. In terms of these angles it can be shown that the total energy of the system of particles is given by

$$H = \sum_{i=1}^{N} \left[\frac{\mathbf{p}_i \cdot \mathbf{p}_i}{2m} + \frac{1}{2} \boldsymbol{\omega}_i^T \cdot \mathbf{I}_p \cdot \boldsymbol{\omega}_i \right] + \Phi \tag{8.5}$$

where $\mathbf{I}_p$ is the *principle inertia tensor* and $\boldsymbol{\omega}_i = (\omega_{1i}, \omega_{2i}, \omega_{3i})$ with ω_{ki} the angular velocity of the ith particle about its kth principle axis [13]. The angular velocities are given in terms of the Euler angles and their time derivatives [13]

$$\begin{aligned} \omega_x &= \dot{\psi}_1 \sin \psi_2 \sin \psi_3 + \dot{\psi}_2 \cos \psi_3 \\ \omega_y &= \dot{\psi}_1 \sin \psi_2 \cos \psi_3 - \dot{\psi}_2 \sin \psi_3 \\ \omega_z &= \dot{\psi}_1 \cos \psi_2 + \dot{\psi}_3 \end{aligned} \tag{8.6}$$

We see at once that the relation between the rotational velocity components and the Euler angles is considerably more complex than that between the translational velocity components and the spatial coordinates. As far as the statistical mechanics is concerned, we are interested in the changes brought about when we integrate over the angular velocities. The integral in the expression for the grand partition function (see eq. 3.33) is of the form

$$Q_N = h^{-6N} \int \exp \frac{-H}{kT} \, d\mathbf{r}_1 \cdots d\mathbf{p}_N \, d\boldsymbol{\psi}_1 \cdots d\boldsymbol{\psi}_N \, d\mathbf{q}_1 \cdots d\mathbf{q}_N \tag{8.7}$$

where the term h^{-6N} comes from the fact that there are $6N$ position and orientational coordinates needed to define the configuration and where the generalized momenta, $\mathbf{q}_i$, and the Euler angles, $\boldsymbol{\psi}_i$, are *conjugate coordinates* [13]. That is, they satisfy the equation

$$q_{ki} = \frac{\partial L}{\partial \dot{\psi}_{ki}} = \frac{\partial L}{\partial \omega_{xi}} \frac{\partial \omega_{xi}}{\partial \dot{\psi}_{ki}} + \frac{\partial L}{\partial \omega_{yi}} \frac{\partial \omega_{yi}}{\partial \dot{\psi}_{ki}} + \frac{\partial L}{\partial \omega_{zi}} \frac{\partial \omega_{zi}}{\partial \dot{\psi}_{ki}}$$

where L, the Lagrangian, is the difference between the kinetic and potential energies, $L = H - 2\Phi$ for our problem. As the potential energy of the system, Φ, is independent of the quantities $\dot{\psi}_{ki}$, it follows from eqs. 8.5 and 8.6 and the definition of q_{ki} that

$$\begin{aligned} q_{1i} &= I_1 \omega_{xi} \sin \psi_{2i} \sin \psi_{3i} + I_2 \omega_{yi} \sin \psi_{2i} \cos \psi_{3i} + I_3 \omega_{zi} \cos \psi_{2i} \\ q_{2i} &= I_1 \omega_{xi} \cos \psi_{3i} - I_2 \omega_{yi} \sin \psi_{3i} \\ q_{3i} &= I_3 \omega_{zi} \end{aligned} \tag{8.8}$$

where I_1, I_2, I_3 are the diagonal components of $\mathbf{I}_p$.

Obviously, $q_{ij} \neq \omega_{ij}$ and hence $\boldsymbol{\psi}_i$ and $\boldsymbol{\omega}_i$ are not conjugate coordinates. Thus the integrals over $\mathbf{q}_i$, and hence $\boldsymbol{\omega}_i$, in eq. 8.7 are not of the form $\int_{-\infty}^{\infty} \exp(-ax^2)\, dx$ found for the integration over translational momenta (see eq. 3.32). To convert eq. 8.7 into an equation containing angular velocity integrations of this form, we have to transform coordinates of integration from $\mathbf{q}_i$ to $\boldsymbol{\omega}_i$:

$$Q_N = h^{-6N} \int \exp \frac{-H}{kT}\, d\mathbf{r}_1 \cdots d\mathbf{p}_N \prod_{i=1}^{N} J_i\, d\boldsymbol{\psi}_i\, d\boldsymbol{\omega}_i \tag{8.9}$$

where J_i is the Jacobian of the transformation, $J_i = |\partial \mathbf{q}_i / \partial \boldsymbol{\omega}_i|$. Using eq. 8.8 to evaluate the Jacobian we obtain

$$J_i = I_1 I_2 I_3 \sin \psi_{2i} \tag{8.10}$$

and the integral takes the form

$$Q_N = h^{-6N} \int \exp \frac{-H}{kT}\, d\mathbf{r}_1 \cdots d\mathbf{p}_N \prod_{i=1}^{N} (I_1 I_2 I_3 \sin \psi_{2i}\, d\boldsymbol{\psi}_i\, d\boldsymbol{\omega}_i) \tag{8.11}$$

Looking at eq. 8.5 again we see that both the $\mathbf{p}$ integrals and the $\boldsymbol{\omega}$ integrals are now of identical mathematical form, $\int_{-\infty}^{\infty} \exp(-ax^2)\, dx$, and so as in the derivation of eq. 3.32 we can write

$$Q_N = \lambda_{tr}^{-3N} \lambda_{rot}^{-3N} \int \exp \frac{-\Phi(\mathbf{r}_1, \ldots \boldsymbol{\psi}_N)}{kT} \prod_{i=1}^{N} (\sin \psi_{2i}\, d\mathbf{r}_i\, d\boldsymbol{\psi}_i) \tag{8.12}$$

where $\lambda_{tr} = (h^2/2\pi mkT)^{1/2}$ as before and $\lambda_{rot} = [h^2/2\pi(I_1 I_2 I_3)^{1/3} kT]^{1/2}$.

We can now derive expressions for the virial coefficients by expanding the grand partition function and number density as in Section 3.7, using as the expansion parameter $z = \exp[\mu/kT]\lambda_{tr}^{-3}\lambda_{rot}^{-3}$. The result for the second virial coefficient is

$$B_2(T) = \frac{-1}{16\pi^2} \int d\mathbf{r} \int_0^{2\pi} d\psi_1 \int_0^{\pi} \sin \psi_2\, d\psi_2 \int_0^{2\pi} d\psi_3 \exp \frac{-\phi(\mathbf{r}, \boldsymbol{\psi})}{kT} \tag{8.13}$$

where $\phi(\mathbf{r}, \boldsymbol{\psi})$ is the pair potential and depends in general on three relative spatial coordinates and three Euler angles specifying the relative orientation of the two molecules.

Two methods have been used to compute the integrals in eq. 8.13. The first is applicable to molecules that have a spherically symmetric repulsive core and a weakly anisotropic tail; the second is applicable to molecules that are strongly asymmetric. An example of the first approach is to calculate the second virial coefficient of the hard sphere potential with embedded point dipole, eq. 8.1. This potential depends on only three angles and the distance between the molecules, so that the second virial coefficient is given by [2]

$$B_2(T) = -\tfrac{1}{4}\int_0^{\infty} r^2\,dr \int_0^{2\pi} d\,(\phi_1-\phi_2) \int_0^{\pi} \sin\theta_1\,d\theta_1 \int_0^{\pi} \sin\theta_2\,d\theta_2 \times \left\{ \exp \frac{-[\phi_{\mathrm{HS}}(r) - \mu^2/r^3\,D(\theta_1,\theta_2,\phi_1-\phi_2)]}{kT} - 1 \right\} \quad (8.14)$$

If the diameter of the hard sphere is σ, this expression may be integrated by expanding the term exp $(\mu^2 D/r^3 kT)$ in a power series in $1/kT$, where it is found that [2, 14]

$$B_2(T) = \frac{2\pi\sigma^3}{3}\left[1 - \sum_{n=1}^{\infty} \frac{D_n}{(2n)!\,(2n-1)}\left(\frac{\mu^2}{\sigma^3 kT}\right)^{2n}\right] \quad (8.15)$$

where

$$D_n = \frac{1}{8\pi}\int_0^{2\pi} d(\phi_1-\phi_2) \int_0^{\pi} \sin\theta_1\,d\theta_1 \int_0^{\pi} \sin\theta_2\,d\theta_2\, D^{2n}(\theta_1,\theta_2,\phi_1-\phi_2)$$

A similar procedure can be used to calculate $B_2(T)$ for the Stockmayer potential, eq. 8.3, for potentials containing higher multipole moments, and for higher virial coefficients [15].

The second method for calculating the virial coefficients for nonspherical potentials is to use multidimensional numerical integration. Although it is in principle possible to evaluate integrals such as that given in eq. 8.13 by successive application of one-dimensional formulae, such as Simpson's rule, in practice this is often not economical. For example, if the range of each variable were divided into, say, 10 regions, the number of function evaluations would be 10 (in one-dimension), 100 (in two dimensions), 1000 (in three dimensions) and 10^N in general. As most integrands of interest in virial coefficient calculations are rapidly varying it is unlikely that 10 points in every dimension would be sufficient. Fortunately, it is possible to use the so-called nonproduct integration methods [16] to reduce the number of function evaluations needed to compute virial coefficients. Using such formulae it is possible to evaluate an N-dimensional cubic polynomial in $2N$ function evaluations rather than with the 3^N evaluations required using Simpson's rule. Nonproduct rules have been used to compute the second and higher virial coefficients of complicated interaction potentials such as those used for water [17, 18] and benzene [19].

8.2 Equilibrium Properties of Dense Fluids

The radial distribution function, $g(r)$, introduced in Section 3.5 as a convenient measure of short range order or correlations in fluids, can be generalized to include angular dependence. Such an *angular correlation function* can be written $g(\mathbf{r}_{12}, \boldsymbol{\psi}_{12})$, where $\mathbf{r}_{12}$ is the vector along the line of molecular centers and $\boldsymbol{\psi}_{12}$ denotes the Euler angles giving the relative orientation of the two molecules. Certain equilibrium properties can be related to the simpler *center correlation function* $g(r)$ defined by

$$g(r) = \langle g(\mathbf{r}_{12}, \boldsymbol{\psi}_{12}) \rangle_{\Omega} \tag{8.16}$$

where the angle brackets denote an unweighted average over the relative orientations of the two molecules including polar coordinates and Euler angles. One example is the isothermal compressibility which has the same form as for the spherically symmetric case (eq. 3.63)

$$kT\left(\frac{\partial \rho}{\partial p}\right)_T = 1 + 4\pi\rho\int_0^\infty r^2[g(r)-1]\,dr \tag{8.17}$$

There are very few exact results for nonspherical molecules, even for relatively idealized systems. Nevertheless, results for fluids composed of molecules with orientationally dependent forces have been reported using simulation studies, integral equation approximations, and perturbation theories.

Both Monte Carlo and molecular dynamics methods have been used to calculate the properties of nonspherical molecules [20–30]. The orientation dependence of the intermolecular forces increases the computation time severalfold over the corresponding calculations for simple fluids. An additional complication is that the anisotropic component of the potential may be long ranged, as are the electrostatic dipole-dipole terms. Machine calculations employ a potential that is set to zero outside a distance of a few molecular diameters, and so contributions from interactions beyond this distance have to be included by treating the outside region as a continuum (see Chapter 9) or by using Ewald summation [28] (see Chapter 10).

Machine calculations using spherically symmetric potentials have also played an important role in the analysis of nonspherical molecular interactions. Results for the spherical potential are used as reference systems for perturbation theories; thus the anisotropic component of the potential is added as a correction to the essentially "exact" simulation results. This approach is particularly attractive if the molecules are nearly spherical so that the perturbation expansion is rapidly convergent.

One of the first molecular dynamics studies of nonspherical molecules was reported by Harp and Berne [7, 20], who used the Stockmayer potential with parameters appropriate to nitrogen and carbon monoxide to compute bulk properties such as the pressure and internal energy.

They obtained reasonable agreement with experiment and also found that contributions to the thermodynamic properties from anisotropic interactions tended to be small. Similar observations were made by Wang et al. [23, 24] who used the Stockmayer potential with both dipole and quadrupole interactions, and with a short range anisotropic term representing overlap forces, in Monte Carlo studies of the pair correlation function. Contributions to $g(r)$ from the long range multipole interactions are much smaller than those arising from anisotropic short range interactions. The main effect of the multipole terms is to shift the peaks in $g(r)$ to slightly smaller distances, whereas the anisotropic overlap forces have a much greater effect on $g(r)$, changing its shape markedly. In an explicit test of this observation, Barojas et al. [21] simulated a system of 500 nitrogen molecules interacting through Lennard-Jones 12-6 potentials centered on each nitrogen atom. They found that the computed pressure and internal energy were within a few percent of experimental values at densities near the triple point of nitrogen. It is interesting to note that the center correlation function $g(r)$ given in eq. 8.16 resembles the pair correlation function for a monatomic gas in a similar thermodynamic state. However, the angle averaged pair distribution function measuring the correlation between an atom on one molecule and a second atom on another molecule differs considerably as can be seen from Fig. 8.2. The first peak in the distribution function is lower, broader, and shows two maxima, corresponding to the atoms of the nearest neighbor molecule.

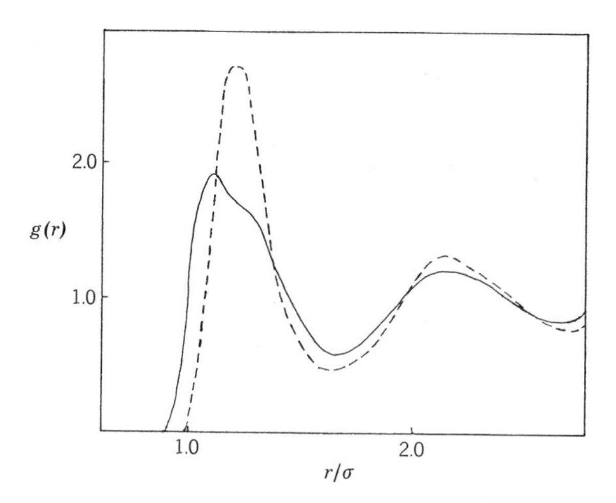

Figure 8.2 Radial distribution functions for liquid nitrogen at $\rho\sigma^3 = 0.696$. —— atom-atom correlation function at $kT/\epsilon = 1.83$; - - - - center-center correlation function at $kT/\varepsilon = 1.51$. (Adapted with permission from *Phys. Rev.*, A**2**, 1092, 1973.)

Barojas et al. [21] computed the static structure factor $S(q)$ using the atom-atom potential described above and their results are compared with experiment in Fig. 8.3. The position of the first and second maxima in $S(q)$ are in excellent agreement with experiment, but the heights of the peaks differ by several percent at both densities for which a comparison is reported. The work of Barojas et al. for liquid nitrogen was extended by Gibbons and Klein [30] who used Monte Carlo methods to study solid nitrogen. Their results for the thermodynamic properties of the crystal favored a softer Lennard-Jones 9-6 potential rather than the 12-6 form.

Other models have been proposed as interaction potentials for liquids composed of linear molecules [26, 29]. For example, Vieillard-Baron [26] computed the equation of state for hard spherocylinders (cylinders with a hemisphere at each end) using Monte Carlo methods. He found that the packing fraction (number density times the volume of one spherocylinder) is an increasing function of the length to breadth ratio. Although models such as this are rather unrealistic, they are able to give a qualitative description of phenomena associated with long rodlike molecules.

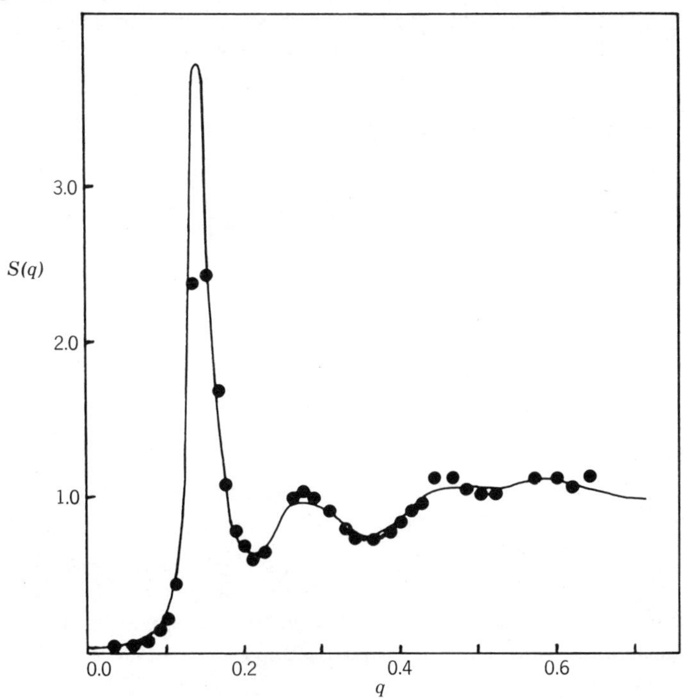

Figure 8.3 Static structure factor for nitrogen at $\rho\sigma^3 = 0.696$ and $kT/\varepsilon = 1.46$. ●● molecular dynamics results; —— experimental data. Units of q are $\text{Å}^{-1}/2\pi$. (Adapted with permission from *Phys. Rev.*, **A2,** 1092, 1973.)

Evans and Watts [31] have reported Monte Carlo studies for liquid benzene using a six-center Lennard-Jones 12-6 potential for each molecule. With this model, the total interaction between two molecules is given by

$$\phi(\mathbf{r}_1, \mathbf{r}_2) = \sum_{i=1}^{6} \sum_{j=1}^{6} 4\varepsilon\left[\left(\frac{\sigma}{R_{ij}}\right)^{12} - \left(\frac{\sigma}{R_{ij}}\right)^{6}\right] \tag{8.18}$$

where the distance R_{ij} is the distance between group i on molecule 1 and group j on molecule 2. The six sites on each molecule were on the vertices of a regular hexagon of side 1.756 Å and the Lennard-Jones parameters had the values $\varepsilon/k = 77$ K and $\sigma = 3.5$ Å. These values had been chosen previously on the basis of a fit to second virial coefficients and solid state data [32]. Evans and Watts [31] computed several structural properties of liquid benzene at two temperatures and densities. Properties computed included the three radial distribution functions $g_{S-S}(r)$, $g_{S-C}(r)$, and $g_{C-C}(r)$ and the three angular distribution functions $g_{u_i-r}(r)$, $g_{u_j-r}(r)$, and $g_{u_i-u_j}(r)$. These distribution functions have the following interpretation. The radial functions measure the relative distributions of site-site distances (g_{S-S}), of site-center of mass distances (g_{S-C}), and of the center of mass-center of mass distribution (g_{C-C}). Thus these three functions correspond to generalizations of the usual radial distribution functions. The angular distributions give a measure of the relative orientation of the molecules. Let $\mathbf{u}_i$ be a unit vector perpendicular to the plane of the ith molecule and let $\mathbf{r}_{ij} = \mathbf{r}_j - \mathbf{r}_i$ be the vector joining the centers of mass of molecules i and j. Then

$$g_{u_i-r}(r) = \frac{\langle \mathbf{u}_i \cdot \mathbf{r}_{ij} \rangle_\delta}{4\pi r^3 g(r)\, dr}$$

$$g_{u_j-r}(r) = \frac{\langle \mathbf{u}_j \cdot \mathbf{r}_{ij} \rangle_\delta}{4\pi r^3 g(r)\, dr}$$

$$g_{u_i-u_j}(r) = \frac{\langle \mathbf{u}_i \cdot \mathbf{u}_j \rangle_\delta}{4\pi r^2 g(r)\, dr}$$

where $\langle \mathbf{a} \cdot \mathbf{b} \rangle_\delta$ implies that the average was taken over the interval $(r_{ij} - \delta/2,\ r_{ij} + \delta/2)$. Note that the functions g_{u_i-r} and g_{u_j-r} are not necessarily the same, as the molecule at one end of the line joining the centers will, in general, have a different orientation to that at the other. To make this clear, the direction of the vectors, $\mathbf{u}_i$, $\mathbf{u}_j$, $\mathbf{r}_{ij}$ were chosen so that $0 \leq \mathbf{u}_i \cdot \mathbf{r}_{ij} < \mathbf{u}_j \cdot \mathbf{r}_{ij}$.

Figure 8.4 shows the three functions g_{S-S}, g_{S-C}, and g_{C-C} as a function of distance at a temperature of 328 K and the experimental number density at a pressure of one atmosphere, $\rho = 0.00633$ molec Å^{-3}. The

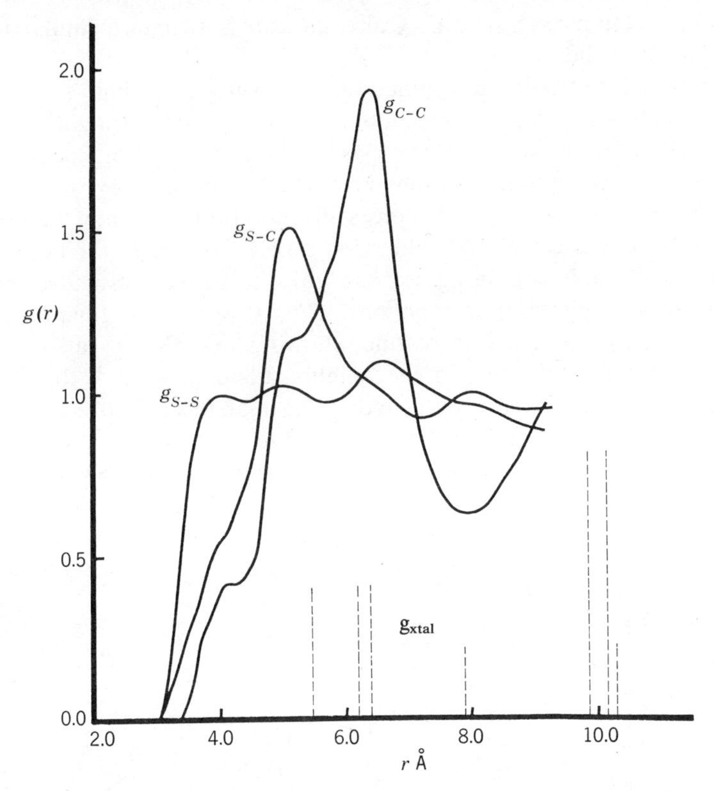

Figure 8.4 Radial distribution functions for liquid benzene at $T = 328$ K and $\rho = 0.00633$ molec Å^{-3}.

center-of-mass–center-of-mass distribution, $g_{C-C}(r)$, shows more detail than is found for simpler liquids. Evans and Watts [31] related this behavior to the structure of the benzene crystal. Their interpretation of the structure is indicated on the figure and recognizes three contributions. First, there is a shoulder at about 4 Å arising from a configuration not found in the crystal, the so-called "stacked" configuration. In this relative orientation one benzene molecule is found on top of another, as in a pile of dinner plates. The second shoulder on the first peak can be correlated with the nearest neighbor position, and relative orientation, found in the crystal. In this configuration the nearest neighbor molecules have the symmetry vector $\mathbf{u}$ perpendicular to that of the central molecule. The main peak arises primarily from second nearest neighbors, although there may also be a contribution from nearest neighbor molecules lying in the same plane. Figure 8.4 includes the corresponding distribution function for the solid after scaling to the liquid number density, and it can be seen that the functions g_{S-C} and g_{S-S} also indicate a structure similar to that found in the solid.

The angular distribution functions are shown in Fig. 8.5 for the conditions $T = 298$ K and $\rho = 0.00662$ molec Å^{-3}. It is straightforward to show that if the relative orientations of the molecules are uncorrelated the angular distribution functions have the values $g_{u_i-r} = \frac{1}{3}$, $g_{u_j-r} = \frac{2}{3}$, and $g_{u_i-u_j} = \frac{1}{2}$ [31, 33]. At longer distances all three functions shown in Fig. 8.5 approach these uncorrelated values. At shorter distances, the behavior of the various functions is in agreement with the general structure inferred from the radial distribution functions. Thus at $r \sim 4$Å the functions $g_{u_i-u_j}$, g_{u_i-r}, and g_{u_j-r} are all close to one, showing that the so-called stacked configuration is preferred. In the neighborhood of $r = 5$ Å the function g_{u_j-r} has a value of about 0.95, indicating that the axis of symmetry of molecule j is parallel to the line joining the centers of mass. However, both g_{u_i-r} and $g_{u_i-u_j}$ are much smaller, showing that in this region molecule i is inclined at an angle to the line joining the centers of mass. Such a configuration is typical of that found in the solid and the distribution function $g_{C-C}(r)$ shows that a substantial number of the nearest

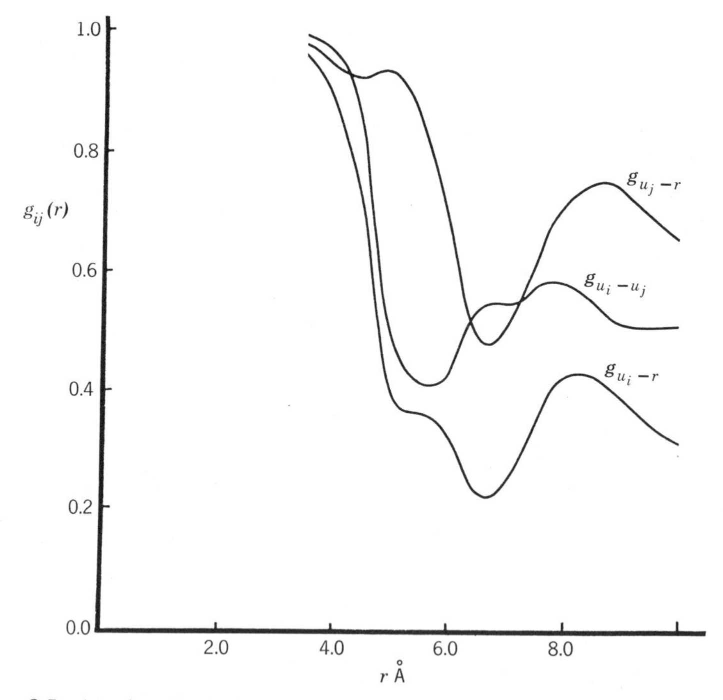

Figure 8.5 Angular distribution functions for liquid benzene at $T = 298$ K and $\rho = 0.00622$ molec Å^{-3}.

neighbor molecules are in this configuration rather than in the stacked configuration. Evans and Watts [31, 33] also discussed the structure of the liquid at distances greater than 5.5 Å in terms of Figs. 8.4 and 8.5, and concluded that the six-center Lennard-Jones model gave a good description of liquid benzene.

8.3 Approximate Theories for Nonspherical Molecules

Both integral equation approximations and perturbation theory have been used to estimate the equilibrium properties of fluids composed of nonspherical molecules. These approaches have met with rather limited success, primarily because of the computational difficulties inherent in dealing with shape dependent potentials. Nevertheless, a significant amount of work has been reported in this field.

As we said earlier, the position and orientation of an arbitrary rigid molecule is determined by the three position coordinates of its center-of-mass and the three Euler angles describing its orientation in some laboratory frame of reference. For this case the Ornstein-Zernike equation of Section 3.8 can be written

$$h(\mathbf{x}_1) = c(\mathbf{x}_1) + \frac{\rho}{8\pi^2}\int h(\mathbf{x}_2)c(\mathbf{x}_2 - \mathbf{x}_1)\, d\mathbf{x}_2 \tag{8.19}$$

where

$$\int d\mathbf{x} = \int_{-\infty}^{\infty} dx_1 \int_{-\infty}^{\infty} dx_2 \int_{-\infty}^{\infty} dx_3 \int_0^{2\pi} d\psi_1 \int_0^{\pi} \sin\psi_2\, d\psi_2 \int_0^{2\pi} d\psi_3$$

with $\mathbf{x}$ representing both the cartesian coordinates and the Euler angles. The generalized Ornstein-Zernike equation is the starting point for most integral equation studies of nonspherical molecules. As in the study of simple fluids, it is customary to assume some approximate relation between $h(\mathbf{x})$ and $c(\mathbf{x})$ to close the above equation. Even if this is done, we still need to specify $h(\mathbf{x})$ for all six coordinates both to solve the integral equation and to give a complete description of the system. For a numerical integration involving N grid points for each coordinate, we would need to have values of $h(\mathbf{x})$ at N^6 points. Consequently, any integral equation approach must include some method for reducing the amount of computation. The calculations can be simplified somewhat by restricting investigations to the properties of linear molecules. Most studies do this although Blum and Torruella [34] have proposed an invariant expansion for two-body correlation functions which reduces the angle dependent Ornstein-Zernike equation to a more manageable form.

For linear molecules, Chen and Steele [35, 36] decomposed the generalized Ornstein-Zernike equation into a set of coupled integral equations by expanding the correlation functions in a series of spherical harmonics. After Fourier transforming the equations and truncating the hierarchy, the set of equations was solved using an iterative procedure. They investigated the thermodynamic and statistical mechanical properties of various mildly nonspherical molecules and found that the pair correlation function is given by the first few terms in the series of spherical harmonics. For such slightly nonspherical molecules the equation of state and virial coefficients do not show any great sensitivity to molecular shape. The expansion given by Chen and Steele [35, 36] has been used by Morrison [37] to calculate the radial distribution function of chlorine molecules using a two-center Lennard-Jones potential.

One of the few exact results reported for nonspherical molecules using integral equation methods is that of Wertheim [38] for a fluid composed of hard spheres with embedded dipoles. This problem was solved for an infinite fluid using the Mean Spherical Model (MSM) of Lebowitz and Percus [39]. For polar fluids, this model supplements the Ornstein-Zernike equation with two additional approximations. The first assumption $g(r)=0$ for $r<\sigma$ is exact for hard core potentials, if σ is the hard core radius. The second relation is an approximation to the direct correlation function and takes the form

$$c(\mathbf{x})=\frac{-\mu^2}{kTr^3}D(\theta_1, \theta_2, \phi_1-\phi_2)$$

where μ is the dipole moment and $D(\theta_1, \theta_2, \phi_1-\phi_2)$ is the function defined in eq. 8.2. Andersen and Chandler [40] have shown that the MSM approximation is the lowest in an hierarchy of approximations which they call the optimized cluster expansion. The solutions to the approximation are expressed in terms of the solution to the Percus-Yevick equation for hard spheres [41, 42] and are useful both for comparison with machine calculations and perturbation theories and for empirical theories of properties such as the dielectric constant of water [4]. Blum [43] has extended Wertheim's solution to include interactions between higher multipole moments while other workers [44] have generalized the MSM approximation to, and obtained analytic solutions for, mixtures.

In the above model the basic shape of the molecules remains spherical at short distances. Nienhuis and Deutsch [45] developed a molecular theory for strong polar dielectric fluids containing particles with arbitrary shape that may or may not be embedded in a dielectric continuum. The result of their investigation was an explicit expression for the two-body correlation function at large intermolecular separations in terms of the short range correlations. Although in principle any short range anisotropic interactions can be described by this theory, the contribution from short range correlations is expressed as a complicated graphical expansion that is not always convenient for calculations.

Perhaps the most computationally convenient integral equation theory is that investigated by Andersen, Chandler, and Lowden [46–49] and known as the Reference Interaction Site Model (RISM). The RISM integral equation resembles the Fourier transformed Ornstein-Zernike equation and was derived within the framework of the optimized cluster expansion of Andersen and Chandler [40, 50–53]. Strong shape dependent interactions appear explicitly in this model. Although approximate, the RISM theory has successfully predicted the properties of a number of dense molecular fluids. Results have been obtained from this theory for the structure of liquid nitrogen [48] and are found to be in quite good agreement with the high density molecular dynamics calculations of Barojas et al. [21]. More recently calculations using the RISM equation with a hard sphere site-site interaction have been extended to other molecular liquids such as carbon tetrachloride and benzene [49]. There is reasonably good qualitative agreement between the RISM radial distribution functions for benzene and the Monte Carlo results of Evans and Watts [31, 33]. No doubt the agreement between the integral equation approximations and machine results for the more complicated molecules will improve as more realistic interaction potentials are introduced into the former methods.

A good deal of progress has been made in recent years toward an understanding of the structure of liquids with noncentral interactions by using statistical mechanical perturbation theory [54]. Many such theories were first derived with the intention of using spherically symmetric molecules as a reference system to calculate corrections to thermodynamic properties of fluids made up of slightly nonspherical particles. Although the importance of perturbation theory to fluids containing nonspherical molecules was recognized some time ago [55], detailed studies were restricted by the absence of accurate results for the reference systems. Consequently, studies were usually limited either to fluids at low densities [12, 56] or to relatively straightforward analyses of neutron diffraction data for molecular fluids [57].

In most cases the perturbation theory begins with a simple intermolecular potential ϕ_0, usually spherically symmetrical, and generates a series expansion about ϕ_0 for a more general anisotropic potential function, ϕ. Usually one expands ϕ (or $\exp[-\phi/kT]$) in a Taylor series about ϕ_0 (or $\exp[-\phi_0/kT]$) retaining the first one or two nonzero terms in the series. As in the simple perturbation theories discussed in Section 3.9, the pair potential for the actual system is written in two parts

$$\phi(\mathbf{r}_{12}, \boldsymbol{\psi}_{12}) = \phi_0(r_{12}) + \lambda\phi_1(\mathbf{r}_{12}, \boldsymbol{\psi}_{12})$$

where r_{12} is the distance between the centers-of-mass of the molecules, $\boldsymbol{\psi}_{12}$ describes the relative orientations of the molecules, and λ is an expansion parameter (for example, it may be the square of the dipole moment). The reference potential is sometimes taken to be the angle averaged potential

$$\phi_0(\mathbf{r}_{12}) = \int W(\boldsymbol{\chi}_{12})\phi(\mathbf{r}_{12}, \boldsymbol{\psi}_{12})\, d\boldsymbol{\chi}_{12}$$

where $W(\boldsymbol{\chi}_{12})$ is some weight function, usually $\exp[-\phi(\mathbf{r}_{12}, \boldsymbol{\psi}_{12})/kT]$, and $\boldsymbol{\chi}$ is the vector $\boldsymbol{\psi}_{12}$ together with the two polar angles.

There have been several perturbation studies which have used either the hard sphere or Lennard-Jones interactions as reference potentials. Gubbins and Gray [58] have generalized the Barker-Henderson theory [59] to angle dependent potentials and obtained an expansion for $g(\mathbf{r}_{12}, \boldsymbol{\psi}_{12})$. Their formal expressions were applied to the calculation of several thermodynamic properties of simple molecules including the isothermal compressibility, pressure, configurational energy, entropy, and specific heat, both along and away from the liquid-vapor coexistence curve [60]. They found that these thermodynamic quantities differ in their sensitivity to the long and short range noncentral interactions. For example, the specific heat is more sensitive to long range anisotropic forces than are the other properties, whereas the pressure is particularly sensitive to the short range anisotropic forces. The isothermal compressibility was only slightly affected by the anisotropic component of the intermolecular potential, even in the dense fluid region. The Gubbins-Gray theory gave excellent agreement with the experimental results for methane, oxygen, and nitrogen in the liquid state. It is important to remember that the reference potential for these calculations was spherical and that the multipole moments and weak short range anisotropic terms were added as a perturbation. The inclusion of strongly anisotropic interactions at short distances apparently leads to poorer convergence of the perturbation series [61].

Stell and co-workers [4, 62] have considered the thermodynamic properties of simple polar fluids interacting through the Stockmayer potential and modifications of this model. A perturbation expansion was extended to include terms of order μ^6, where μ is the dipole moment, and the effects of higher order terms were approximated using a simple but accurate Padé extrapolation. They used their expressions to compute the contributions to the free energy of HCl, CO_2, and N_2 from the dipole, quadrupole, and octupole moments. Their results for the free energy of dense fluids were in good agreement with the Monte Carlo calculations of Verlet and Weis [25]. At the same time, Verlet and Weis proposed another, equally successful perturbation theory based on a partial resummation of a perturbation expansion of Lebowitz, Stell, and Baer [63]. Their theory resembles the Andersen-Chandler-Weeks integral approximations discussed earlier [53].

One approach to the problem of including strongly repulsive forces in the perturbation theory of nonspherical molecules was reported by Sung and Chandler [64] using the "blip function" expansion of Andersen, Chandler, and Weeks [65]. This expansion is a method for describing the equilibrium structure and thermodynamic properties of systems interacting through repulsive forces in terms of the properties of a hard sphere fluid. The theory neglects the slowly varying attractive forces and in this sense corresponds to a high temperature approximation. Despite the errors inherent in the blip function expansion, the results of Sung and Chandler were nevertheless in good agreement with molecular dynamics results for the radial distribution function.

Sandler [66] has reported important modifications to the Helmholtz free energy expansion when noncentral interactions are included in the reference potential. A similar approach, which had been considered earlier by Perram and White [67], ensures that the strong noncentral forces associated with short range repulsions are properly taken into account. Sandler's analysis showed that the lowest order terms in the perturbation expansion are proportional to the second power of the dipole and quadrupole moments and not to the fourth power as had been found previously using a spherical reference potential function [62]. Initial results, using first order corrections only, were most encouraging for the chlorine and hydrogen chloride molecules.

The apparently diverse approaches to perturbation theories of fluids discussed in this section have been neatly summarized by Smith [54]. He showed that all the expansions could be derived from a functional Taylor expansion [68] of the Helmholtz free energy and radial distribution function. As we have seen, most perturbation theories include all of the nonspherical contributions to the interactions between molecules as a perturbation term. This approach appears to be sufficient for nearly spherical molecules or for those with relatively small multipole moments. However, for more anisotropic molecules this approach is unsatisfactory, and it is likely that theories that use a noncentral interaction as the reference potential will be both more accurate and more rapidly convergent than perturbation theories based on spherical reference potential functions [69]. In addition, it is likely that any adequate treatment of polar fluids will include the dominant three-body contributions to the

Helmholtz free energy [70]. However we omit this problem from our discussions in this chapter.

8.4 Transport Phenomena in Dilute Polyatomic Gases

The kinetic theory of dilute monatomic gases was discussed briefly in Section 2.4 where we gave formulae relating the transport coefficients to the interatomic potential through the collision integrals $\Omega^{(l,s)}$ (see eqs. 2.45–2.49). Many textbooks have given full descriptions of the derivation of these formulae and have discussed their application to simple systems [2, 71]. Briefly, the theory leading to these formulae is based on the assumption that the gas is so dilute that the only collisions of importance involve no more than two particles. The extension of kinetic theory to dilute polyatomic gases is nontrivial, and very few calculations of the transport coefficients of such systems have been reported. Wang Chang and Uhlenbeck (WCU) [72] published an extension of the standard kinetic theory in which the internal degrees of freedom of the molecules were treated quantum mechanically, although it was assumed that the translational degrees of freedom could be handled classically. This theory was based on a number of assumptions, including the existence of *inverse collisions.*

To understand an inverse collision, consider an example where two molecules have positions and velocities $(\mathbf{r}_1, \mathbf{v}_1)$ and $(\mathbf{r}_2, \mathbf{v}_2)$ prior to the collision and $(\mathbf{r}_1', \mathbf{v}_1')$ and $(\mathbf{r}_2', \mathbf{v}_2')$ after the collision. The inverse collision is the collision where the initial state for particle 1 is $(\mathbf{r}_2, \mathbf{v}_1')$, that for particle 2 is $(\mathbf{r}_1, \mathbf{v}_2')$ and where the final states are $(\mathbf{r}_2', \mathbf{v}_1)$ and $(\mathbf{r}_1', \mathbf{v}_2)$, respectively. Figure 8.6 demonstrates that *in general the inverse collision does not exist.* The concept of inverse collisions and associated concepts such as a closed cycle of corresponding collisions is fully discussed by Tolman [73], and it is not possible to give a detailed account here. A fundamental assumption underlying the derivation of the Boltzmann equation, and hence expressions for the transport coefficients for monatomic systems, is that inverse collisions exist.

In addition to assuming the existence of inverse collisions, the WCU theory also ignores the conservation of angular momentum in polyatomic fluids which can be serious for many systems. However, the WCU theory leads to expressions for the transport coefficients in terms of more general collision integrals and these can be used in studies of transport processes in dilute molecular gases [33].

Subsequent to the WCU theory Taxman [74] published a derivation of dilute gas transport coefficients for molecules with internal rotational and

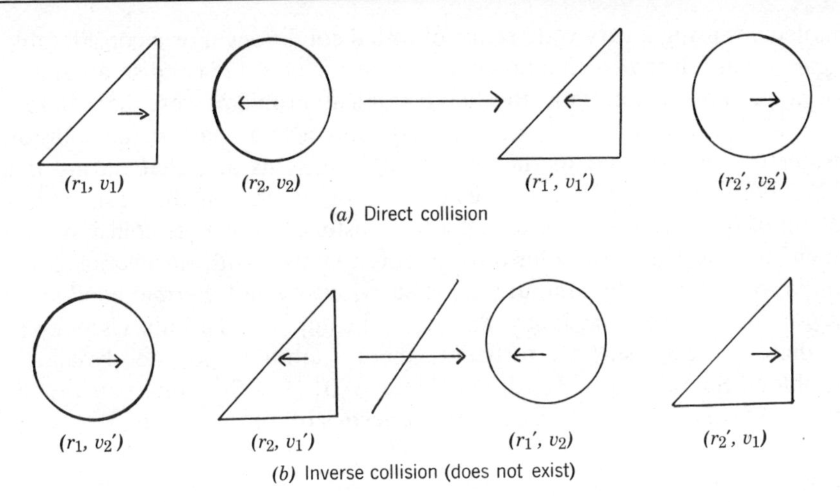

Figure 8.6 Demonstration that the inverse collision does not always exist.

vibrational states wholly in terms of classical mechanics. This theory is likely to be reasonably accurate for systems containing molecules with relatively large moments of inertia and whose internal vibrational states are largely uncoupled from the translational and rotational degrees of freedom. Taxman derived his theory without assuming the existence of inverse collisions but did ignore the conservation of total angular momentum of the system.

Both the WCU theory and the Taxman theory are based on the assumption that the mean time between collisions in the dilute gas is much greater than the duration of any given collision. This assumption is also fundamental to the corresponding monatomic theory [2, 71] and apparently is accurate for these systems at sufficiently low densities [75]. In systems of spherical particles there is only a relatively narrow range of impact parameters and energies in which orbiting collisions occur (see Section 2.4). The corresponding ranges are much greater for molecular collisions, where the angular momentum of the system about its center of mass can be absorbed in internal degrees of freedom. Thus as two molecules collide it is possible for translational kinetic energy to be converted into intramolecular rotational kinetic energy, possibly lengthening the duration of the collision. A number of papers have been written extending the WCU and Taxman theories [76], and there have been solutions to the equations reported for rather idealized molecular systems such as rigid convex bodies [77].

In order to compute the collision integrals for the WCU and Taxman theories it is necessary to solve the equations of motion for two colliding molecules using a very wide range of initial conditions. For a considerable time it was felt that such a task was almost hopeless [78] and so a number of approximate solutions to the dynamical problem were considered. Mason and Monchick [78] assumed that only a small part of the collision trajectory contributed to the transport coefficients and that during this part of the trajectory the molecules did not rotate. On the basis of this assumption they were able to show the existence of inverse collisions and then use the standard collision integrals together with an average over initial orientations to compute the shear viscosity and thermal conductivity of simple molecular gases. Parker [79] computed the bulk viscosity of a diatomic gas using perturbation theory solutions to the dynamical problem. Subsequently, Lordi and Mates [80] used the same pair potential as Parker to compute transport properties of nitrogen using the WCU theory. Similarly, Clarke and Smith [81] used the WCU theory to compute the transport properties of a system of hard spheres with embedded dipoles. These authors assumed that the collision process occurred in a single plane and ignored some aspects of the dependence of the collision integrals in the WCU theory on the angular momenta.

Probably the most extensive analysis of the WCU and Taxman theories of transport in dilute molecular gases was reported by Evans [33, 82], who computed the shear and bulk viscosity coefficients and the thermal conductivity of benzene vapor over a fairly wide range of temperatures. Evans used the six-centered Lennard-Jones interaction developed for liquid state studies [31, 33] (discussed in Section 8.2), obtaining good agreement with experimental measurements. He showed that if the dynamical problem of two colliding benzene molecules was computed accurately, and adequate thermal averaging over initial molecular configurations performed, then the transport coefficients could be computed with errors of a few percent. Small but systematic differences were found between transport coefficients computed using the WCU and Taxman theories. Although these differences were not very much larger than the numerical errors in the calculations, they suggested that values predicted by the Taxman theory are consistently greater than those obtained from the WCU theory. Figure 8.7 shows the shear viscosity coefficient for benzene vapor predicted using the six-center potential and the WCU and Taxman theories. Also included in the figure are experimental results [83] and some indication of the numerical errors in the calculations. It can be seen that the agreement with experiment is reasonably good. Similarly, accurate results for the thermal conductivity were obtained, provided certain assumptions about the effect of internal vibrational states of the molecules were used.

Evans [33, 82] also examined the nature of the collision process in

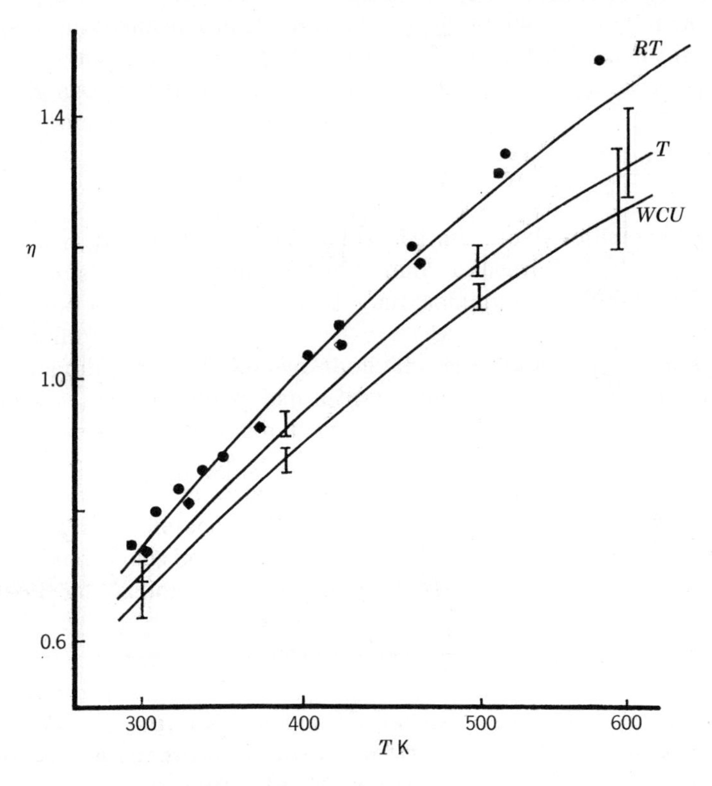

Figure 8.7 Shear viscosity for benzene vapor in units of 10^{-4} poise. Curves T and WCU are from the Taxman and WCU theories using the potential reported in Section 8.2. The curve RT comes from a similar potential fitted to the experimental results using the Taxman theory. ● experimental data.

some detail in an attempt to examine the usefulness of the Mason-Monchick assumption of fixed molecular orientation during a collision. This assumption ignored out-of-plane scattering by setting all torques acting between the molecules to zero. Figure 8.8 gives the contributions to the total energy of a colliding pair of benzene molecules as a function of time, assuming that the molecules began the collision with initial positions, velocities, orientations, and angular velocities given by

$$\mathbf{r}_2 = -\mathbf{r}_1; \qquad \mathbf{v}_2 = -\mathbf{v}_1; \qquad \boldsymbol{\omega}_2 = \boldsymbol{\omega}_1 = 0$$
$$\mathbf{r}_1 = (0.758, -8.0, 1.46)\ \text{Å}$$
$$\mathbf{v}_1 = (0.0, 5.74 \times 10^4, 0.0)\ \text{cm s}^{-1}$$
$$\boldsymbol{\psi}_1 = (1.571, -0.246, 0.0)\ \text{rad}$$
$$\boldsymbol{\psi}_2 = (0.0, 0.0, 0.0)\ \text{rad}$$

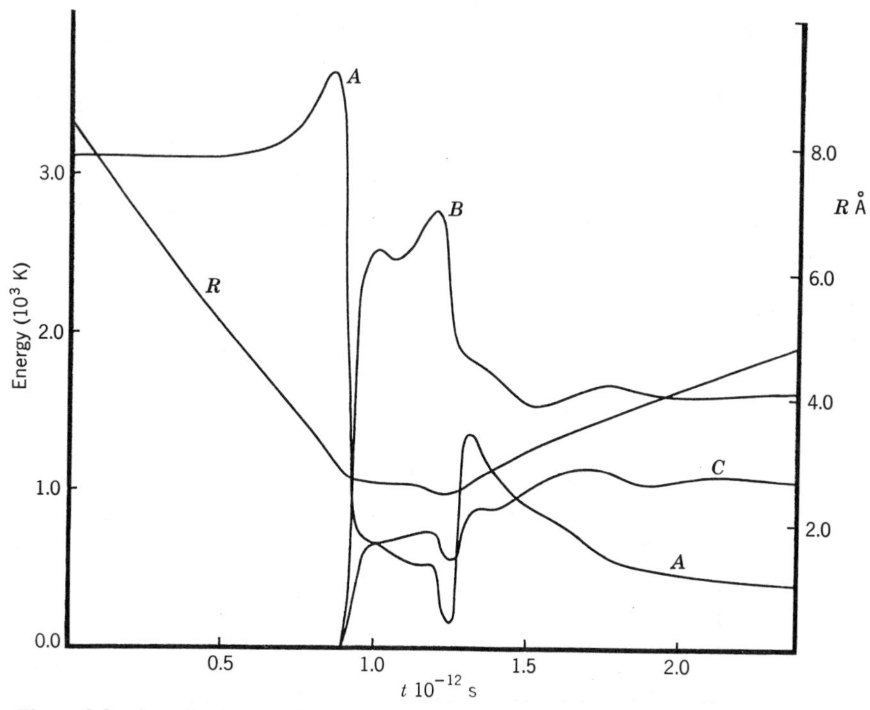

Figure 8.8 Contributions to the total energy of a colliding pair of benzene molecules. *A*, total translational kinetic energy; *B*, rotational kinetic energy of molecule 2; *C*, rotational kinetic energy of molecule 1; *R*, separation of the molecules.

Quantities represented in Fig. 8.8 include the separation of the molecules, total translational and rotational kinetic energies, and individual rotational kinetic energies of the two molecules. We see that there is an initial increase in the total translational kinetic energy to compensate the decreasing potential energy as the two molecules fall into the region of the potential well. As the repulsive cores of the molecules come into contact, the translational kinetic energy decreases extremely rapidly. At the same time the molecules, whose initial angular velocities were zero, begin to tumble. For a period of about 5×10^{-13} s the molecular centers of mass remain almost stationary as the molecules oscillate, first colliding on one side of the ring and then on the other. The molecules then separate with much reduced translational energy but with fairly substantial rotational energies due to tumbling motions. This collision demonstrates very well the way in which energy can pass between translational states and rotational states when nonspherical molecules collide. Other effects, such as out-of-plane scattering and very long lived collisions, can also be examined using the methods discussed by Evans.

8.5 Time Dependent Phenomena in Dense Molecular Liquids

Studies of transport phenomena in dense molecular fluids have taken two forms. First, there have been several attempts to extend the linear response theory described in Chapter 4 to include nonspherical particles [84], and second, there have been some molecular dynamics studies of simple diatomics [7, 85] and of water [10, 86]. The extension of linear response theory is complicated and involves considering flux-flux correlations from a number of sources. For example, the shear and bulk viscosities that give a measure of the transport of linear momentum through a fluid are paralleled by rotational viscosity coefficients measuring the transport of angular momentum [84]. Similarly, the self-diffusion coefficient, which measures the transport of mass by translational motion, is complemented by a rotational diffusion coefficient measuring the effects of molecular reorientation. We do not consider these effects in any detail.

The molecular dynamics studies of diatomic fluids have been largely restricted to nitrogen. There have been several experimental studies of the properties of this substance, including measurements of the self-diffusion coefficient [87] and neutron scattering cross-sections [88]. Theoretical studies of time dependent phenomena include those of Harp and Berne [7] and of Cheung and Powles [85]. The former authors used the Stockmayer potential with quadrupole-quadrupole interactions (given in eq. 8.4) to model liquid nitrogen, and Cheung and Powles used a two-centered Lennard-Jones potential to predict properties of the same substance. It was found that the velocity autocorrelation function for liquid nitrogen was very similar to that found for monatomic systems (discussed in Section 5.5). The self-diffusion coefficient computed from the two-centered Lennard-Jones model was in good agreement with the experimental measurements for nitrogen [87]. Cheung and Powles [85] reported quantitative differences between the computed and measured angular momentum correlation functions and attributed these to inadequacies in the pair potential at intermediate distances. However, the overall qualitative agreement between theory and experiment for liquid nitrogen is good, and there is no doubt that future molecular dynamics studies of molecular liquids will be important. In the next chapter we describe studies of time dependent phenomena in liquid water.

References

1. P. J. Flory, 1969, *Statistical Mechanics of Chain Molecules*, Interscience, New York.
2. J. O. Hirschfelder, C. F. Curtiss, and R. B. Bird, 1954, *Molecular Theory of Gases and Liquids*, Wiley, New York.

3. R. M. Gibbons, 1969, *Mol. Phys.*, **17,** 81; A. B. Ritchie, 1967, *J. Chem. Phys.*, **46,** 618; T. Kihara, 1953, *Rev. Mod. Phys.*, **25,** 831.
4. G. S. Rushbrooke, G. Stell, and J. S. Høye, 1973, *Mol. Phys.*, **26,** 1199; V. M. Jansoone and E. U. Franck, 1972, *Ber. Bunsenges. Phys. Chem.*, **76,** 943.
5. W. H. Stockmayer, 1941, *J. Chem. Phys.*, **9,** 398, 863.
6. J. S. Rowlinson, 1949, *Trans. Faraday Soc.*, **45,** 974.
7. G. D. Harp and B. J. Berne, 1970, *Phys. Rev.*, A**2**, 975.
8. J. M. Prausnitz, 1969, *Molecular Thermodynamics of Fluid Phase Equilibrium*, Prentice-Hall, New Jersey.
9. B. C. Freasier, 1975, *Chem. Phys. Lett.*, **35,** 280; Y. D. Chen and W. A. Steele, 1968, *J. Chem. Phys.*, **50,** 1428; M. Rigby, 1970, *ibid.*, **53,** 1021; J. W. Perram and F. Kohler, 1976, to be published.
10. A. Rahman and F. H. Stillinger, 1971, *J. Chem. Phys.*, **55,** 3336; A. Ben-Naim and F. H. Stillinger, 1972, in *Structure and Transport Processes in Water and Aqueous Solutions*, R. A. Horne, Ed., p. 295, Interscience, New York.
11. A. I. Kitaigorodskii, 1966, *J. Chem. Phys.*, **63,** 9; 1960, *Tetrahedron*, **9,** 183; D. E. Williams, 1966, *J. Chim. Phys.*, **45,** 3770.
12. J. R. Sweet and W. A. Steele, 1967, *J. Chem. Phys.*, **47,** 3022, 3029.
13. H. Goldstein, 1953, *Classical Mechanics*, Addison-Wesley, Cambridge, Mass.
14. W. H. Keesom, 1912, *Comm. Phys. Lab. Leiden, Suppl. 24b*, Section 6.
15. J. S. Rowlinson, 1951, *J. Chem. Phys.*, **19,** 827.
16. A. H. Stroud, 1971, *Approximate Calculation of Multiple Integrals*, Prentice-Hall, New Jersey; C. B. Haselgrove, 1961, *Math. Comp.*, **15,** 323; D. J. Evans, 1975, in *Computational Methods in Mathematical Physics*, R. S. Anderssen and R. O. Watts, Eds., p. 29, University of Queensland Press.
17. D. J. Evans and R. O. Watts, 1974, *Mol. Phys.*, **28,** 1283; R. O. Watts, 1972, *ibid.*, **23,** 445.
18. C. H. J. Johnson and T. H. Spurling, 1971, *Aust. J. Chem.*, **24,** 1567.
19. D. J. Evans and R. O. Watts, 1975, *Mol. Phys.*, **29,** 777.
20. B. J. Berne and G. D. Harp, 1970, *Adv. Chem. Phys.*, **17,** 63.
21. J. Barojas, D. Levesque, and B. Quentrec, 1973, *Phys. Rev.*, **A7,** 1092.
22. G. N. Patey and J. P. Valleau, 1973, *Chem. Phys. Lett.*, **21,** 297.
23. S. S. Wang, P. A. Egelstaff, C. G. Gray, and K. E. Gubbins, 1974, *Chem. Phys. Lett.*, **24,** 453.
24. S. S. Wang, C. G. Gray, P. A. Egelstaff, and K. E. Gubbins, 1973, *Chem. Phys. Lett.*, **21,** 123.
25. L. Verlet and J. J. Weis, 1974, *Mol. Phys.*, **28,** 665.
26. J. Vieillard-Baron, 1974, *Mol. Phys.*, **28,** 809.
27. V. M. Jansoone, 1974, *Chemical Physics*, **3,** 78.
28. D. J. Adams and I. R. McDonald, 1974, *J. Phys., C (London)*, **7,** 2761; P. P. Ewald, 1921, *Ann. Phys. (Leipzig)*, **64,** 253.
29. G. A. Few and M. Rigby, 1973, *Chem. Phys. Lett.*, **20,** 433; B. J. Berne and P. Pechukas, 1972, *J. Chem. Phys.*, **56,** 4213; D. Levesque, D. Schiff, and J. Vieillard-Baron, 1969, *ibid.*, **51,** 3625; J. Vieillard-Baron, 1972, *ibid.*, **56,** 4729.
30. T. G. Gibbons and M. L. Klein, 1974, *Chem. Phys. Lett.*, **29,** 463.
31. D. J. Evans and R. O. Watts, 1976, *Mol. Phys.*, in press.
32. D. J. Evans and R. O. Watts, 1976, *Mol. Phys.*, **31,** 83.
33. D. J. Evans, 1976, Ph.D. thesis, Australian National University.
34. L. Blum and A. Torruella, 1972, *J. Chem. Phys.*, **56,** 303.
35. Y. D. Chen and W. A. Steele, 1971, *J. Chem. Phys.*, **54,** 703.

36. Y. D. Chen and W. A. Steele, 1970, *J. Chem. Phys.*, **52,** 5284.
37. P. F. Morrison, 1972, Ph.D. thesis, California Institute of Technology; P. F. Morrison and C. J. Pings, 1974, *J. Chem. Phys.*, **60,** 2323.
38. M. S. Wertheim, 1971, *J. Chem. Phys.*, **55,** 4291.
39. J. L. Lebowitz and J. K. Percus, 1966, *Phys. Rev.*, **144,** 251.
40. H. C. Andersen and D. Chandler, 1972, *J. Chem. Phys.*, **57,** 1918.
41. M. S. Wertheim, 1963, *Phys. Rev. Lett.*, **10,** 321.
42. E. Thiele, 1963, *J. Chem. Phys.*, **39,** 474.
43. L. Blum, 1972, *J. Chem. Phys.*, **57,** 1862; 1973, *ibid.*, **58,** 3295.
44. S. A. Adelman and J. M. Deutsch, 1973, *J. Chem. Phys.*, **59,** 3971; E. Waisman, 1973, *ibid.*, **59,** 495.
45. G. Nienhuis and J. M. Deutsch, 1971, *J. Chem. Phys.*, **55,** 4213; 1972, *ibid.*, **56,** 235; 1972, *ibid.*, **56,** 5511; J. M. Deutsch, 1973, *Ann. Rev. Phys. Chem.*, **24,** 301; J. W. Sutherland, G. Nienhuis, and J. M. Deutsch, 1974, *Mol. Phys.*, **27,** 721.
46. D. Chandler and H. C. Andersen, 1972, *J. Chem. Phys.*, **51,** 1930.
47. D. Chandler, 1973, *J. Chem. Phys.*, **59,** 2742.
48. L. J. Lowden and D. Chandler, 1973, *J. Chem. Phys.*, **59,** 6587.
49. L. J. Lowden and D. Chandler, 1974, *J. Chem. Phys.*, **61,** 5228.
50. H. C. Andersen and D. Chandler, 1970, *J. Chem. Phys.*, **53,** 547.
51. D. Chandler and H. C. Andersen, 1971, *J. Chem. Phys.*, **54,** 26.
52. H. C. Andersen and D. Chandler, 1971, *J. Chem. Phys.*, **55,** 1497.
53. H. C. Andersen, D. Chandler, and J. D. Weeks, 1972, *J. Chem. Phys.*, **56,** 3812.
54. W. R. Smith, 1974, *Can. J. Phys.*, **52,** 2022; W. R. Smith, 1973, Chapter 2 in *Specialist Periodical Reports: Statistical Mechanics*, K. Singer, Ed., Vol. 1, Chemical Society, London.
55. J. A. Pople, 1954, *Proc. Roy. Soc. (London), Ser. A*, **221,** 498, 508; R. W. Zwanzig, 1955, *J. Chem. Phys.*, **23,** 1915; A. D. Buckingham, 1967, *Discuss. Faraday Soc.*, **43,** 205.
56. J. R. Sweet and W. A. Steele, 1969, *J. Chem. Phys.*, **50,** 668; T. H. Spurling and E. A. Mason, 1967, *ibid.*, **46,** 322; T. H. Spurling, A. G. De Rocco, and T. S. Storvick, 1968, *ibid.*, **48,** 1006; H. P. Hung and T. H. Spurling, 1970, *Aust. J. Chem.*, **23,** 377; K. K. Datta and Y. Singh, 1971, *J. Chem. Phys.*, **55,** 3541.
57. V. F. Sears, 1966, *Can. J. Phys.*, **44,** 1279; 1967, *ibid.*, **45,** 237; P. A. Egelstaff, D. I. Page, and J. G. Powles, 1971, *Mol. Phys.*, **20,** 881; K. E. Gubbins, C. G. Gray, P. A. Egelstaff, and M. S. Ananth, 1973, *ibid.*, **25,** 1353.
58. K. E. Gubbins and C. G. Gray, 1972, *Mol. Phys.*, **23,** 187.
59. J. A. Barker and D. Henderson, 1967, *J. Chem. Phys.*, **47,** 4714; W. R. Smith, D. Henderson, and J. A. Barker, 1970, *ibid.*, **53,** 508.
60. M. S. Ananth, K. E. Gubbins, and C. G. Gray, 1974, *Mol. Phys.*, **28,** 1005.
61. J. W. Perram and L. R. White, 1974, *Mol. Phys.*, **28,** 527.
62. G. Stell, J. C. Rasaiah, and H. Narang, 1974, *Mol. Phys.*, **27,** 1393; 1972, *ibid.*, **23,** 393.
63. J. L. Lebowitz, G. Stell, and S. Baer, 1965, *J. Math. Phys.*, **6,** 1282.
64. S. Sung and D. Chandler, 1972, *J. Chem. Phys.*, **56,** 4989.
65. H. C. Andersen, J. D. Weeks, and D. Chandler, 1971, *Phys. Rev.*,**A4,** 1597.
66. S. I. Sandler, 1974, *Mol. Phys.*, **28,** 1207.
67. J. W. Perram and L. R. White, 1972, *Mol. Phys.*, **24,** 1133.
68. V. Volterra, 1959, *Theory of Functionals*, Dover, New York.
69. P. J. Stiles, 1975, *Chem. Phys. Lett.*, **30,** 126.
70. J. C. Rasaiah and G. Stell, 1974, *Chem. Phys. Lett.*, **25,** 519.

71. S. Chapman and T. G. Cowling, 1939, *The Mathematical Theory of Nonuniform Gases*, Cambridge University Press; J. H. Ferziger and H. G. Kaper, 1972, *Mathematical Theory of Transport Processes in Gases*, North Holland, Amsterdam.
72. C. S. Wang Chang and G. E. Uhlenbeck, 1951, "Transport Phenomena in Polyatomic Molecules," Michigan University Engineering Research Institute Report CM-681; C. S. Wang Chang, G. E. Uhlenbeck, and J. de Boer, 1964, in *Studies in Statistical Mechanics*, J. de Boer and G. E. Uhlenbeck, Eds., Vol. 3, Part c, North Holland, Amsterdam.
73. R. C. Tolman, 1948, *The Principles of Statistical Mechanics*, Oxford University Press.
74. N. Taxman, 1957, Ph.D. thesis, Northwestern University; 1958, *Phys. Rev.*, **110,** 1235.
75. J. A. Barker, M. V. Bobetic, and A. Pompe, 1971, *Mol. Phys.*, **20,** 347.
76. F. J. McCormack, 1968, *Phys. Fluids*, **11,** 2533; F. R. W. McCourt and R. F. Snider, 1965, *J. Chem. Phys.*, **43,** 2276; L. Waldman, 1967, *Z. Naturforsch.*, *a***12,** 660.
77. C. F. Curtiss, 1956, *J. Chem. Phys.*, **24,** 225.
78. E. A. Mason and L. Monchick, 1962, *J. Chem. Phys.*, **36,** 1622.
79. J. G. Parker, 1959, *Phys. Fluids*, **2,** 449.
80. J. A. Lordi and R. E. Mates, 1970, *Phys. Fluids*, **13,** 291.
81. A. G. Clarke and E. B. Smith, 1970, *J. Chem. Phys.*, **53,** 1235.
82. D. J. Evans and R. O. Watts, 1976, *Mol. Phys.*, in press.
83. A. K. Pal and A. K. Barua, 1968, *J. Chem. Phys.*, **48,** 872; P. M. Craven and J. D. Lambert, 1951, *Proc. Roy. Soc.* (*London*), *Ser. A*, **205,** 439; Landolt and Bornstein, 1951, *Zahlenwerte und Funktionen*, J. Bartels, Ed., Part 3, p. 105, Springer-Verlag, Berlin; T. Titani, 1933, *Bull. Chem. Soc. Japan*, **8,** 255.
84. R. F. Snider and K. S. Lewchuk, 1967, *J. Chem. Phys.*, **46,** 3163; N. D. Gershon and I. Oppenheim, 1972, *Physica*, **62,** 198; N. K. Ailawadi, B. J. Berne, and D. Forster, 1971, *Phys. Rev.*, *A***3**, 1462.
85. P. S. Y. Cheung and J. G. Powles, 1976, *Mol. Phys.*, to be published.
86. F. H. Stillinger and A. Rahman, 1972, *J. Chem. Phys.*, **57,** 1281; 1974, *ibid.*, **60,** 1545.
87. K. Krynicki, E. J. Rakhamaa, and J. G. Powles, 1974, *Mol. Phys.*, **28,** 853.
88. J. C. Dore, G. Walford, and D. I. Page, 1975, *Mol. Phys.*, **29,** 565.

Nine

Water and Ionic Solutions

One of the most interesting applications of the theory developed in previous chapters is to the structure and properties of liquid water and ionic solutions. Although it is true that a great deal of work remains to be completed on the properties of aqueous fluids, recent studies have given a microscopic description of such systems that gives a good qualitative account of both structural and bulk properties. In all the other chapters we have measured energies in terms of degrees Kelvin, by dividing the appropriate quantities in MKS units by Boltzmann's constant. It is standard practice in physical chemistry to discuss bond strengths in water and aqueous solutions in terms of kcal mole^{-1} and we have followed this procedure in this chapter. To enable the reader to compare results given here with those presented elsewhere in the book, the conversion factor is given by

$$1 \text{ kcal mole}^{-1} \equiv 503.22 \text{ K}$$

Thus the strength of the hydrogen bond, $\sim$5 kcal mole^{-1}, is $\sim$2516 K or approximately 17.7 times as strong as the binding energy of two argon atoms in their minimum energy configuration.

9.1 Mixture and Continuum Models: A Summary

We have seen in previous chapters, particularly that devoted to the inert gases (Chapter 7), that a rigorous theory of the liquid state depends on the solution to two problems. First, there must exist an accurate interaction potential for the molecules in the liquid, and second, it must be possible to evaluate the statistical mechanics with high precision. Several methods have been used to tackle the first problem, and computer simulation methods provide a partial answer to the second problem. In addition to the recent work using machine simulation methods, many less fundamental approaches have been tried. Eisenberg and Kauzmann [1] give a comprehensive account of such models and refer to many earlier reviews. More recently, articles in a book edited by Horne [2] include reviews of continuum [3] and structural theories [4] and these can be consulted by readers interested in a more detailed account of the less fundamental theories.

Most empirical theories of water postulate a model for the liquid structure using some experimental properties together with a significant helping of intuition. On the basis of this model a relatively simple partition function is established containing several parameters that are subsequently adjusted to fit a range of experimental properties. Two main types of empirical models exist—the mixture models, which postulate the simultaneous existence of two or more relatively long lived structures in the liquid, and the continuum models, which are based on the assumption that the structure relaxes on a time scale that is similar to that observed in other liquids. On closer examination, it is found that the structural and continuum models differ primarily in their assumptions about the nature of the interactions between water molecules.

X-Ray studies of various phases of ice, together with less direct evidence, have established that there is a very strong interaction between water molecules in certain configurations. In all forms of ice the protons lie along lines joining neighboring oxygen atoms, so that each oxygen atom is surrounded by four hydrogen atoms, two of which are covalently bonded to it and two of which are less tightly held. The interaction that favors the collinear O—H · · · O configuration is usually called the *hydrogen bond.* The strongly directional character of the hydrogen bond interaction is undoubtedly retained in the liquid state, and the essential difference between continuum and structural theories of liquid water is a disagreement about the orientational dependence of the water-water interaction. The controversy on the character of this interaction will not be resolved until an accurate interaction potential including both intramolecular terms and many body effects is available.

Structural theories of liquid water depend on the assumption that the hydrogen bond is so strongly directional that it can be considered either as "made" or "broken." That is, structural theories assume that in many respects the hydrogen bond is similar to a chemical bond. On the basis of this assumption, any water molecule in the liquid can be in one of five states, that is, it can be bonded to 0, 1, 2, 3, or 4 other water molecules. The numbers of molecules in these categories, or alternatively the number of formed hydrogen bonds, are frequently associated with parameters in structural theories. Many such theories have been propounded and all give a satisfactory description of one or more properties of liquid water over a limited range of temperature and density. It is worth remarking that the theories fall into three groups—simple two-state models, interstitial models, and models associated with partition functions. A typical example from the first group is the model of Davis and Litovitz [5], who assume an open icelike arrangement of hexagonal rings mixed with a close packed structure. Probably the best known interstitial model is that of Pauling [6], who assumed that there was a hydrogen bonded clathrate framework with interstitial water molecules. Némethy and Scheraga [7] developed a partition function model by considering the fraction of water molecules in each of the five hydrogen bonded states mentioned earlier. Eisenberg and Kauzmann [1] give an excellent table

Table 9.1 Estimates of broken hydrogen bonds in water. The table is adapted from that given by Falk and Ford [8]. References are given at the end of the chapter

Experimental basis	T(K)	% Broken	Reference
Dielectric constant	273	9	8*a*
Heat of melting	273	15	8*b*
Heat capacity	273	66	8*c*
Infrared spectrum	273	60	8*d*
		46	8*e*
		38	8*f*
		32	8*g*
		9–16	8*h*
Model of structure	273	47	8*i*
		28.5	8*j*
Raman spectroscopy	299	50	8*k*
	273	10.5	8*l*
Ultrasonic absorption	273	71.5	8*m*
		30	8*n*
	293	17.5	8*o*
Ultraviolet spectroscopy	298	0.1	8*p*
Volumetric data	273	57.5	8*q*
		56	8*r*
		18	8*s*
		18	8*t*
		2.5	8*u*
Viscosity measurements	273	46	8*v*
X-Ray diffraction	298	20	8*w*

summarizing the principle characteristics of structural models (Table 5.1 in their book).

As was stated earlier, all structural models are capable of giving a reasonable account of the bulk properties of liquid water. This is not too surprising as they are generally fitted directly to bulk data. Although a considerable amount of intuition is built into the model at an early stage, the consequences of this intuition remain qualitative until very late in the model development. Consequently, the parameters to be estimated are usually fitted to experimental data *of the type to be predicted* after the model is fully developed. It is unlikely that the results from such models give useful information about the microscopic behavior of the liquid. Falk and Ford [8] pointed out the inconsistencies between various theories of liquid water in 1966 when they published the table reproduced here as Table 9.1. In this table the fraction of broken hydrogen bonds in liquid water is given for a wide range of structural theories. Apparently there is no agreement between the several theories at the microscopic level.

The other type of empirical theory, continuum theory, is based on the assumption that there exists a range of angles about the hydrogen bonding configuration over which the strength of the water-water interaction changes by a relatively small amount. Figure 9.1 shows schematically the difference between the assumptions of the two approaches. In this

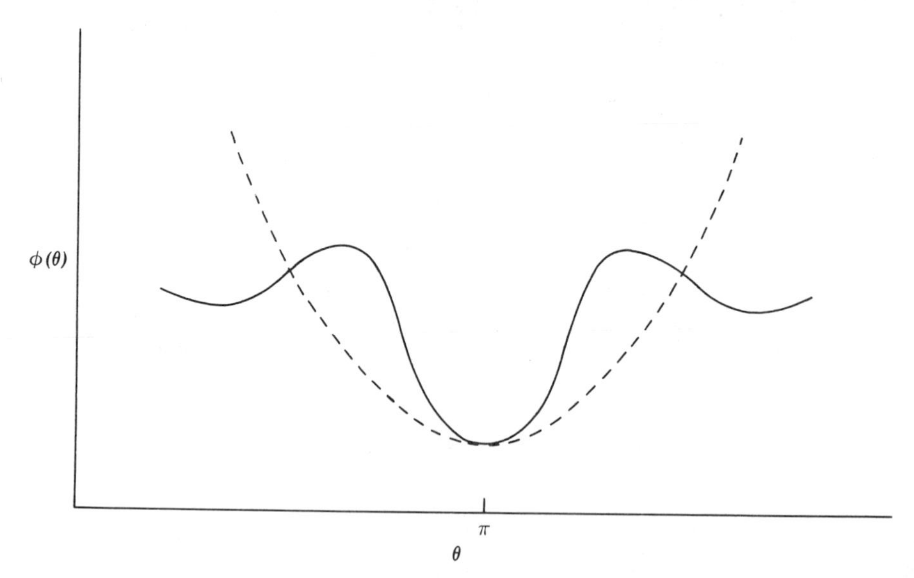

Figure 9.1 Interpretation of the water-water interaction as a function of O—H · · · O angle according to structural and continuum theories. ——— structural theories; - - - - - continuum theories.

figure the intermolecular potential is given as a function of the angle between the O—H · · · O directions in a given bond. It can be seen that the structural theories replace the rather broad distribution of energies with potential barriers, and once the O—H · · · O angle is such that the energy falls outside the central well the bond is termed broken.

The best known continuum theory of water is that of Pople [9], who developed the distorted hydrogen bond model. A similar model based on a random network of hydrogen bonds was described later by Bernal [10]. Both workers demonstrated that their models were consistent with available data on the radial distribution function of water. Pople [9] also showed that his model gave a good account of the dielectric constant of water and the decrease in volume of ice on melting. Eisenberg and Kauzmann suggest that the Pople model can also account for the heat capacity and thermal energy of water [11]. As we show in the next section, currently available water-water interaction potentials favor continuum theories over structural theories. However, a number of questions remain to be answered before any definite conclusion can be reached.

9.2 Interaction Potentials

In Section 2.3 the Rowlinson model [12] for the pair interaction potential of water was introduced. A similar model was described by Ben-Naim and Stillinger [13] (BNS). Both these models were empirical and contain a number of parameters that were adjusted to agree with certain experimental data. The parameters in the Rowlinson model were chosen to fit the static lattice energy of ice, the second virial coefficient of steam and the dipole moment of the isolated water molecule, whereas those of Ben-Naim and Stillinger were chosen in a somewhat arbitrary fashion to obtain a model that was applicable to liquid water. An alternative to using empirical interaction models is to use accurate quantum mechanical calculations. A number of such calculations have been reported [14–16] for water-water pair interactions, and some have been used in liquid state calculations [17, 18].

Both the Rowlinson [12] and BNS [13] models consist of a Lennard-Jones interaction with embedded point charges and can be described in terms of a molecular coordinate system with the oxygen atom at the origin [19]. We begin by defining two orthogonal unit vectors $\mathbf{a}$ and $\mathbf{b}$ originating at the oxygen atom and a third vector $\mathbf{c}=\mathbf{a}\wedge\mathbf{b}$ that is orthogonal to the plane ($\mathbf{a}$, $\mathbf{b}$). This gives us a molecule based frame of reference whose orientation in the laboratory is determined by the components of $\mathbf{a}$ and $\mathbf{b}$. Let the positions of the positive charges for both models be given by $\mathbf{s}_1$ and $\mathbf{s}_2$, the positions of the negative charges be given by $\mathbf{t}_1$ and $\mathbf{t}_2$, and the Lennard-Jones interaction be centered at $\mathbf{u}$, where for each molecule $\mathbf{s}_i$, $\mathbf{t}_i$, and $\mathbf{u}$ are in the molecular frame of reference. Then in terms of $\mathbf{a}$, $\mathbf{b}$, and $\mathbf{c}$ we can write

$$\begin{aligned}
\mathbf{s}_1 &= \alpha^+\mathbf{a} + \beta^+\mathbf{b} \\
\mathbf{s}_2 &= \alpha^+\mathbf{a} - \beta^+\mathbf{b} \\
\mathbf{t}_1 &= \alpha^-\mathbf{a} + \beta^-\mathbf{c} \\
\mathbf{t}_2 &= \alpha^-\mathbf{a} - \beta^-\mathbf{c} \\
\mathbf{u} &= \gamma\mathbf{a}
\end{aligned} \tag{9.1}$$

The parameters α^+, α^-, β^+, β^-, γ together with the size of the charges, q, the Lennard-Jones parameters ε and σ, and the values of their dipole moments are given for both models in Table 9.2. Both models assume that the positive charges are on the hydrogen atoms.

To avoid complications arising at short distances, when two unlike charges could be coincident, Ben-Naim and Stillinger [13] multiplied the total Coulombic interaction by a switching function

$$S(r) = \begin{cases} 0 & r < R_1 \\ \dfrac{(r-R_1)^2(3R_2-R_1-2r)}{(R_2-R_1)^3} & R_1 \leqslant r < R_2 \\ 1 & R_2 \leqslant r \end{cases} \tag{9.2}$$

where r is the oxygen-oxygen distance, $R_1 = 2.0379$ Å and $R_2 = 3.1877$ Å. This function varies smoothly from zero to one over the range (R_1, R_2). Although similar overlap problems can arise for the Rowlinson

Table 9.2 Parameters for the Rowlinson and BNS models of the water-water interaction

	Rowlinson	BNS
ε(kcal mole^{-1})	0.707	0.0721
σ(Å)	2.725	2.82
q	0.3278	0.19
α^+(Å)	0.5844	0.5774
α^-(Å)	0.0	−0.5774
β^+(Å)	0.7616	0.8165
β^-(Å)	0.2539	0.8165
γ(Å)	0.2922	0.0
μ(D)	1.84	2.10

potential, it does not contain a switching function and most calculations using the model [19–21] have included a hard sphere term of diameter 2 Å acting between oxygen atoms.

Accurate quantum mechanical calculations of the full water-water interaction potential have not been reported. However, a number of workers [14–16, 22, 23] have given approximate results based on the Hartree-Fock self-consistent field and related methods. The most extensive, and probably the most accurate, of these calculations have been reported by Clementi and co-workers [16, 17] who claim that their results are close to the Hartree-Fock limit. Clementi and his collaborators have calculated the interaction potential of two water molecules in over 200 different configurations and have fitted the results to a multiparametric model potential. This model, which can be described using the local coordinate system constructed for the Rowlinson and BNS potentials, has the following properties. Exponential repulsions act between all the atoms on one molecule and all the atoms on the second molecule. In addition, Coulombic interactions act between various point charges distributed in a manner similar to that used for the BNS and Rowlinson models. Details are given in the paper by Popkie et al. [17]. It is important to remember that this model was determined from quantum mechanical calculations alone and contains no empirically determined parameters. The Hartree-Fock method ignores contributions to the interaction potential from electron correlation terms. Kistenmacher et al. [18] have suggested several approximations for these terms. The dominant term in the correlation energy correction at longer distances is the induced dipole-induced dipole interaction described in Chapter 2. Two useful approximate expressions for the coefficient C_6 in this term have been given by London [24] and by Kirkwood and Müller [25], taking the form

$$C_6 = 676.5 \text{ kcal mole}^{-1}\text{ Å}^6 \qquad \text{(London)} \tag{9.3}$$

$$C_6 = 1222.2 \text{ kcal mole}^{-1}\text{ Å}^6 \qquad \text{(Kirkwood-Müller)} \tag{9.4}$$

These terms were included in the Hartree-Fock water-water potential by adding the interaction $-C_6/r^6$ where r is the oxygen-oxygen distance. At shorter distances Kistenmacher et al. [18] included correlation energy corrections using an approximation suggested by Wigner [26] for the electron gas in metals. The Wigner correction term takes the form

$$\phi_w(r) = \int\left[\frac{a_1\rho_d^{4/3}}{(a_2+\rho_d^{1/3})} - \frac{2a_1\rho_m^{4/3}}{(a_2+\rho_m^{1/3})}\right] d\mathbf{r} \tag{9.5}$$

where ρ_d is the water dimer electron density and ρ_m is the water monomer electron density, both determined from the Hartree-Fock calculations, and a_1 and a_2 are two parameters determined so that the exact correlation energy of the monomer is obtained [27, 28]. It was found that the Wigner correlation term is described by

$$\phi_w(r) = A_w \exp(-\alpha_w r) \tag{9.6}$$

where r is the oxygen-oxygen distance, $A_w = 64.62$ kcal mole^{-1} and $\alpha_w = 0.9752$ Å^{-1}.

Kistenmacher et al. [18] constructed a number of pair potentials for water using various combinations of the Hartree-Fock potential and the several approximations for the correlation energy. Figure 9.2 gives the

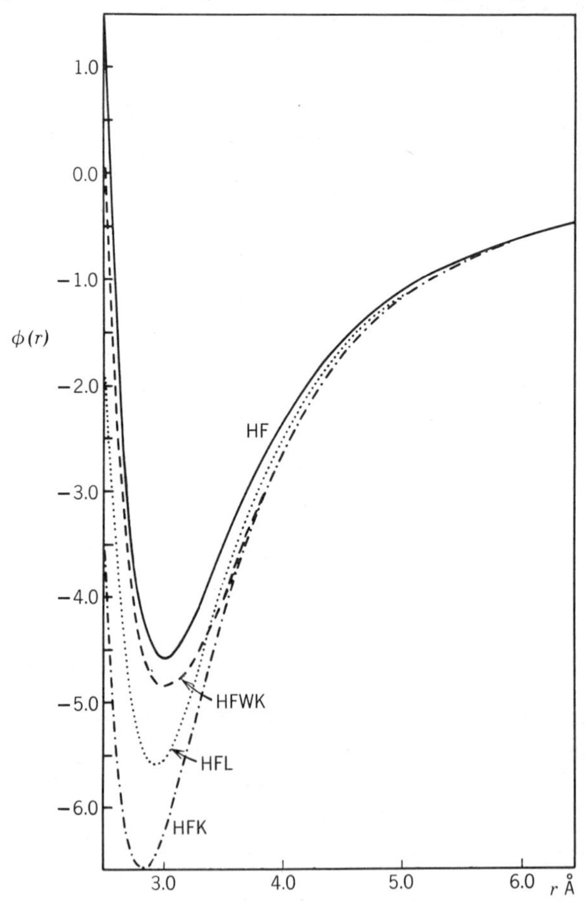

Figure 9.2 Interaction potential in kcal mole^{-1} of two water molecules in the hydrogen bonding orientation as a function of distance. Results are given for the Hartree-Fock potential with several correction terms added. (Adapted with permission from *J. Chem. Phys.*, **60**, 4455, 1974).

interaction energy of two water molecules as a function of distance, with the relative orientations of the molecules being held in the hydrogen bonding configuration. Results are given for several models, including the pure Hartree-Fock calculation, the Hartree-Fock potential with the Kirkwood-Müller and London corrections and the Hartree-Fock and Kirkwood potential with the Wigner correction. Figure 9.2 shows that all models have a minimum in the interaction energy of between −4.0 and −7.0 kcal mole^{-1} at distances in the range 2.8–3.0 Å. There is a great deal of experimental evidence [1] that supports a hydrogen bond length of about 2 Å, and it is known that the O—H bond length in a water molecule is just under 1 Å. Consequently, the minima in all the model interactions are in the correct region. Experimental evidence [1] also favors a hydrogen bond energy of about -6 ± 3 kcal mole^{-1}, and if this bond strength is associated with the binding energy of a water dimer in the hydrogen bonding configuration, then the depth of the minima in Fig. 9.2 is also acceptable [29].

Although Fig. 9.2 shows that the various interaction models have minima of the correct depth and at the correct position for dimers in the hydrogen bonding configuration, they do not demonstrate that these are local minima in the potential surface. This question can be resolved by examining potential surfaces such as that found in Fig. 9.3.

Define the (X, Y) plane to be that containing the three atoms on one molecule, with the Z axis passing through the oxygen atom and calculate the pair potential as follows. A plane parallel to the (X, Y) plane and containing the oxygen atom of the second molecule is considered. Holding the second oxygen atom at a distance $R = (X^2 + Y^2)^{1/2}$ from the Z axis, the second molecule is rotated until the minimum energy configuration is found. By repeating this procedure for many points in the (X, Y) plane the contours of constant energy shown in Fig. 9.3 may be constructed [18, 29].

Figure 9.3 gives a cross-section of the energy surface for the Hartree-Fock potential, taken through the plane $Z = 0$ and also indicates the relative configuration of the two molecules. The potential surface in the upper half plane is the mirror image of that given in the lower half plane. It can be seen from the figure that the interaction potential shows three minima, two of which correspond to the water dimer with the O—H bond on the central molecule collinear with the O—O direction. These two configurations correspond to the hydrogen bonding positions examined in Fig. 9.2 and demonstrate that such orientations are indeed local minima in the potential surface. The third minimum, occurring in a configuration in which one hydrogen atom on the second molecule is directed toward the oxygen atom on the central molecule, is associated with the hydrogen

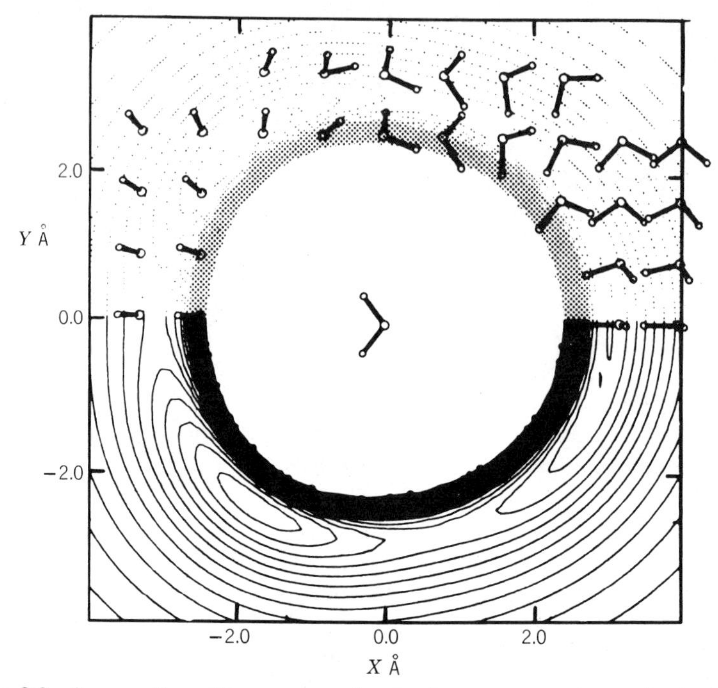

Figure 9.3 Cross-section of the energy surface for the water-water interaction in the plane $Z=0$. The energy of the lowest contour is -4.39 kcal mole^{-1} and the contour interval is 0.31 kcal mole^{-1}. (Adapted with permission trom *J. Chem. Phys.*, **60,** 4455, 1974.)

bond formed when an O—H bond on the second molecule is collinear with the O—O direction. This bond gives a minimum value for the interation potential in planes above and below that shown in Fig. 9.3, as can be seen in Fig. 9.4. This figure, representing a cross-section of the potential surface for the plane $Z=1.5$ Å, shows a single minimum in the hydrogen bonding configuration. At larger distances most of the structure in the potential surface vanishes and it is found that the potential is spherically symmetric [18]. The symmetry of the interaction potentials discussed in this section demands that the surfaces for positive values of Z are mirrored for the transformation $Z \rightarrow -Z$. Consequently, we can use pairwise additivity of the interactions in water to account for the observed tetrahedral structure of dense systems of water molecules.

Returning to Fig. 9.3, it is instructive to consider the effect of rotating the second oxygen atom about the first, keeping the hydrogen atom directed towards the oxygen. If the potential energy were plotted as a function of the angle, it is apparent that only minima corresponding to the hydrogen bonding configurations would be found. The shape of the

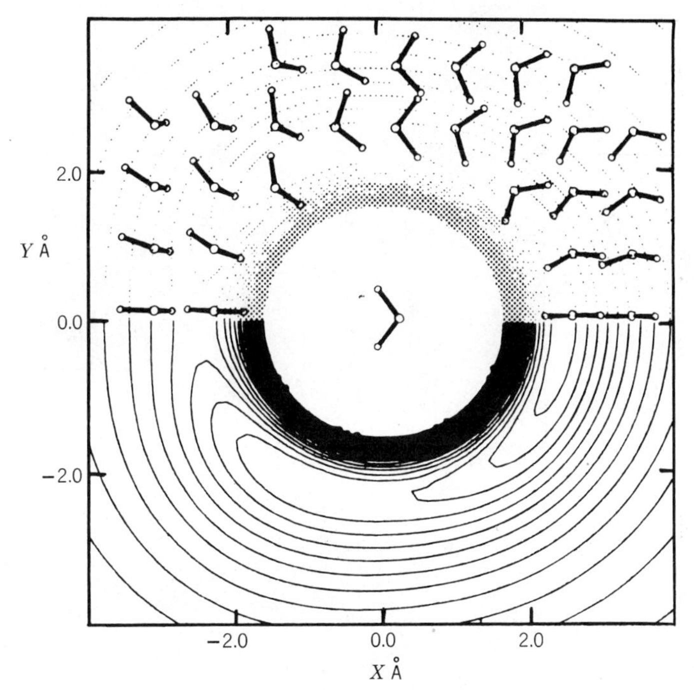

Figure 9.4 Cross-section of the energy surface for the water-water interaction in the plane $Z = 1.5$ Å. Conditions for the energy contours are the same as for Fig. 9.3. (Adapted with permission from *J. Chem. Phys.*, **60,** 4455, 1974.)

resulting curve would resemble that envisaged by proponents of the continuum theories of water (see Fig. 9.1). It appears that both *a priori* and empirical models of the pair interaction favor the continuum theory of water rather than structural theories.

The accuracy of the pair potentials can be examined to some extent by computing the second virial coefficient of steam and comparing the results with experimental measurements. It is a necessary (but certainly not sufficient!) condition for a pair potential to be accurate that it reproduce experimental measurements of the second virial coefficient. Evans and Watts [30] have calculated the second virial coefficient of steam using the various *a priori* interaction models given by Kistenmacher et al. [18]. Their results are reproduced in Fig. 9.5 together with a collection of experimental data [31–34]. The pure Hartree-Fock potential gives results that are in reasonable agreement with experiment over the temperature range considered. If the potential is modified by adding either the London or the Kirkwood-Müller dispersion energy terms the agreement is very poor. Apparently, neither of these modifications is a

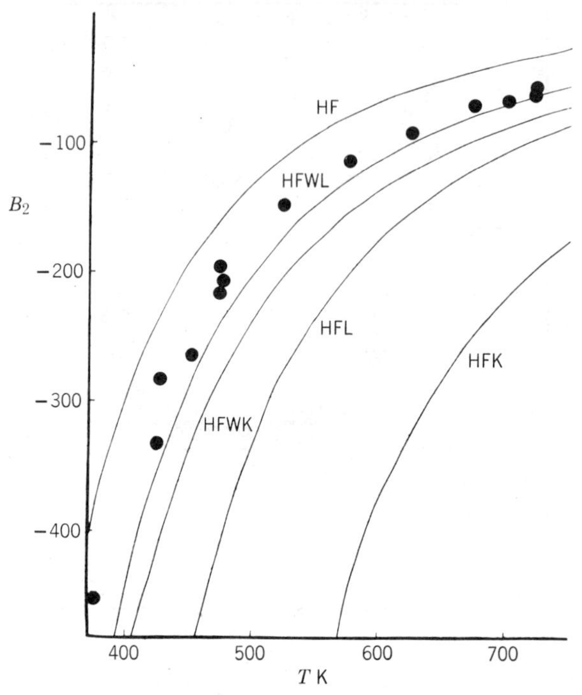

Figure 9.5 Second virial coefficient of steam in $cm^3\,mole^{-1}$ for several Hartree-Fock models compared with experiment. (Adapted with permission from *Mol. Phys.*, **28,** 1233, 1974).

good representation of the electron correlation energy at short distances. The best fit to experiment is found when the Wigner correction is used at short distances and the London term at longer distances as in the model HFWL (see Fig. 9.5). This model was constructed by switching smoothly from the Wigner correction at short distances to the London correction at longer distances using the formula

$$\phi(\mathbf{r}) = \phi_{\mathrm{HF}}(\mathbf{r}) + [1 - S(r)]\phi_{\mathrm{w}}(r) - \frac{S(r)C_6}{r^6} \tag{9.7}$$

where ϕ_{HF} is the Hartree-Fock potential, C_6 is given by eq. 9.3, and $\phi_{\mathrm{w}}(r)$ by eq. 9.6 and $S(r)$ by eq. 9.2 with $R_1 = 3.0$ Å and $R_2 = 4.0$ Å. Evans and Watts [30] also compared the Hartree-Fock results with values computed from the Rowlinson and BNS models. They concluded that the HFWL model was no worse than either of the empirical models *as a pair potential.*

There is no doubt that many body forces exist between clusters of water molecules but it is difficult to obtain an accurate estimate of their importance. Probably the most significant many body contributions arise from three-body interactions, and some results have been reported for such terms. Hankins, Moskowitz, and Stillinger [15] reported self-consistent field calculations for pairs and triplets of water molecules. They showed that three-body terms are large in magnitude and vary in sign according to the relative orientations of the three water molecules. Del Bene and Pople [23] also reported self-consistent field calculations of the interactions between water molecules and found qualitatively similar results. They extended their calculations to chains of up to six molecules and found that polymers of the type OH · · · OH · · · OH · · · OH · · · were preferred. Again, the nonadditive effects were quite large. The three-body terms tended to favor the tetrahedral coordination found in ice crystals, and so could be interpreted in terms of nonadditive hydrogen bond formation. Popkie and Clementi [35] have carried out an extensive analysis of three-body effects and report that the Hartree-Fock three-body term accounts for 15–20% of the binding energy of the water trimer. In addition to the three-body terms available from Hartree-Fock calculations, there will be contributions arising from the electron correlation energy. At present no quantitative results are available for such terms.

In the study of simple ionic solutions several interaction potentials are of importance, namely, the water-water interaction, the water-anion interaction, the water-cation interaction, and the three types of ion-ion interaction. We have seen that a number of models exist for water-water interactions, and similarly there are several competing empirical models for use in pure ionic systems [36]. Accurate Hartree-Fock pair potentials exist [37–39] for water interacting with the alkali metal ions Li^+, Na^+, K^+, and for the halide ions F^- and Cl^-. In addition, the Hartree-Fock calculations have been extended to include three-body interactions for the trimer H_2O-Li^+-F^- [40]. As with the pure water interactions, Clementi et al. [37–40] have fitted the numerical data for the ion-water interactions to an interpolatory formula [41].

Given the lack of detailed information on the importance of many body terms in liquid water and ionic solutions and the inaccuracies of the several pair potentials proposed, some care must be given to the interpretation of results from liquid state calculations. It is best to treat the several models as representing plausible effective pair potentials and to use results from several models before reaching any conclusions.

9.3 Computer Simulation Methods

Both Monte Carlo (MC) [19–21, 42, 43] and molecular dynamics (MD) [44–46] calculations have been reported for pure water, and MC studies have been reported for ion-water systems [47–49]. Basic computational details for both methods have been given in Section 3.10. There is one problem that is important in water calculations that does not occur for many simpler systems. At large distances the dominant term in the water-water potential arises from the interactions between the permanent dipoles on the molecules. The contribution of this term to the potential energy of two water molecules depends on the inverse third power of the distance between them, providing this distance is sufficiently large, and in bulk water the total size of these contributions depends on the shape of the system. As computer simulation calculations are carried out for an infinite (although periodic) system the long range interactions should be included. It has been assumed in all MD calculations [44–46] and in some of the MC studies [20, 42, 43] that if the system is fairly large the long range effects are small and can be included at a later stage. However, it has been shown that for dielectric properties the proper freatment of long range dipole interactions is important [19, 21]. Consequently, this section gives a brief account of how such effects can be included.

The obvious way to include the dipole-dipole terms at large distances is to make use of the periodic character of the structural model used in simulation studies. In solid state physics, particularly in the study of ionic solids [50], it is important to sum Coulombic interactions over the whole periodic crystal. In computer simulation studies the basic cell is reproduced an infinite number of times, and so the model resembles a crystalline solid with large unit cell. Consequently, methods developed to sum over periodic systems, such as Ewald's method [51], can be used to calculate the contribution to the energy and forces acting on a molecule due to long range interactions. Although such calculations are possible for infinite systems of dipoles [52, 53], they are computationally expensive and have not been introduced into studies of liquid water.

An alternative method to Ewald summation has been reported by Barker and Watts [19, 21]. During simulation calculations the interactions between a given molecule and its neighbors are put equal to zero if the distance between the centers of the molecules is greater than some distance R_c. If no correction is made for long range effects, then to a good first approximation a given water molecule is effectively surrounded by a sphere of water molecules, of radius R_c, sitting in a vacuum. If R_c is sufficiently large some contribution from long range interactions can be included provided it is assumed that the molecules surrounding the sphere form a continuum of dielectric constant ε. At any instant during the calculation the total dipole moment of the sphere surrounding the central molecule, i, is given by

$$\mathbf{M}_i = \sum_j{}' \boldsymbol{\mu}_j + \boldsymbol{\mu}_i \tag{9.8}$$

where the prime on the summation indicates that all molecules j whose centers lie within a radius R_c of the central molecule are included. Suppose the dielectric constant of the material in the sphere is ε_1, then the following phenomenon occurs. The dipole moment $\mathbf{M}_i$ induces surface charges on the sphere whose effect is to oppose the polarization of the sphere. The strength and direction of this *depolarizing field* can be calculated [54] and is given by

$$\mathbf{P}_i = \frac{-2(\varepsilon_1 - 1)\mathbf{M}_i}{(2\varepsilon_1 + 1)R_c^3} \tag{9.9}$$

If it is now assumed that the molecules surrounding the sphere form a dielectric continuum, of dielectric constant ε_2, then $\mathbf{M}_i$ interacts with this medium to produce the Onsager reaction field [54, 55], given by

$$\mathbf{R}_i = \frac{2(\varepsilon_2 - 1)\mathbf{M}_i}{(2\varepsilon_2 + 1)R_c^3} \tag{9.10}$$

Obviously, if the dielectric constants inside and outside the sphere are equal, $\varepsilon_1 = \varepsilon_2 = \varepsilon$, as they are for pure water, we have

$$\mathbf{R}_i = -\mathbf{P}_i \tag{9.11}$$

and the depolarizing field is canceled by the reaction field. The Onsager reaction field corresponds to an indirect interaction between particle i and its neighbors, transmitted through the surrounding medium. It can be shown that these interactions contribute to the free energy of the system [19, 54]

$$A_i = \frac{-(\varepsilon - 1)}{(2\varepsilon + 1)R_c^3}\boldsymbol{\mu}_i \cdot \mathbf{M}_i \tag{9.12}$$

This contribution to the free energy should be taken into account in the computer simulation calculations. It has been shown that its effect is small for many properties of water, but that the effect of these long range interactions on the dielectric properties and on some structural properties is significant [19].

9.4 Static Structure of Liquid Water

We have seen in earlier chapters that the structure of a liquid is best described in terms of the radial distribution function. In liquid water three atom-atom distribution functions can be identified—$g_{OO}(r)$ measuring the distribution of oxygen-oxygen separations, $g_{OH}(r)$ measuring the oxygen-hydrogen distribution, and $g_{HH}(r)$ measuring the hydrogen-hydrogen distribution. In principle, these distributions can be obtained experimentally using a combination of X-ray and neutron scattering data, and some measurements have been reported [56, 57]. The only distribution function that is known with any accuracy, however, is the oxygen-oxygen distribution [56]; the others (g_{OH} and g_{HH}) have been obtained using a model of the liquid to aid the analysis of experimental data [57]. Experimental neutron scattering data for heavy water has been reported by Page and Powles [58] who combined their results with X-ray scattering data in an attempt to obtain more insight into the structure of the liquid. Page and Powles did not obtain atom-atom distribution functions but compared their results for the structure factor with results predicted by various models. Several features in the scattering data have also been found in other molecular liquids, and they concluded that the structure of liquid water was neither unique nor unusual.

Although there is a lack of good experimental data on the atom-atom distribution functions, it is straightforward to compute these functions using computer simulation methods. Such results have been reported by several groups [18–21, 42–46] and all show similar qualitative behavior. Figure 9.6 gives the oxygen-oxygen distribution function calculated using

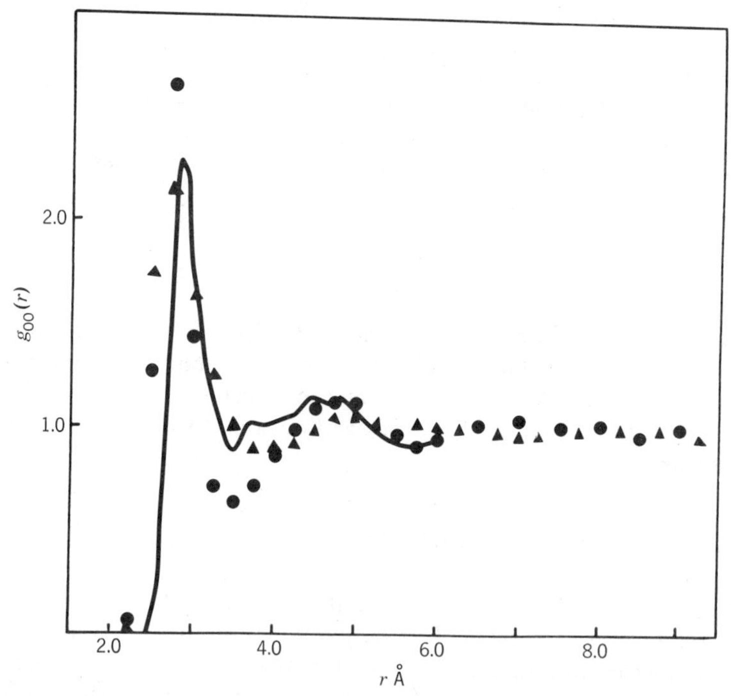

Figure 9.6 Oxygen-oxygen distribution function at $T = 298$ K and $\rho = 0.0334$ molec Å^{-3}. —— experimental results; ● BNS potential; ▲ Rowlinson potential. (Adapted with permission from *Mol. Phys.*, **28,** 1069, 1974.)

the Rowlinson [12] and BNS [13] empirical interaction potentials together with the X-ray diffraction data of Narten et al. [56]. Although the computed results are taken from the work of Watts [19], the result for the BNS model is in agreement with that published by Rahman and Stillinger [44]. There are distinct differences both between the two models and between the models and the experimental results. In particular, the BNS interaction produces an oxygen-oxygen distribution function that contains too much structure, and the Rowlinson model underemphasizes the structure. However, the qualitative agreement between calculations and experiment is good. The first peaks of both models are in the correct position, at around 2.8 Å, although on close examination they both give a rather steeper initial slope. Furthermore, both models predict that the second maximum in $g_{OO}(r)$ should be in the range 4.5–5.0 Å. We have seen in Chapter 5 that simple spherical interactions, such as are found between noble gas atoms, predict that the second peak occurs at twice the distance to the first peak, in this case 5.6 Å. It is possible to calculate the number of molecules whose centers lie between the center of one molecule and the position of the first minimum. This corresponds to the number of nearest neighbors and is obtained by integrating $g_{OO}(r)$. It is found that the models predict 5.5–6.0 neighbors and that experiment gives about 5.0. As the corresponding coordination number in the inert gases is about 12, we again have a quantitative difference. Consequently, we can feel satisfied that the models are qualitatively correct. Kistenmacher et al. [18] reported similar results for the Hartree-Fock interaction together with the several corrections for the correlation energy discussed in Section 9.2. Their results for the Hartree-Fock potential with Kirkwood correction were in good agreement with those of the Rowlinson model.

Results for the oxygen-hydrogen distribution calculated using the two empirical potentials are given in Fig. 9.7. The figure shows that both models predict two strong peaks, at about 1.8 Å and at about 3.2 Å, together with less significant maxima at longer distances. These results are expected on the basis of the strong tendency for water molecules to form tetrahedrally coordinated structures. Given an oxygen-oxygen nearest neighbor distance of about 2.8 Å and an oxygen-hydrogen bond length of about 1.0 Å, we can expect the nearest neighbor O—H distance to be about 1.8 Å. With strong tetrahedral coordination the second neighbor maximum will correspond to the distance between the oxygen atom on one molecule and the nonbonded hydrogen atom on the second molecule. Using simple trigonometry, we expect the second neighbor peak to occur at around 3.2 Å. It is clear that both models are qualitatively accurate. Variations between the models are largely confined to the relative heights

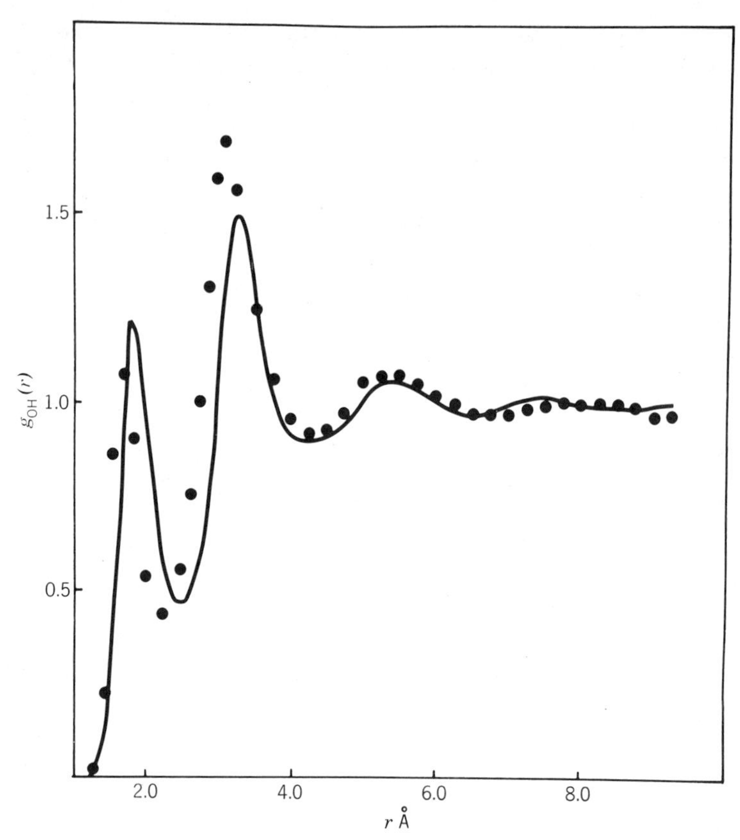

Figure 9.7 Oxygen-hydrogen distribution function at $T = 298$ K and $\rho = 0.0334$ molec Å^{-3}. ——— BNS potential; ● Rowlinson potential. (Adapted with permission from *Mol. Phys.*, **28,** 1069, 1974.)

of the various maxima, and can be interpreted in terms of the different emphasis given to the electrostatic terms in the two models. The results of Narten [57], obtained using a combination of experimental data and a model of the liquid structure, also show maxima in the region of 1.8 Å and 3.2 Å. However, the widths of the first two maxima are much greater than those of the Rowlinson and BNS models. Kistenmacher et al. [18] have compared their Hartree-Fock results with the results of Narten and find similar differences. The Hartree-Fock calculations, particularly when correlation energy corrections are added, resemble those obtained from the empirical models. There is little doubt that the results reported by Narten are very dependent on his model of liquid structure, and it is more likely that when accurate experimental data for g_{OH} become available they will resemble the simulation results.

The third atom-atom distribution function, $g_{HH}(r)$, is given for the Rowlinson and BNS models in Fig. 9.8. Once again, the positions of the first two maxima predicted by these model interactions agree reasonably well. There are also significant differences in the heights of the Rowlinson and BNS models. In particular, the Rowlinson potential gives a much stronger peak at about 3.5 Å than does the BNS potential, and at longer distances the two distributions are out of phase. Both models show strong peaks at about 2.4 Å; in addition, the BNS result shows a small but strong shoulder at 1.5 Å. This shoulder has been attributed [44] to a configuration in which a hydrogen atom lies between two negative charges. If the potential energy surface for the BNS model is examined [17] this configuration can be identified as a local minimum. The first neighbor peak can be explained qualitatively in terms of known oxygen-hydrogen bond lengths together with the general tetrahedral coordination found in liquid water.

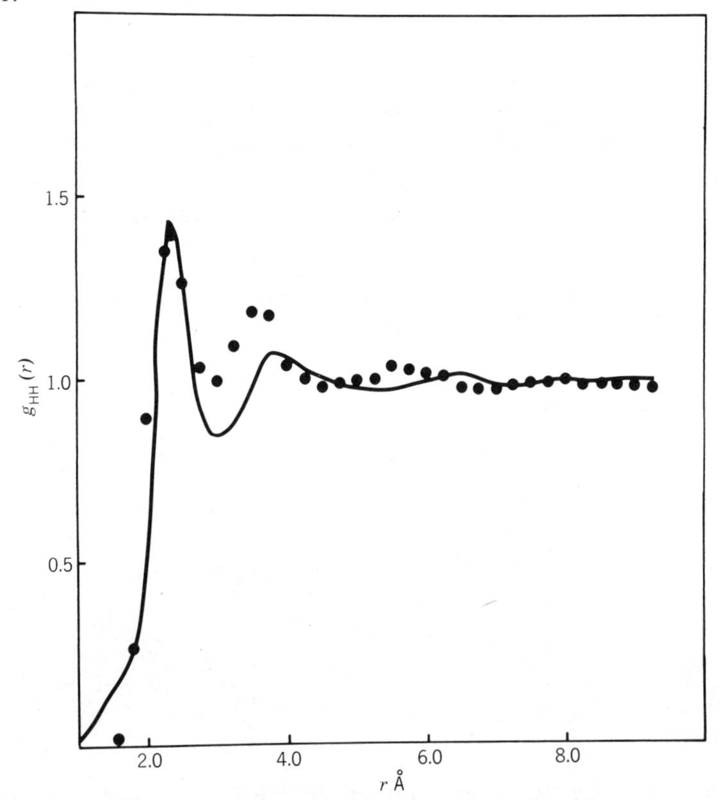

Figure 9.8 Hydrogen-hydrogen distribution function. Conditions are as for Fig. 9.7. (Adapted with permission from *Mol. Phys.*, **28,** 1069, 1974.)

Rahman and Stillinger [44, 46] have examined the temperature dependence of the atom-atom distribution functions and their results for $g_{HH}(r)$ are given for $T = 265$ K, 307.5 K, and 588 K in Fig. 9.9. The figure shows that increasing the temperature at fixed density causes a significant decrease in the local molecular correlation rather similar to that displayed for simple fluids in Chapter 5. Similar results were obtained for the other atom-atom distribution functions. Rahman and Stillinger also analyzed their calculations to obtain explicit information about the molecular orientations. Their analysis showed that at low temperatures the liquid structure had very many icelike characteristics whereas at higher temperatures these structural properties diminish. Sarkisov et al. [43] analyzed their Monte Carlo calculations in terms of hydrogen bonding, defining a hydrogen bond as a pair configuration in which the distance between the oxygen atoms does not exceed 3.2 Å and in which the O—H · · · O angle is not less than 125°. They showed that it was not possible to distinguish any clear division into hydrogen bonded and nonhydrogen bonded configurations, although the number of molecules not involved in some form of hydrogen bonding is small. Even though the definition of hydrogen bonding given by Sarkisov et al. is quite arbitrary, their results did not support any of the usual structural models of water [4].

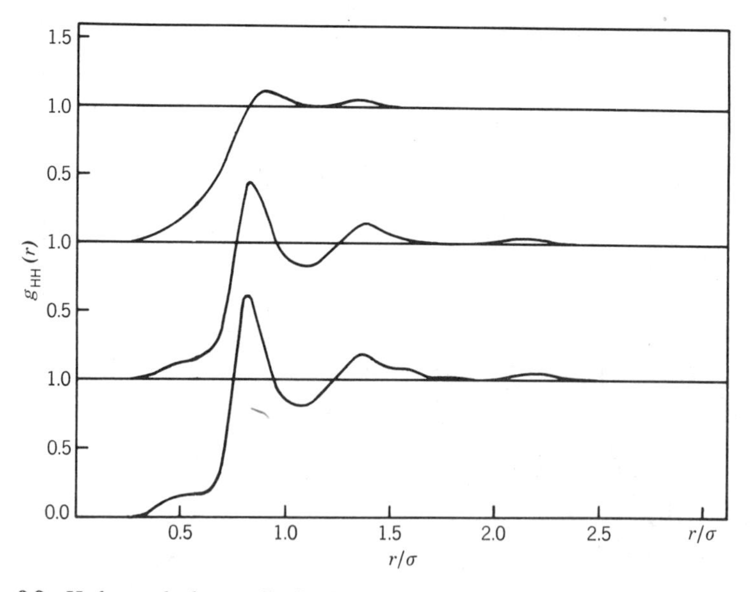

Figure 9.9 Hydrogen-hydrogen distribution function for BNS model calculated at temperatures $T = 265$ K, $T = 307.5$ K, and $T = 588$ K. (Adapted with permission from *J. Chem. Phys.*, **57,** 1281, 1972.)

9.5 Thermodynamic Properties and Bond Energy Distributions

Although the qualitative details of water structure are reproduced by all the models considered in the previous section, there are several obvious quantitative differences. A similar situation exists for the thermodynamic properties, including the internal energy, U, and specific heat, C_v. Table 9.3 collects data for these two quantities predicted by the various models and compares them with experimental results. It can be seen that the overall agreement with both internal energy and specific heat is good. It is particularly impressive to notice that the unusually high value of C_v is given by all the models. The range of values for U predicted by the Hartree-Fock based potentials is rather large and the Kirkwood correlation energy correction is obviously too large. However, it is apparent that all the models are reasonably good as effective pair potentials for water. Probably the most disappointing results for thermodynamic properties are those for the equation of state. In general pV/NkT is either very large and positive or very large and negative [19, 59]. These discrepancies can be related to the relative sensitivity of the equation of state to details of the structure, particularly the shape of the radial distribution function in the region of the first maximum.

In addition to predicting thermodynamic properties, the simulation calculations can also be used to examine bond energy distributions. Stillinger and Rahman [45] have given results for the BNS potential that throw some light on the hydrogen bond "making" and "breaking"

Table 9.3 Internal energy and specific heat at constant volume of water calculated for several models using Monte Carlo methods

Potential	U(kcal mole^{-1})	C_v(cal K^{-1} mole^{-1})	Number of molecules
Rowlinson	−8.4	21.8	64
	−8.4	18.5	216
BNS	−7.8	19.7	64
	−7.5	17.1	216
	−6.6	20.0	500
HF	−6.9	18.0	64
HF + London	−9.2	15.0	64
HF + Kirkwood	−12.2	19.0	64
HF + London + Wigner	−8.1	—	64
HF + London + Kirkwood	−8.6	—	64
Experiment	−8.12	18.0	—

controversy [3, 4]. They defined $\rho(\phi)$ to be the average number of neighbors of a given molecule whose interaction energy with that molecule is ϕ. Their results for the temperatures 265 K, 307.5 K, and 588 K are given in Fig. 9.10. Obviously there are no interactions whose energies are less than -6.5 kcal mole^{-1}, the minimum in the BNS potential. Also, the number of pairs whose interaction is essentially zero is very large, corresponding to the very large number of molecules found at large separations. There are a significant number of repulsive interactions corresponding to molecules that are very close to the central molecule. The most noticeable features in the diagram are the local maximum in the region of -5 kcal mole^{-1} and the near invariance of the energy density for $\phi \sim -3.5$ kcal mole^{-1}. Stillinger and Rahman note that this invariant point is reminiscent of the isosbestic point found in the Raman spectrum of water at different temperatures [60]. On the basis of this observation they proposed that both isosbestic points could be related to the presence of a single excitation mechanism. Using heuristic arguments, they suggested that if the hydrogen bond was associated with the local maximum in $\rho(\phi)$, then on average the energy required to break a

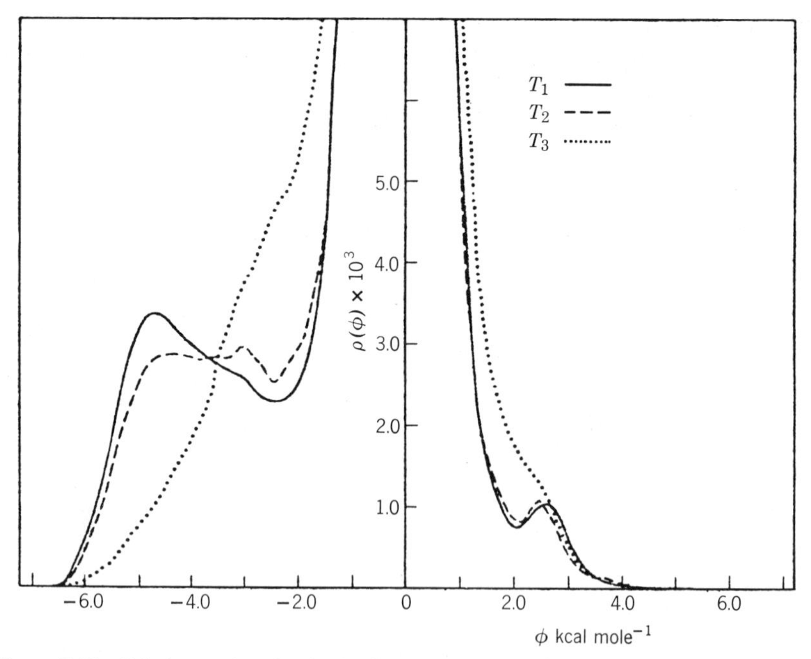

Figure 9.10 Pair interaction density in liquid water from BNS potential at temperatures $T = 265$ K, $T = 307.5$ K, and $T = 588$ K. (Adapted with permission from *J. Chem. Phys.*, **57,** 1281, 1972.)

hydrogen bond would be 2.54 kcal mole^{-1}. This figure is close to the value of 2.55 kcal mole^{-1} suggested by Walrafen [61] on the basis of Raman spectra. Similar values have been suggested by Worley and Klotz [62] on the basis of infrared studies (2.4 kcal mole^{-1}) and by Bucaro and Litovitz [63] on the basis of depolarized light scattering measurements. After using the isosbestic point as a reference from which to define hydrogen bonded and nonhydrogen bonded water molecules, Stillinger and Rahman [45] then discuss the structure of the liquid in similar terms to those used by Sarkisov et al. [43], which are reported in the preceding section. Their conclusions are much the same as those of Sarkisov et al., and Stillinger and Rahman remark that if their definition of hydrogen bonding is adopted (or any other for that matter), then the questionable two-state theories [57, 64–66] of water structure would be supplanted.

9.6 Time Dependent Phenomena

As has been pointed out earlier, if time dependent phenomena are to be studied using simulation methods, then the molecular dynamics method must be used. Consequently, only Rahman and Stillinger [44–46] have reported results for such properties of water. They have examined the self-diffusion coefficient, the translational velocity and angular momentum autocorrelation functions, and the van Hove [67] self-correlation function. They have also examined some properties concerned with dielectric relaxation. These are discussed in the following section.

As was shown in Section 4.2, the self-diffusion coefficient can be calculated using either the mean square displacement of a molecule as a function of time

$$D = \lim_{t\to\infty} \frac{\langle [\mathbf{r}_1(t) - \mathbf{r}_1(0)]^2 \rangle}{6t} \tag{9.13}$$

or by integrating the velocity autocorrelation function

$$D = \tfrac{1}{3}\int_0^\infty \langle \mathbf{v}_1(t) \cdot \mathbf{v}_1(0) \rangle \, dt \tag{9.14}$$

Rahman and Stillinger found that [44, 45] their results from the two formulae differed, primarily because of the small time interval over which information was available. Several possibilities exist for the value of $\mathbf{r}_1$ in eq. 9.13—for example, the oxygen position, the hydrogen positions, and the position of the center of mass. In the limit of long times these three displacements will lead to identical self-diffusion coefficients as the BNS potential models a rigid molecule. At short times, there will be considerable interference between the rotational and translational displacements of the atoms so that the most rapid approach to the limiting value of D can be expected from the center-of-mass displacement. Rahman and Stillinger found that at low temperatures the agreement with experiment was poor. Consequently, they rescaled the BNS potential by multiplying by a scale factor ζ, so that $\phi(\zeta)=\zeta\phi_{BNS}$ where ζ was chosen so that good agreement was obtained with experimental values of D and the internal energy. Given the apparent insensitivity of the self-diffusion coefficient to details of the potential [68], and the fact that $\langle[\mathbf{r}_1(t)-\mathbf{r}_1(0)]^2\rangle/6t$ does not reach its limiting value until about 10^{-11} s [69]—considerably longer than the length of the Rahman and Stillinger calculation—the rescaled potential should be treated with some caution.

The center-of-mass velocity autocorrelation function for water and its Fourier transform, or power spectrum, are qualitatively different to those found for simple liquids, such as argon (discussed in Section 5.5) [44]. In particular, the velocity autocorrelation function shows oscillatory behavior after the first negative region. Rahman and Stillinger attribute this behavior to the greater structural rigidity in water, leading to more persistently oscillatory motions. The power spectrum shows a maximum at about $\omega=10^{13}\,s^{-1}$ and a broad shoulder between $2\text{–}4\times10^{13}\,s^{-1}$. In their paper on the temperature dependence of these properties Stillinger and Rahman [45] conjecture that these features can be identified with the broad intermolecular bands observed by Raman infrared and neutron spectroscopy at 60 and 170 cm^{-1}. They suggest that the higher frequency feature is associated with oscillation of a molecule in a cage formed by its neighbors and that the lower frequency feature is evidence of cooperative structural relaxation phenomena.

Rahman and Stillinger considered several details of the oscillatory motions in water, reporting autocorrelation functions for the three independent angular momenta, for the weighted (by the reciprocal of the moments of inertia about the principal axes) total angular momentum, and for the time dependence of the dipole-dipole correlations. They found that these functions showed a substantial oscillatory character and that all the functions decayed rapidly. The Fourier transforms showed maxima at different frequencies, corresponding to the different components of the inertia tensor, with very little fine structure. In the following section we present evidence suggesting that some details of the structure, particularly orientational properties, are sensitive to the way in which long range interactions are treated in the simulation calculations. Consequently, no detailed discussion of the orientational properties is given here.

9.7 Dielectric Properties

The problem of including long range dipole-dipole terms in simulation calculations was discussed in Section 9.3, where it was suggested that the most convenient approach is to treat the water molecules at large distances from a central molecule as a dielectric continuum. If this approximation is adopted then we have the following situation. There is a sphere of radius R_c surrounding a central molecule, i. All molecules within this sphere are considered to interact with the central molecule through the usual intermolecular potential (Rowlinson or BNS, for example). If the molecules outside this sphere are assumed to behave as a dielectric continuum, then a simple relation can be derived relating the dielectric constant of the continuum, to the mean square polarization of the sphere [54]

$$\varepsilon - 1 = \frac{3\varepsilon}{2\varepsilon + 1} \frac{\langle \mathbf{M}_i^2 \rangle}{kTR_c^3} \tag{9.15}$$

where $\mathbf{M}_i$, the total dipole moment of the sphere around molecule i, is given by eq. 9.8. Alternatively, if all interactions outside the sphere are ignored so that the sphere is apparently in a vacuum, then a similar relation can be obtained for the dielectric constant of the material within the sphere [54]

$$\varepsilon - 1 = \frac{(\varepsilon + 2)}{3kTR_c^3} \langle \mathbf{M}_i^2 \rangle \tag{9.16}$$

Should the size of the sphere be sufficiently large, so that boundary effects can be ignored, it is possible to make the approximation

$$\langle \mathbf{M}_i^2 \rangle = N\langle \boldsymbol{\mu}_i \cdot \boldsymbol{\mu}_j \rangle = N\mu^2 g_K \tag{9.17}$$

where N is the number of particles in the sphere and g_K is known as the Kirkwood factor [70]. It is important to note that the above formulae are only valid for spherical systems and that the Kirkwood factor is only defined for a sufficiently large sphere. It is doubtful if the size of the spheres involved in computer simulation studies (6–12 Å radius) are sufficiently large, and for this reason the Kirkwood factor is not considered in any detail. However, it is apparent that if the experimental value of the dielectric constant of water is inserted into either eqs. 9.15 or 9.16 then experimental estimates of $\langle \mathbf{M}_i^2 \rangle$ may be obtained.

Table 9.4 gives results for the mean square polarization of different sized spheres calculated using both the Rowlinson and BNS potentials [19]. The quantities G_x^2, G_y^2, and G_z^2 give the mean square polarization divided by the square of the dipole moment of the model in the x, y, and

Table 9.4 Mean square polarizations computed for water using the Rowlinson and BNS potentials. Results are given as a function of R_c and of the value of ε used for the Onsager reaction field. Experimental results were obtained from eq. 9.16 with $\varepsilon = 78.5$

Potential	R_c(Å)	ε	G_x^2	G_y^2	G_z^2	G^2	M^2	M^2(experimental)
Rowlinson	6.2	1.0	7.7	6.0	6.5	20.2	68.3	28.2
	6.2	78.5	15.4	25.7	30.8	71.9	243.5	509.7
	9.2	1.0	27.0	25.2	22.3	74.4	251.9	95.6
BNS	6.2	1.0	6.9	7.3	6.4	20.6	90.8	28.3
	6.2	78.5	16.8	76.1	37.5	130.4	574.9	509.7
	9.2	1.0	36.4	33.0	32.1	101.5	447.5	95.6
	12.3	1.0	1842.3	78.2	90.2	2010.7	8867.2	222.1

z directions, respectively, and G^2 is the sum of these terms. For a sufficiently large sphere this quantity is equal to the Kirkwood factor multiplied by the number of particles in the sphere. The term M^2 is the quantity $\langle \mathbf{M}_i^2 \rangle$ found in eqs. 9.15 and 9.16, and is compared with the corresponding experimental values in the last two columns of the table. The results corresponding to $\varepsilon = 1$ represent calculations that did not include the long range correction discussed in Section 9.3. Equation 9.12 shows that the correction requires that the model dielectric constant be known if the correct reaction field is to be applied. To overcome this problem Watts [19] included the reaction field using the experimental dielectric constant of water at 298 K, $\varepsilon = 78.5$; results for these calculations are included in the table.

The table shows clearly that, although neither model is in quantitative agreement with experiment, both reproduce qualitatively the large increase in M that occurs when the reaction field is applied. In addition the BNS model, for the largest system, shows very strong polarization effects and it appears that it may be indicating ferroelectric behavior. It is likely that the BNS potential overemphasizes the electrostatic contributions to the interaction, and the modifications reducing these contributions suggested by Rahman and Stillinger [46] may produce a better account of dielectric behavior. The dielectric constants of the models can be obtained by solving eq. 9.16 for ε given the calculated value of $\langle \mathbf{M}_i^2 \rangle$. Results are disappointing, with $\varepsilon \lesssim 16$, for both models, and it is apparent that the question of calculating dielectric properties requires considerably more work.

Many aspects of the dielectric properties of a system are related to the static and dynamic dipole moment correlations. Barker and Watts [19, 21] have reported results for the dipole-dipole correlation function as follows:

$$g_{\mu}(r)=\frac{\langle \boldsymbol{\mu}(0)\cdot\boldsymbol{\mu}(r)\rangle}{4\pi\rho\mu^2 r^2\, dr} \tag{9.18}$$

This function measures the probability of finding a dipole at a distance r from a given dipole in terms of the expected relative orientation. Figure 9.11 shows results for the BNS potential calculated by truncating the water-water interactions at 6.3 Å, 9.3 Å, and 12.3 Å. Two calculations are included for the 6.3 Å truncation, showing the influence of the Onsager reaction field. The diagram shows clearly that the long range interactions are particularly important, in terms of their influence on

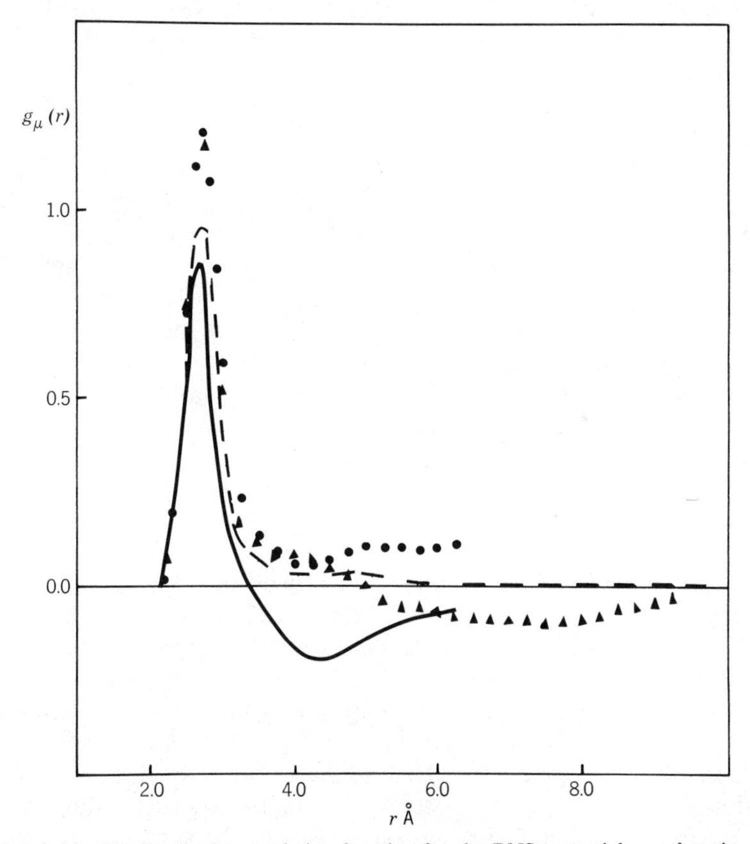

Figure 9.11 Dipole-dipole correlation function for the BNS potential as a function of the potential range. ——— $R_c = 6.2$ Å, no reaction field; ●●● $R_c = 6.2$ Å, with reaction field; ▲▲▲ $R_c = 9.3$ Å, no reaction field; --- $R_c = 12.3$ Å, no reaction field. (Adapted with permission from *Mol. Phys.*, **28,** 1069, 1974.)

water structure, beyond the nearest neighbor shell. All four curves have a strong positive maximum at about 3.0 Å, corresponding to the tetrahedral relative orientations in water. Beyond this peak there are qualitative differences in the structure of the correlation functions. The 6.3 Å result, in the absence of the reaction field, shows relatively strong negative correlations beyond about 3.5 Å. If the reaction field is included, using $\varepsilon = 78.5$ for the dielectric constant, the correlations are positive everywhere. Apparently, using a value of $\varepsilon = 78.5$ overemphasizes the correction term, leading to the positive correlations. If the correct (model) value of ε were used the correlation function should go smoothly to zero with increasing distance. The results for the longer interaction radii show that this does occur, the 12.3 Å results having the expected behavior. Similar results were reported by Watts [19] for the Rowlinson potential, although there were some differences in peak height due to the smaller dipole moment of the Rowlinson model.

Rahman and Stillinger [44–46] have examined the time dependent dipole-dipole self-correlation functions

$$\Gamma_n(t) = \langle P_n[x_1(t)] \rangle \tag{9.19}$$

where the $P_n(x)$ are the Legendre polynomials and $\mu^2 x_1(t) = \boldsymbol{\mu}_1(0) \cdot \boldsymbol{\mu}_1(t)$. They found that both Γ_1 and Γ_2 had very little structure and fell smoothly towards zero with exponential behavior. The function Γ_1 is central to the frequency dependence of the dielectric constant of the liquid and was used by Rahman and Stillinger [44] to obtain Cole-Cole plots [71]. They obtained reasonable qualitative agreement with experiment, but did not consider the influence of long range interactions.

9.8 Theory of Ionic Solutions

Theoretical models of ionic solutions have been largely concerned with the Debye-Hückel model [72] and its extensions [73]. The Debye-Hückel theory treats the ions as point charges in a dielectric continuum and solves the corresponding Poisson equation. Around each ion in a solution there exists on average a number of oppositely charged ions. These ions, together with the solvent molecules in the neighborhood, form a *co-sphere* around the central ion. This co-sphere will be of opposite charge to that of the central ion owning it. When an ionic solution is diluted two terms contribute to the change in free energy of the system. First, there will be a contribution from the change in energy of an ideal solution, and second there will be a change due to dilution of the co-sphere. This second contribution to the free energy change arises primarily from work done against the electrostatic interaction between the ions and their co-spheres. The Debye-Hückel theory enables this free energy change to be calculated for very dilute solutions.

A statistical mechanical treatment of ionic solution theory is often based on a dilute system of charged hard spheres embedded in a dielectric continuum, the so-called *primitive model.* Advances on this model have been made by including the second virial coefficient term in the expansion of thermodynamic properties in powers of electrolyte concentration [74–76]. The formal theory of electrolyte solutions is usually based on the McMillan-Meyer theory [79], which modifies the cluster theory of fluids developed in Sections 3.7 and 3.8 to cover the conditions appropriate to ionic solutions. The main modification is to treat the solvent as a continuum, but the long range of the Coulombic interactions also poses mathematical problems that have to be overcome. In this section we outline a derivation of Debye-Hückel theory based on solving the Poisson equation and then develop the corresponding cluster expansion theory. Following this, a brief review of recent results in the field is given. A full treatment of the physical chemistry of electrolyte solutions can be found in several text books and the reader is referred to them for more detail [77–79].

Assume that at very low concentrations an ion may be treated as a point charge in a continuum of dielectric constant ε. Let ψ be the potential at a point $\mathbf{r}$ due to a positive ion of charge $z_i e$ at the origin together with its co-sphere. Then

$$\psi(r) = \frac{z_i e}{\varepsilon r} + \psi_{\text{atm}}(r) \tag{9.20}$$

where $\psi_{\text{atm}}(r)$ is the electrostatic potential at $\mathbf{r}$ due to the co-sphere. If the concentration of ions in the solution is n_i^0, then the local concentration is given by

$$n_i(r) = n_i^0 \exp \frac{-z_i e \psi(r)}{kT} \tag{9.21}$$

from the Boltzmann distribution. It follows that the charge density at a distance r is given by

$$\rho(r) = \sum_i n_i(r) z_i e \tag{9.22}$$

where the sum is taken over all ion types. Poisson's equation for the electrostatic potential at $\mathbf{r}$ can be written

$$\nabla^2 \psi = -\frac{4\pi}{\varepsilon}\rho(r) = -\frac{4\pi e}{\varepsilon}\sum_i n_i^0 z_i \exp \frac{-z_i e \psi(r)}{kT} \tag{9.23}$$

which is the nonlinear Poisson equation for $\psi(r)$. If the temperature is sufficiently high, so that $z_i e\psi(r) \ll kT$, we can expand the exponential

$$\exp\frac{-z_i e\psi(r)}{kT} \approx 1 - z_i\frac{e\psi(r)}{kT} + \cdots$$

so that the Poisson equation becomes

$$\nabla^2\psi(r) = -\frac{4\pi e}{\varepsilon}\sum_i\left[n_i^0 z_i\left(1 - z_i\frac{e\psi(r)}{kT}\right)\right]$$

which on using the fact that the ionic solution is electrically neutral, $\sum_i n_i^0 z_i = 0$, gives

$$\nabla^2\psi = \kappa^2\psi \tag{9.24}$$

where κ is given by

$$\kappa^2 = \frac{4\pi e^2}{\varepsilon kT}\sum_i n_i^0 z_i^2 \tag{9.25}$$

Note that κ^{-1} has dimensions of length. The co-sphere around a given ion acts as a shield, reducing the electrostatic interaction between the ion and another some distance away. The *Debye length* or *screening parameter*, κ^{-1}, gives the effective radius of the co-sphere. Once eq. 9.24 is transformed to spherical polar coordinates the solution for the radial component is given by

$$\psi(r) = \frac{A\exp(-\kappa r)}{r} + \frac{B\exp(\kappa r)}{r} \tag{9.26}$$

To find the constants A and B we have to use appropriate boundary conditions. We know that at large distances the potential is zero, so that $\psi(\infty) = 0$ and hence $B = 0$. To find A, suppose that the distance of closest approach of two ions is a. Then the field due to the central charge at $r = a$ is given by elementary electrostatics

$$-\left.\frac{\partial\psi}{\partial r}\right|_{r=a} = \frac{z_i e}{\varepsilon a^2} \tag{9.27}$$

using the inverse square law. Differentiating eq. 9.26 and equating to eq. 9.27 gives

$$\frac{A\exp(-\kappa a)}{a^2}(1 + \kappa a) = \frac{z_i e}{\varepsilon a^2}$$

or

$$A = \frac{z_i e\exp(\kappa a)}{\varepsilon(1 + \kappa a)} \tag{9.28}$$

In a very dilute solution the screening parameter will be very small so that $\exp(\kappa a) \approx 1 + \kappa a$, and the electrostatic potential becomes

$$\psi(r) = \frac{z_i e}{\varepsilon} \frac{\exp(-\kappa r)}{r}$$

or the so-called *Debye potential*, and the potential energy of an ion of charge $z_j e$ at position r is

$$V_{ij}(r) = z_j e \psi(r) = \frac{z_i z_j e^2}{\varepsilon r} \exp(-\kappa r) \tag{9.29}$$

We see shortly that if we make certain assumptions in the cluster theory we can derive the Debye potential using statistical mechanics.

Before doing this, it is worth showing how the Debye potential can be used to obtain thermodynamic properties of the electrolyte solution. Inserting the expression for $\psi(r)$ into eq. 9.20 we have

$$\psi_{\text{atm}}(r) = \frac{z_i e}{\varepsilon r} \left\{ \frac{\exp[\kappa(a-r)]}{(1+\kappa a)} - 1 \right\} \tag{9.30}$$

This gives the electrostatic potential at r resulting from the co-sphere around the ion. Note that when r is equal to the ion diameter a, so that in dilute solutions $\kappa a \ll 1$

$$\psi_{\text{atm}}(a) = \frac{z_i e}{\varepsilon a}[1 - \kappa a + \cdots - 1] = -\frac{z_i e}{\varepsilon(1/\kappa)} \tag{9.31}$$

so that the co-sphere acts as a charge at a distance $1/\kappa$ away.

To study the influence of the co-sphere on the thermodynamic properties, we consider the equation for the chemical potential [80]

$$\mu = \mu^0 + kT \log m + kT \log \gamma \tag{9.32}$$

where μ^0 is the chemical potential of the ionic solution at infinite dilution, $kT \log m$ is the chemical potential for an ideal solution, and $kT \log \gamma$ measures the deviation from ideality. The quantity γ is called the activity coefficient. Obviously, for an ionic solution the deviation from ideality is measured by the chemical potential due to the co-sphere. This term can be calculated by recognising that it is just the potential energy due to ψ_{atm} evaluated at $r = a$. This is given by $z_j e \psi_{\text{atm}}(a)$ so that

$$\log \gamma = \frac{z_j z_i e^2}{2\varepsilon kT} \frac{\kappa}{(1+\kappa a)} \tag{9.33}$$

where the factor of 2 is introduced as the chemical potential is shared by both ions. Once the chemical potential is obtained it is possible to calculate other thermodynamic properties of the system. One of the more important properties is the *osmotic pressure*, Π. It was shown by van't Hoff that at low concentrations measurements of the osmotic pressure of a solution as a function of concentration obeyed similar laws to the ideal gas laws [80]. That is, in some ways a dilute solution behaves as a dilute gas. Using this observation, it is possible to show that if p_0 is the vapor pressure of the pure solvent and p is the vapor pressure of the solute, the osmotic pressure is given by

$$\Pi = \rho kT \log \frac{p_0}{p} \tag{9.34}$$

where ρ is the partial molal number density of the solvent constituents in the solution.

To obtain the Debye potential using statistical mechanics we recall from Chapter 3 that the radial distribution function of a one-component fluid can be written in the form (eq. 3.93)

$$g(r) = \exp - \frac{\phi(r)}{kT}\, y(r) \tag{9.35}$$

where $y(r)$ can be represented as a density expansion

$$\log y(r) = \sum_{m \geq 3} \rho^{m-2} g_{m-2} \tag{9.36}$$

with g_{m-2} a sum of irreducible cluster integrals with two root points. We also recall (eq. 3.87) that the potential of mean force for the fluid is defined as

$$W^{(2)}(r) = -kT \log g(r) \tag{9.37}$$

These terms can be generalized to multicomponent systems, and McMillan and Meyer [79, 81] have made the appropriate generalization for ionic solutions. Briefly, the potential of mean force obtained from the cation-cation distribution function is identified with the potential $\psi(r)$ given in eq. 9.20, with similar terms for cation-anion and anion-anion interactions. To obtain the Debye-Hückel theory the summation in eq. 9.36 is carried out for a subclass of the irreducible diagrams, the chain diagrams. Before we investigate this aspect of the theory, however, there is a mathematical problem to overcome.

In the discussion leading up to eq. 9.34 we mentioned that at low concentrations the ionic solution behaves in a similar fashion to an ideal gas. Pursuing this analogy, consider the second virial coefficient of a one-component monatomic ionic gas interacting with potential

$$\phi(r)=\frac{e^2}{r} \tag{9.38}$$

Using eq. 3.77 we have

$$\begin{aligned} B_2 &= \lim_{R\to\infty} 2\pi\int_0^R r^2\left(\exp\frac{-e^2}{rkT}-1\right)dr \\ &= \lim_{R\to\infty} 2\pi\int_0^R r^2\left[\frac{-e^2}{rkT}+\frac{e^4}{2r^2(kT)^2}-\frac{e^6}{6r^3(kT)^3}\cdots\right]dr \end{aligned} \tag{9.39}$$

from which it follows that the first three terms in the expansion for B_2 are divergent as $R\to\infty$. In addition, the remaining terms in the sum diverge at the lower limit, even when R is finite. It can be shown that the integral in B_2 converges [79], and so it follows that the two sets of divergences must cancel. These observations can be extended to the full cluster expansion for thermodynamic properties, and the Debye-Hückel theory is obtained by careful grouping of the divergent terms.

Now consider the expression for the radial distribution function of our monatomic ionic gas, eqs. 9.35–9.37. We can identify some of the cluster diagrams, g_{m-2}, with chain diagrams, so that in terms of these integrals we have

$$\log y(r)\approx\rho\,\bigwedge+\rho^2\,\sqcap+\rho^3\,\bigcap+\cdots \tag{9.40}$$

with ρ the number density. Each of the bonds in this equation represents an f bond

$$f(r)=\exp\frac{-\phi(r)}{kT}-1 \tag{9.41}$$

so that in our case we can write $f(r)=\exp(-e^2/rkT)-1$. Letting $\lambda=4\pi e^2/kT$ and expanding the exponential we have

$$f(r)=-\frac{\lambda}{4\pi r}+\frac{\lambda^2}{2(4\pi r)^2}-\frac{\lambda^3}{6(4\pi r)^3}+\cdots \tag{9.42}$$

Truncating this expansion at the first term and substituting into eqs. 9.35 and 9.40 and using the result $\kappa^2=\lambda\rho$ gives

$$\log g(r_{12})=\frac{\lambda}{\kappa^2}\sum_{m\geq 1}(-\kappa^2)^m p_m(r_{12}) \tag{9.43}$$

where $p_1=1/4\pi r_{12}$ and

$$p_m(r_{12})=(4\pi)^{-m}\int\frac{1}{r_{13}}\cdot\frac{1}{r_{34}}\cdot\frac{1}{r_{45}}\cdots\frac{1}{r_{mm+1}}\cdot\frac{1}{r_{m+12}}\,d\mathbf{r}_3\cdots d\mathbf{r}_m,\qquad m\geq 2 \tag{9.44}$$

The integrals $p_m(r)$ represent extensions of the convolution integral used to define the Ornstein-Zernike equation (see Section 3.8) and can be evaluated in the following way. Define the Fourier transform of a function $f(r)$ as

$$\tilde{f}(s) = \int f(r) \exp(i\mathbf{r} \cdot \mathbf{s})\, d\mathbf{r} = \frac{4\pi}{s} \int_0^\infty r f(r) \sin(sr)\, dr$$

for a spherically symmetric function. Using the convolution theorem then gives [79]

$$\tilde{p}_n(s) = [\tilde{c}(s)]^n \tag{9.45}$$

where $\tilde{c}(s)$ is the Fourier transform of the function $c(r) = 1/4\pi r$:

$$\tilde{c}(s) = \frac{4\pi}{s} \int_0^\infty r \sin(rs) \cdot \frac{1}{4\pi r}\, dr \tag{9.46}$$

To evaluate $\tilde{c}(s)$, we define

$$d(r) = \frac{\exp(-\alpha r)}{4\pi r} \tag{9.47}$$

so that $\lim_{\alpha\to 0} d(r) = c(r)$. Then

$$\tilde{d}(s) = \frac{1}{s} \int_0^\infty \exp[-\alpha r] \sin(rs)\, dr = \frac{1}{\alpha^2 + s^2}$$

and we find that

$$\tilde{c}(s) = \lim_{\alpha\to 0} d(s) = \frac{1}{s^2} \tag{9.48}$$

Using this result, together with eq. 9.45 for $\tilde{p}_n(s)$, in eq. 9.43 gives, with $q(r) = \log g(r)$

$$\tilde{q}(s) = \frac{\lambda}{\kappa^2} \sum_{m=1}^{\infty} \left(\frac{-\kappa^2}{s^2}\right)^m = \frac{-\lambda}{\kappa^2 + s^2} \tag{9.49}$$

and hence

$$q(r) = \log g(r) = \frac{-\lambda}{2\pi^2 r} \int_0^\infty \frac{s \sin(sr)}{\kappa^2 + s^2}\, ds$$

so that from eq. 9.37

$$W^{(2)}(r) = \frac{e^2}{r} \exp(-\kappa r) \tag{9.50}$$

which is the functional form of the Debye potential. Note that if instead of the one-component electron gas we consider a system where the

Coulomb interaction is given by

$$\phi(r) = \frac{e^2}{\varepsilon r}$$

with ε the dielectric constant, and where ρ becomes the mean square charge density, $\rho = \sum_i n_i^0 z_i^2$ as in eq. 9.25, eq. 9.50 becomes equivalent to eq. 9.29. Thus in the statistical mechanical treatment of ionic solutions the Debye potential is given by the potential of mean force.

The McMillan-Meyer theory asserts that if $\phi(r)$ is the potential of mean force at infinite dilution between two ions in solution, then the solution behaves like an ideal gas interacting through this pair potential. We see that the McMillan-Meyer theory is a formal extension of the thermodynamic arguments given by van't Hoff when introducing his relation between osmotic pressure and vapor pressure [80]. The most common model for the potential of mean force at infinite dilution for a system of ions with nonzero radius is the primitive model

$$\phi(r) = \phi_{\mathrm{HS}}(r) + \frac{e_i e_j}{\varepsilon r} \tag{9.51}$$

where $\phi_{\mathrm{HS}}(r)$ is the hard sphere potential of diameter σ_{ij} for ions i and j, and e_i and e_j are the charges on ions i and j. As with the Debye-Hückel theory the solvent is approximated as a dielectric continuum. If the hard sphere diameters are allowed to go to zero and the approximate cluster diagram summation is introduced, the primitive model reduces to the Debye-Hückel model. If it is assumed that the hard sphere diameter is the same for all ions, the model becomes the *restricted primitive model.*

Having noted that the Debye-Hückel theory is obtained by approximating the cluster expansion for the radial distribution function, it is reasonable to look for more complete summations of eq. 9.36. The Percus-Yevick (PY) and Hypernetted Chain (HNC) approximations for the radial distribution function (see Section 3.8) provide such partial summations when modified to account for the long range interactions [82]. Carley [83] and Rasaiah and Friedman [84] have solved the HNC and PY approximations using the restricted primitive model for a system of 1:1 electrolytes. They have reported a number of properties of the electrolyte solution including ion-ion distribution functions, the osmotic pressure, and the internal energy. Waisman and Lebowitz [85–87] have introduced an approximation to electrolyte theory that is similar to the PY approximation. They assumed that the Ornstein-Zernike equation could be closed using the equations

$$c_{ij}(r) = c_{ij}(r)\ (\mathrm{PY}) \qquad r<\sigma$$
$$c_{ij}(r) = \frac{e_i e_j}{\varepsilon r} \qquad r \geqslant \sigma \tag{9.52}$$

where $c_{ij}(r)$ is the direct correlation function for ions i and j. This approximation, known as the Mean Spherical Model (MSM) has also been used in the theory of simple liquids [88] and liquid metals [89]. When the MSM is used with the potential corresponding to the restricted primitive model of electrolyte theory, it can be solved analytically. Having obtained an expression for the distribution functions it is relatively easy to calculate the thermodynamic properties (see Section 3.6).

Perturbation theory has also been used to obtain expressions for the properties of electrolyte solutions, again within the context of the restricted primitive model. In Section 3.9 we showed that if the pair potential could be written in the form

$$\phi(r) = u_0(r) + u_1(r)$$

then the properties of the system could be obtained in terms of those of the reference system, $u_0(r)$, by expanding the canonical partition function in powers of $1/kT$. A similar expansion can be made using the restricted primitive model, with $\phi_{\mathrm{HS}}(r)$ as the reference system. However, if this is done, reference to eq. 3.111 shows that the first correction term is

$$A_1 = 2\pi V \sum_i \sum_j \rho_i \rho_j \int_0^\infty r^2 g_{ij}^0(r) \frac{e_i e_j}{\varepsilon r}\, dr$$

where ρ_i is the number density of ion i and $g_{ij}^0(r)$ is the radial distribution function of the hard sphere fluid corresponding to the term $\phi_{\mathrm{HS}}(r)$ for ion pair (i, j). This integral is divergent. Consequently, to use perturbation theory in this form the Coulomb term in eq. 9.51 must be replaced by the screened interaction given by the Debye-Hückel theory, eq. 9.29, so that the integral is convergent. Alternatively, the method used to obtain the Fourier transform of the potential can be used, together with the electroneutrality property of the solution, to obtain [90]

$$A_1 = 2\pi \frac{V}{\varepsilon} \sum_i \sum_j e_i e_j \rho_i \rho_j \int_{\sigma_{ij}}^\infty r[g_{ij}^0(r) - 1]\, dr \tag{9.53}$$

This expansion was first obtained by Rasaiah and Stell [90], who also showed that for the restricted primitive model, where all the hard sphere diameters are equal, $A_1 = 0$.

Andersen and Chandler [91] have reported a form of perturbation theory known as the mode expansion. Again, they assume that the interaction potential can be written as the sum of a reference term and a perturbing term. If the perturbing term, $u_1(r)$, has Fourier transform $\tilde{u}_1(q)$, then Andersen and Chandler have shown that the free energy is given by

$$\frac{A - A_0}{VkT} = -\frac{\rho^2}{2kT}\left(1 - \frac{1}{N}\right)\tilde{u}_1(0) + \sum_{i=1}^{\infty} a_i$$

where A_0 is the free energy of the reference system and the first term on the right hand side is the mean field or van der Waals contribution to A, and the terms a_i are related to various moments of the density fluctuations. If the reference system is taken to be the ideal gas, that is if $\phi_{HS}(r)$ is assumed to have $\sigma_{ij} = 0$, then truncating this expansion at the term a_1 leads to the Debye-Hückel theory. Other expansions for use in electrolyte theory have been given by Lebowitz and Stell [92–94].

The primitive model, if used with any of the theories just discussed, can be used to obtain thermodynamic properties of electrolyte solutions at quite high concentrations. At concentrations significantly greater than infinite dilution the approximation of regarding the solvent as a dielectric continuum becomes questionable. In fact, Stokes has shown [95] that the primitive model theories can be brought into agreement with both experiment and with other more empirical theories [96, 97] if the hard sphere diameter is treated as an adjustable parameter. Consequently, the results from these theories are better tested against Monte Carlo calculations for the primitive model rather than against experimental results. If this is done, both theory and simulation results can be obtained using the same interaction potential, thus providing a more rigorous test of theory.

Monte Carlo results for the restricted primitive model have been reported by Card and Valleau [98] and by Vorontsov-Vel'yaminov and co-workers [99, 100]. Most of the approximate theories have been evaluated numerically, results being given for the HNC approximation by Carley [83], for the PY and HNC approximations by Rasaiah and Friedman [84], for the mode expansion by Chandler and Andersen [101], and for the MSM approximation by Waisman and Lebowitz [87]. Other numerical results have been given by Rasaiah [102–104] and by Vorontsov-Vel'yaminov et al. [105]. Most of these results are for the restricted primitive model, although Rasaiah and Friedman have also given results for a square well potential with Coulombic interactions [106].

If the various results for 1:1 electrolytes are compared, it is found that the HNC approximation is most useful, and this conclusion also follows from studies of 2:1, 3:1, and 3:2 electrolytes at concentrations up to 2

M litre^{-1}. The PY approximation is not particularly good, but the MSM results, and particularly the mode expansion results, agree well with the machine calculations. These results contain a contribution from the hard sphere interactions, and all the approximations give different values for these terms. If the hard sphere terms are removed all the approximations are in reasonably good agreement with the simulation predictions. This suggests that the short range structure is important in determining the properties of the electrolyte solution, and thus raises doubt about the validity of treating the solvent as a continuum. Stell [107] has extended the concept of the primitive model to include solvent structure, but he has not reported any numerical results.

9.9 Structure of Ion-Water Clusters

Although the formal theory of ionic solutions has not been extended to take full account of solvent-solvent and solvent-solute interactions some work has been reported on the structure of alkali halide-water systems. Kistenmacher et al. [108, 109] and Watts and co-workers [47–49] have used Hartree-Fock calculations of the ion-water interactions [41] to examine the structure of small clusters of water molecules around single ions [108, 109] and ion pairs [47–49]. Kistenmacher et al. [108] studied polymeric clusters of the type $Li^+(H_2O)_n$, $K^+(H_2O)_n$, $F^-(H_2O)_n$, and $Cl^-(H_2O)_n$, for $n = 2, 3, \ldots 10$, optimizing the structure of the clusters so that the minimum energy configuration was adopted. For $n < 5$, the minimum energy configuration was the symmetric configuration, but with higher values of n some of the water molecules formed a second solvation layer. They found that the coordination numbers for ion-water clusters were about 4 for Li^+, between 5 and 6 for Na^+, between 4 and 6 for F^-, and between 6 and 7 for Cl^-. Kistenmacher et al. [108] also computed the ion-atom distribution functions for a cluster containing 27 water molecules at room temperature using a Monte Carlo technique. They found that there were strong nearest neighbor ion-oxygen peaks at about 2.0 Å for Li^+, 2.3 Å for Na^+, 2.8 Å for K^+ and F^-, and 3.4 Å for Cl^-. After examining the three-dimensional structure of the clusters using perspective drawings Kistenmacher et al. [108, 109] found that on average the oxygen atoms were closer to the cations than were the hydrogen atoms, with the reverse being true for the anions.

Watts et al. [47–49] have reported Monte Carlo studies of clusters of 50 and 200 water molecules around the ion pair LiF [47, 48] and KF and LiCl [49], again using the Hartree-Fock potentials discussed in Section 9.2. The Monte Carlo program used to simulate liquid water [19] was modified to simulate the system of water molecules around an ion pair held at a fixed distance apart. Calculations for LiF ranged from ion-ion distances of 2 Å up to separations of 10 Å and at temperatures of 298 K and 500 K. It was shown that at 298 K the water-ion and water-water interactions were sufficiently strong to hold the cluster in a configuration whose number density was similar to that found in liquid water. At 500 K the cluster tended to evaporate although the ions were left with a strongly bound nearest neighbor shell.

The structure of the clusters can be discussed in terms of the density of oxygen and hydrogen atoms at various places in the cluster. Define a cylindrical coordinate system such that the origin lies midway between the cation and anion, with the x axis passing through the centers of the ions. The density of atoms of a particular type will be symmetric about the x axis, but there is no reason to suppose that the density near an anion is the same as that near the cation. Consequently, the structure of the clusters can be examined in terms of the distribution function

$$f_i(x, r) = \frac{\langle N_i[(x, x+\delta x),(r, r+\delta r)]\rangle}{2\pi r\,\delta r\,\delta x}$$

where the quantity averaged is the number of atoms of type i having their x coordinate in the range $(x, x+\delta x)$ and being within a distance $(r, r+\delta r)$ of the x axis, where $r^2 = y^2 + z^2$.

A typical result is given in Fig. 9.12. This figure shows the distribution of oxygen atoms (Fig. 9.12a) and hydrogen atoms (Fig. 9.12b) at 298 K around the ion pair $Li^+ - F^-$, with the ion-ion separation being 4 Å. It can be seen that both the Li^+ ion and the F^- ion have a strongly bound single shell of oxygen atoms surrounding them. There is also some evidence of a second neighbor shell, although the structure is more diffuse. The center of the nearest neighbor oxygen shell is about 2 Å from the lithium ion and about 2.7 Å from the fluoride ion. Examining the hydrogen atom distribution, it is apparent that there is also a strongly bound nearest neighbor shell around both ions and a strong second neighbor shell around the F^- ion. The hydrogen atoms lie closer to the F^- ion (~1.7 Å) than to the Li^+ ion (~3.0 Å). The differences between Figs. 9.12a and 9.12b are related to the polarity and structure of the water molecule. Any given water molecule will orient so that the oxygen atom is towards the cation and at least one hydrogen atom towards the anion. The second neighbor shell of hydrogen atoms is best explained by examining the number of atoms of a given type within a sphere of radius r around the ions. This function is shown as a function of distance, for the cluster represented by Fig. 9.12, in Fig. 9.13. It can be seen that there is a strong plateau in the number of oxygen atoms around both ions for $N=4$. Thus

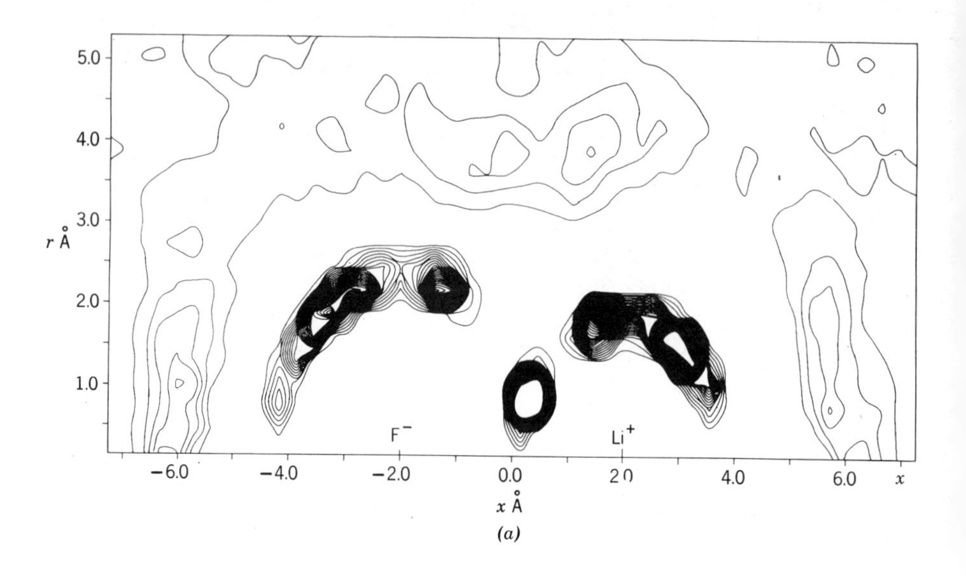

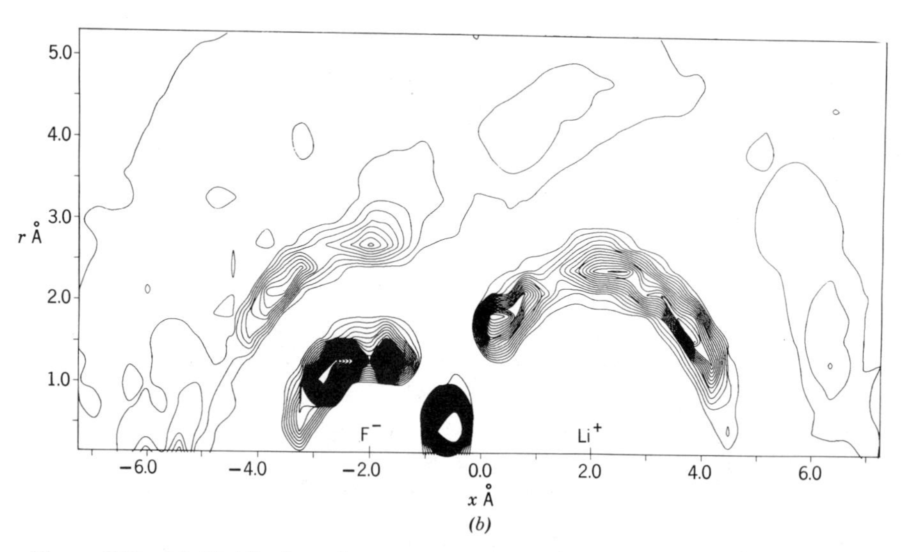

Figure 9.12 (*a*) Distribution of oxygen atoms around the ion pair LiF for an ionic separation of 4.0 Å at 298 K; (*b*) Distribution of hydrogen atoms around the ion pair LiF for an ionic separation of 4.0 Å at 298 K. (Adapted with permission from *J. Chem. Phys.*, **61,** 2550 1974.)

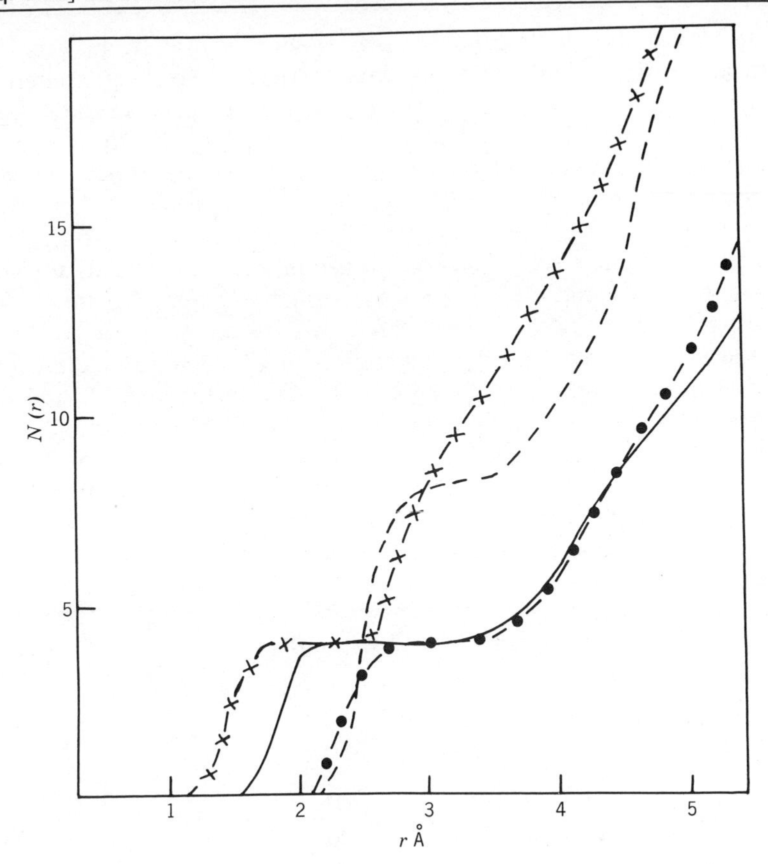

Figure 9.13 Number of atoms within a distance r of Li^+ and F^- ions. ——— Li^+—O; - - - - - - Li^+—H; —●—●— F^-—O; —×—×— F^-—H. (Adapted with permission form *J. Chem. Phys.*, **61,** 2550, 1974.)

both ions have a first hydration layer containing four molecules. Around the anion, the hydrogen atoms are oriented such that one atom from each water molecule is directed toward the ion and the other is directed away from the ion. This property shows that the ion has four nearest neighbor atoms. The set of hydrogen atoms directed away from the anion is primarily responsible for the pronounced second neighbor shell seen in Fig. 9.12*b*. Examining the distribution of atoms around the cation, we see that there are four oxygen atoms in the nearest neighbor shell. Each of these oxygen atoms has two hydrogens attached to it, and as they are directed away from the cation they appear as a plateau having the value $N(Li^+ - H) = 8$ in Fig. 9.13. Watts, Clementi, and Fromm [47] suggested that the strong four-coordination exhibited by the Li^+ and F^- ions was responsible for the identification of these ions as "structure makers" in empirical theories of ionic solutions. After a study of the effect of altering the interionic distance and the number of water molecules in the cluster they found that the first coordination shell was always tightly bound to the ions and that the density of the cluster became more or less uniform beyond the second shell. That is, beyond about two coordination layers the water structure no longer recognized the existence of the ion pair.

Watts [49] examined the effect of ion size on the water structure using calculations for KF and LiCl. The ion-ion separations were fixed at 4 Å so that his results could be compared directly with those reproduced as Figs. 9.12 and 9.13. He reported that although such graphs were qualitatively similar to those found for LiF, the increased sizes of the ions Cl^- and K^+ enlarged the volume of the first coordination layer. After examining the number of atoms of each type within a given distance of the ions, Watts reported that the proximity of the larger ions increased the number of water molecules in the first coordination shell compared with the number around the small ions Li^+ and F^-. Thus the Cl^- ion distorted the water

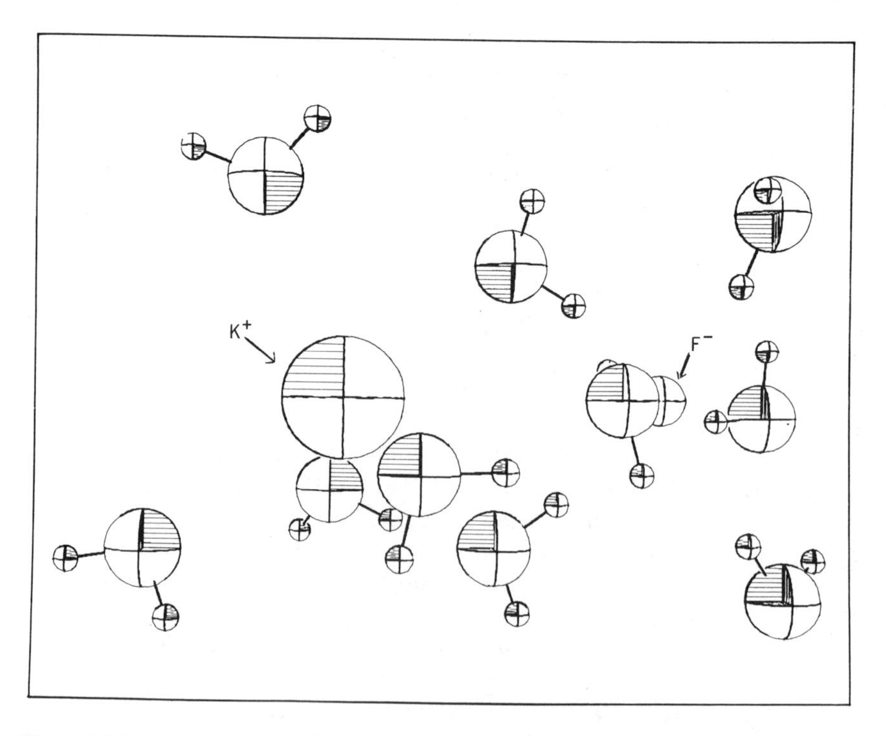

Figure 9.14 Perspective drawing showing the relative orientation of water molecules in the nearest neighbor shell around KF.

structure so that the plateau in the graph of $N(\mathrm{Li}^{+}\text{–O})$ had a value of 5 rather than 4 found in Fig. 9.13. Similarly, the K^{+} ion disturbed the water structure so that more hydrogen atoms were packed around the F^{-} ion. It would appear from Watts' results that when alkali halide ions are close together (e.g., at reasonably high concentrations) it is not possible to consider their solvation shells as being independent. Figure 9.14 gives a perspective drawing of the positions and relative orientations of the water molecules in the first coordination layer around the KF ion pair. The figure shows that in the neighborhood of the cation the oxygen atoms are close to the ion and that near the anion the hydrogen atoms are directed towards the ion. It is also possible to see from the figure that the water molecules tend to be in hydrogen bonding configurations.

The results discussed in this section represent the first attempts to apply methods used to understand simple liquids to the study of ionic solutions. Future work will be directed towards examining both the dynamic and static properties of ionic solutions and to understanding the forces acting between the various entities in these systems. Ultimately, we can expect agreement between fundamental theory, simulation methods, and experiment to be as good as that existing for the condensed inert gases.

References

1. D. Eisenberg and W. Kauzmann, 1969, *Structure and Properties of Water,* Oxford University Press.
2. R. A. Horne, 1972, *Structure and Transport Processes in Water and Aqueous Solutions,* Interscience, New York.
3. G. S. Kell, 1972, "Continuum Theories of Liquid Water," Chapter 9 in reference 2.
4. C. M. Davis and J. Jarzgnski, 1972, "Mixture Models of Water," Chapter 10 in reference 2.
5. C. M. Davis, Jr. and T. A. Litovitz, 1965, *J. Chem. Phys.,* **42,** 2563.
6. L. Pauling, 1959, in *Hydrogen Bonding,* D. Hadzi, Ed., Pergamon, London.
7. A. Némethy and H. A. Scheraga, 1962, *J. Chem. Phys.,* **36,** 3382.
8. M. Falk and T. A. Ford, 1966, *Can. J. Chem.,* **44,** 1699.

   a. G. H. Haggis, J. B. Hasted, and T. J. Buchanan, 1952, *J. Chem. Phys.,* **20,** 1452.
   b. L. Pauling, 1948, *The Nature of the Chemical Bond,* 2nd ed., p. 304, Cornell University Press.
   c. A. Eucken, 1948, *Z. Elektrochem.,* **69,** 2720.
   d. J. J. Fox and A. E. Martin, 1940, *Proc. Roy. Soc. (London), Ser. A,* **174,** 234.
   e. K. Buijs and G. R. Choppin, 1963, *J. Chem. Phys.,* **39,** 2035.
   f. M. R. Thomas, H. A. Scheraga, and E. E. Shrier, 1965, *J. Phys. Chem.,* **69,** 3722.
   g. R. Goldstein and S. S. Penner, 1964, *J. Quant. Spectrosc. Radiative Transfer,* **4,** 441.
   h. W. Luck, 1963, *Ber Bunsenges. Phys. Chem.,* **67,** 186.
   i. G. Némethy and H. A. Scheraga, 1962, *J. Chem. Phys.,* **36,** 3382.
   j. L. Pauling, 1959, in *Hydrogen Bonding,* D. Hadzi, Ed., Pergamon, New York.
   k. P. C. Cross, J. Burnham, and P. A. Leighton, 1937, *J. Am. Chem. Soc.,* **59,** 1134.

l. G. E. Walrafen, 1964, *J. Chem. Phys.*, **40,** 3249.
m. T. A. Litovitz and E. H. Carnevale, 1955, *J. Appl. Phys.*, **26,** 816.
n. L. Hall, 1948, *Phys. Rev.*, **73,** 775.
o. O. Nomoto, 1956, *J. Phys. Soc. Japan*, **11,** 1146.
p. D. P. Stevenson, 1965, *J. Phys. Chem.*, **69,** 2145.
q. G. Wada, 1961, *Bull. Chem. Soc. Japan*, **34,** 955.
r. K. Grjotheim and J. Krogh-Moe, 1954, *Acta. Chem. Scand.*, **8,** 1193.
s. H. S. Frank and A. S. Quist, 1961, *J. Chem. Phys.*, **34,** 604.
t. C. M. Davis and T. A. Litovitz, 1965, *J. Chem. Phys.* **42,** 2563.
u. R. P. Marchi and H. Eyring, 1964, *J. Chem. Phys.*, **68,** 221.
v. R. H. Ewell and H. Eyring, 1937, *J. Chem. Phys.*, **5,** 726.
w. M. D. Danford and H. A. Levy, 1962, *J. Am. Chem. Soc.*, **84,** 3965.

9. J. A. Pople, 1951, *Proc. Roy. Soc.* (*London*), *Ser. A*, **205,** 163.
10. J. D. Bernal, 1964, *Proc. Roy. Soc.* (*London*), *Ser. A*, **280,** 299.
11. Reference 2, Section 4.3(*c*).
12. J. S. Rowlinson, 1951, *Trans. Faraday Soc.*, **47,** 120.
13. A. Ben-Naim and F. H. Stillinger, 1972, "Aspects of the Statistical Mechanics of Water," p. 295 in reference 2.
14. G. F. H. Diercksen, 1971, *Theo. Chim. Acta.*, **21,** 335.
15. D. Hankins, J. W. Moskowitz, and F. H. Stillinger, 1970, *J. Chem. Phys.*, **53,** 4544.
16. E. Clementi and H. Popkie, 1972, *J. Chem. Phys.*, **57,** 1077.
17. H. Popkie, H. Kistenmacher, and E. Clementi, 1973, *J. Chem. Phys.*, **59,** 1325.
18. H. Kistenmacher, H. Popkie, E. Clementi, and R. O. Watts, 1974, *J. Chem. Phys.*, **60,** 4455.
19. R. O. Watts, 1974, *Mol. Phys.*, **28,** 1069.
20. J. A. Barker and R. O. Watts, 1969, *Chem. Phys. Lett.*, **3,** 144.
21. J. A. Barker and R. O. Watts, 1973, *Mol. Phys.*, **26,** 789.
22. K. Morokuma and L. Pedersen, 1968, *J. Chem. Phys.*, **48,** 3275.
23. J. Del Bene and J. A. Pople, 1970, *J. Chem. Phys.*, **52,** 4858.
24. F. London, 1937, *Trans. Faraday Soc.*, **33,** 8.
25. J. G. Kirkwood, 1932, *Phys. Z.*, **33,** 57; A. Müller, 1936, *Proc. Roy. Soc.* (*London*), *Ser. A*, **154,** 624.
26. E. P. Wigner, 1934, *Phys. Rev.*, **46,** 1002.
27. E. Clementi, 1963, *J. Chem. Phys.*, **38,** 2248; 1963, *ibid.*, **39,** 175.
28. E. Clementi, 1970, Chapter 1 in *The Chemistry of the Cyano-group*, J. Rappoport, Ed., Interscience, New York.
29. H. Kistenmacher, G. C. Lie, H. Popkie, and E. Clementi, 1974, *J. Chem. Phys.*, **61,** 547.
30. D. J. Evans and R. O. Watts, 1974, *Mol. Phys.*, **28,** 1233.
31. F. G. Keyes, L. B. Smith, and H. T. Gerry, 1936, *Proc. Am. Acad. Arts Sci.*, **70,** 319.
32. G. S. Kell, G. E. McLaurin, and E. Whalley, 1965, *Advances in Thermophysical Properties at Extreme Temperatures and Pressures*, p. 104, American Society of Mechanical Engineers.
33. M. P. Vukalovich, M. S. Trakhtengerts, and G. A. Spiridonov, 1967, *Teploenergetika*, **14,** 65.
34. G. S. Kell, G. E. McLaurin, and E. Whalley, 1968, *J. Chem. Phys.*, **48,** 3805.
35. G. C. Lie and E. Clementi, 1974, *J. Chem. Phys.*, **60,** 1275, 1288.
36. A fairly extensive set of references to ion-ion potentials is given in Chapter 10.
37. E. Clementi and H. Popkie, 1972, *J. Chem. Phys.*, **57,** 1077.
38. H. Kistenmacher, H. Popkie, and E. Clementi, 1973, *J. Chem. Phys.*, **58,** 1689.

39. H. Kistenmacher, H. Popkie, and E. Clementi, 1973, *J. Chem. Phys.*, **58,** 5627.
40. J. W. Kress, E. Clementi, J. J. Kozak, and M. E. Schwartz, in press.
41. H. Kistenmacher, H. Popkie, and E. Clementi, 1973, *J. Chem. Phys.*, **59,** 5842.
42. G. N. Sarkisov and V. G. Dashevsky, 1972, *Zh. Strukt. Khimii*, **13,** 799.
43. G. N. Sarkisov, V. G. Dashevsky, and G. G. Malenkov, 1974, *Mol. Phys.*, **27,** 1249.
44. A. Rahman and F. H. Stillinger, 1971, *J. Chem. Phys.*, **55,** 3336.
45. F. H. Stillinger and A. Rahman, 1972, *J. Chem. Phys.*, **57,** 1281.
46. F. H. Stillinger and A. Rahman, 1974, *J. Chem. Phys.*, **60,** 1545.
47. R. O. Watts, E. Clementi, and J. Fromm, 1974, *J. Chem. Phys.*, **61,** 2550.
48. J. Fromm, E. Clementi, and R. O. Watts, 1975, *J. Chem. Phys.*, **62,** 1388.
49. R. O. Watts, 1976, *Mol. Phys.*, to appear.
50. W. Cochran, 1971, "Lattice Dynamics of Ionic and Covalent Crystals," in *CRC Critical Reviews in Solid State Sciences.*
51. P. P. Ewald, 1921, *Ann. Phys. (Leipzig)*, **64,** 253.
52. Report of Summer Workshop on Molecular Dynamics and Monte Carlo Calculations on Water, CECAM, Paris, 1972, (issued by C. Moser, unpublished).
53. E. R. Smith and J. W. Perram, 1974, *Phys. Lett.*, **50A**, 294; 1975, *Mol. Phys.*, **30,** 31; 1975, *Phys. Lett.*, **53A**, 121.
54. H. Fröhlich, 1958, *Theory of Dielectrics*, Oxford University Press.
55. L. Onsager, 1936, *J. Am. Chem. Soc.*, **58,** 1486.
56. A. H. Narten, M. D. Danford, and H. A. Levy, 1967, *Discuss. Faraday Soc.*, **43,** 97.
57. A. H. Narten, 1972, *J. Chem. Phys.*, **56,** 5681.
58. D. I. Page and J. G. Powles, 1971, *Mol. Phys.*, **21,** 901.
59. R. O. Watts, unpublished data.
60. G. E. Walrafen, 1968, *J. Chem. Phys.*, **48,** 244.
61. G. E. Walrafen, 1968, in *Hydrogen Bonded Solvent Systems*, A. K. Covington and P. Jones, Eds., Taylor and Francis, London.
62. J. D. Worley and I. M. Klotz, 1966, *J. Chem. Phys.*, **45,** 2868.
63. J. A. Bucaro and T. A. Litovitz, 1972, referred to in reference 45.
64. M. D. Danford and H. A. Levy, 1962, *J. Am. Chem. Soc.* **84,** 3965.
65. O. Ya Samoilov, 1965, *Structure of Aqueous Electrolyte Solutions and the Hydration of Ions*, Consultants Bureau, New York.
66. J. W. Perram, 1971, *Mol. Phys.*, **20,** 1077.
67. L. van Hove, 1953, *Phys. Rev.*, **89,** 1189.
68. P. Schofield, 1974, in *Molecular Motions in Liquids*, J. Lascombe, Ed., p. 15, Reidel, Boston.
69. R. O. Watts, 1972, *Faraday Soc. Symp.*, **6,** 166; R. A. Fisher and R. O. Watts, 1972, *Aust. J. Phys.*, **25,** 529.
70. J. G. Kirkwood, 1939, *J. Chem. Phys.*, **7,** 911.
71. K. S. Cole and R. H. Cole, 1941, *J. Chem. Phys.*, **9,** 341.
72. P. Debye and E. Hückel, 1923, *Z. Physik*, **24,** 195, 305.
73. C. W. Outhwaite, 1971, *Mol. Phys.*, **20,** 705.
74. J. C. Poirier, 1953, *J. Chem. Phys.*, **21,** 972.
75. E. Meeron, 1957, *J. Chem. Phys.*, **26,** 804.
76. R. Abe, 1959, *Progr. Theor. Phys. (Kyoto)*, **22,** 213.
77. H. S. Harned and B. B. Owen, 1958, *The Physical Chemistry of Electrolyte Solutions*, Reinhold, New York.
78. R. A. Robinson and R. H. Stokes, 1959, *Electrolyte solutions*, 2nd. ed., Butterworths, London.
79. H. L. Friedman, 1962, *Ionic Solution Theory*, Interscience, New York.

80. S. Glasstone and D. Lewis, 1960, *Elements of Physical Chemistry*, Macmillan and Co., London.
81. W. G. McMillan and J. E. Mayer, 1945, *J. Chem. Phys.*, **13,** 276.
82. A. R. Allnatt, 1964, *Mol. Phys.*, **8,** 533.
83. D. D. Carley, 1967, *J. Chem. Phys.*, **46,** 3783.
84. J. C. Rasaiah and H. L. Friedman, 1968, *J. Chem. Phys.*, **48,** 2742.
85. E. Waisman and J. L. Lebowitz, 1970, *J. Chem. Phys.*, **52,** 4307.
86. E. Waisman and J. L. Lebowitz, 1972, *J. Chem. Phys.*, **56,** 3086.
87. E. Waisman and J. L. Lebowitz, 1972, *J. Chem. Phys.*, **56,** 3093.
88. R. O. Watts, D. Henderson, and J. A. Barker, 1972, *J. Chem. Phys.*, **57,** 5391.
89. L. Blum and A. H. Narten, 1972, *J. Chem. Phys.*, **56,** 5197.
90. J. C. Rasaiah and G. Stell, 1970, *Mol. Phys.*, **18,** 249.
91. H. C. Andersen and D. Chandler, 1970, *J. Chem. Phys.*, **53,** 547.
92. J. L. Lebowitz, G. Stell, and S. Baer, 1965, *J. Math. and Phys.*, **6,** 1281.
93. G. Stell, 1971, *J. Chem. Phys.*, **55,** 1485.
94. G. Stell and J. L. Lebowitz, 1968, *J. Chem. Phys.*, **49,** 3706.
95. R. H. Stokes, 1972, *J. Chem. Phys.*, **56,** 3382.
96. R. H. Stokes and R. A. Robinson, 1948, *J. Am. Chem. Soc.*, **70,** 1870.
97. E. Glueckaaf, 1953, *Trans. Faraday Soc.*, **51,** 1235.
98. D. N. Card and J. P. Valleau, 1970, *J. Chem. Phys.*, **52,** 6232.
99. P. N. Vorontsov-Vel'yaminov, A. M. El'yashevich, and A. K. Kron, 1966, *Electrokhimiya*, **2,** 708.
100. P. N. Vorontsov-Vel'yaminov, and A. M. El'yashevich, 1968, *Electrokhimiya*, **4,** 1430.
101. D. Chandler and H. C. Andersen, 1971, *J. Chem. Phys.*, **54,** 26.
102. J. C. Rasaiah, 1970, *J. Chem. Phys.*, **52,** 704.
103. J. C. Rasaiah, 1972, *J. Chem. Phys.*, **56,** 3071.
104. J. C. Rasaiah, D. N. Card, and J. P. Valleau, 1972, *J. Chem. Phys.*, **56,** 248.
105. P. N. Vorontsov-Vel'yaminov, A. M. El'yashevich, J. C. Rasaiah, and H. L. Friedman, 1970, *J. Chem. Phys.*, **52,** 1013.
106. J. C. Rasaiah and H. L. Friedman, 1970, *J. Chem. Phys.*, **50,** 7965.
107. G. Stell, 1974, *J. Chem. Phys.*, **59,** 3926.
108. H. Kistenmacher, H. Popkie, and E. Clementi, 1974, *J. Chem. Phys.*, **61,** 799.
109. H. Kistenmacher, unpublished results.

Ten

Liquid Metals and Molten Salts

In this chapter we conclude our discussion of the role of intermolecular forces in dense fluids by considering liquid metals and molten salts. The analysis of these liquids is complicated by the presence of many body forces acting between the atoms and ions. Fortunately, it is possible to construct a theoretical framework that defines effective two-body potentials that are in good agreement with experimental data for the liquid phase. As might be expected, the behavior of these liquids near their melting points is not too different from that of the corresponding solids, and we see that the parameters appearing in model potentials for the liquid state are often derived from solid state data. Much of the experimental data used to investigate properties of the liquid come from X-ray and neutron scattering studies [1]. We discussed the relation between the structure factors obtained from scattering data and the effective pair potential and radial distribution function in Section 5.1 and extend this discussion in Section 10.3.

For the inert gases, the interaction potential is characterized by a short range repulsion and a smooth monotonic increasing long range attraction that vanishes rapidly at long distances. For the molten salts, a similar short range interaction is used but this has to be augmented by the dominant Coulomb potential term at longer distances. By contrast, the long range forces between atoms in liquid metals are very different from the force law appropriate to the same atoms in free space. Each atom in the solid or liquid metal contributes one electron to a delocalized molecular orbital system that extends over the whole sample, so that the substance resembles an electron gas confined by an essentially periodic lattice of ions. These electrons form the so-called conduction band, responsible for such phenomena as the very high electrical conductivity of metals. The interionic forces in the metal must reflect the high density of conduction electrons which surround the ions. In the next section, we show that the forces in metals can be described to a good first approximation by an effective ion-electron interaction called the pseudopotential. This is later combined with the direct ion-ion interaction to obtain an effective pair potential for the metal.

10.1 Effective Pair Interactions from Pseudopotential Theory

An electron near an isolated metal ion experiences a strong electrostatic potential from the nucleus

$$\phi(r) = \frac{Ze^2}{r} \tag{10.1}$$

where e is the electronic charge, Z the atomic number of the metal, and r the distance between electron and ion. In a free atom, the electrons in the outermost atomic orbitals have this attractive field screened considerably by the presence of other electrons. In a metal in the condensed state the conduction electrons tend to distribute themselves in the form of a screening charge which opposes the attractive interactions from the ionic charge. The Coulomb attraction between an ion and an electron is counterbalanced by the repulsion due to the Pauli exclusion principle, which precludes electrons from occupying the same quantum state. Near the ion core this repulsion in fact dominates the interaction and conduction electrons are thus excluded from the ion core region. The ion in a metal, taken together with its screening charge, is often called a *pseudoatom* and the relatively weak interaction between the ions and the conduction electrons is called a *pseudopotential.*

The role of the core electrons is vastly simplified in the pseudopotential model. Their presence excludes the conduction electrons from the region of the ion core, although the latter electrons can roam freely through the rest of the metallic ion array. Quantum mechanically, the absence of conduction electrons in the ion core region can be explained in terms of *orthogonalized plane waves* [2]. In this method the wave function for a conduction electron is forced to be orthogonal to any electronic core state. One can see the effect of this construction in the following simplified picture. First, the wave function for a single conduction band electron, ψ_k, is assumed to satisfy the one electron Schrödinger equation

$$H\psi_k = E_k\psi_k \tag{10.2}$$

where H is the Hamiltonian operator, containing the kinetic energy operator and the potential energy of the conduction electron in the combined fields of the ions and the screening electrons $v(r)$. The orthogonality of the wave functions for the conduction electrons is expressed by

$$\langle \psi_k \mid \psi_{k'} \rangle = \delta_{k,k'} \tag{10.3}$$

where δ_{ij} is the Kronecker delta, $\delta_{ij} = 1,\ i = j; = 0\ i \neq j$.

The notation used in this equation is the "bra" and "ket" convention of Dirac [3]. Thus the symbol $|i\rangle$ represents some wave function with identifier i (possibly a set of quantum numbers), $\langle i|$ represents its complex conjugate together with an integral operator, so that

$$\langle i \mid j \rangle \equiv \int \psi_i^* \psi_j \, d\tau$$

where ψ_i and ψ_j are the wave functions and the integral is over all the coordinates of both wave functions. On occasions the second wave function is operated on by some operator **P** so that, for example

$$\langle i | \mathbf{P} | j \rangle = \int \psi_i^* \mathbf{P} \psi_j \, d\tau$$

A single core electron wave function is represented by $|\alpha\rangle$, where α enumerates all the core electrons around all the ion sites. These wave functions must be orthogonal both to each other and to the conduction electron wave functions

$$\begin{aligned} \langle \alpha \mid \alpha' \rangle &= \delta_{\alpha,\alpha'} \\ \langle \psi_k \mid \alpha \rangle &= 0 \end{aligned} \tag{10.4}$$

We have said earlier that the conduction band electrons are free to move over the whole system and the wave function ψ_k must reflect this freedom. The requisite form for ψ_k can be ensured by writing it as a linear combination of orthogonal plane waves (OPW), u_q. We construct an OPW from a free electron plane wave function $\exp[i\mathbf{q} \cdot \mathbf{r}]$, denoted by $|q\rangle$, by subtracting off core state components so that it is orthogonal to the states $|\alpha\rangle$

$$u_q = |q\rangle - \sum_\alpha \langle \alpha \mid q \rangle \, |\alpha\rangle \tag{10.5}$$

In this construction the coefficients $\langle \alpha \mid q \rangle$ are chosen to make u_q (and hence any linear combination of them) orthogonal to the core states

$$\langle \alpha' \mid u_q \rangle = \langle \alpha' \mid q \rangle - \sum_\alpha \langle \alpha \mid q \rangle \, \langle \alpha' \mid \alpha \rangle = 0 \tag{10.6}$$

Equation 10.5 may alternatively be written using a projection operator **P** acting on the core states $|\alpha\rangle$

$$u_q = (1 - \mathbf{P})|q\rangle \tag{10.7}$$

where

$$\mathbf{P} = \sum_{\alpha} |\alpha\rangle\langle\alpha| \tag{10.8}$$

Notice that $\langle\alpha|$ can be regarded as an integral operator since $\langle\alpha| Q = \int \psi_\alpha^* Q \, d\boldsymbol{\tau}$ for any function Q.

The wave function for a conduction electron is a linear combination of such OPWs:

$$\psi_k = \sum_q a_q u_{k+q} = \sum_q a_q(|k+q\rangle - \sum_\alpha \langle\alpha \mid k+q\rangle \mid \alpha\rangle)$$

$$= \sum_q a_q(1 - \mathbf{P}) \mid k+q\rangle = (1 - \mathbf{P})\chi_k \tag{10.9}$$

The new function χ_k in eq. 10.9 is usually called the *pseudowave function.* Combining eqs. 10.2 and 10.9 we obtain

$$H(1 - \mathbf{P})\chi_k = E_k(1 - \mathbf{P})\chi_k \tag{10.10}$$

or

$$[H + (E_k - H)\mathbf{P}]\chi_k = E_k\chi_k \tag{10.11}$$

This is equivalent to the one-electron Schrödinger equation with the potential energy $v(r)$ replaced by

$$W = v(r) + (E_k - H)\mathbf{P} \tag{10.12}$$

where the function W is known as the pseudopotential. At this point W is an integral operator and is also to some extent arbitrary since any linear combination of core functions $|\alpha\rangle$ can be added to χ_k in eq. 10.10 for the same eigenenergy E_k. Inside the core, $(E_k - H)\mathbf{P}$ is large and positive since H acting on **P** produces core energies E_α which in general will be much smaller than the conduction band energies E_k. On the other hand, $v(r)$ is large and negative in the region of the core due to the unscreened ion charge. Consequently, W arises from the small difference between two large contributions and so can be used in perturbation theory solutions to the Schrödinger equation.

In many cases it is convenient to make the reasonable assumption that W is not an integral operator but a real function of the position variable, replacing the second term in eq. 10.12 by a nonoperator term v_p:

$$W(r) = v(r) + v_p(r) \tag{10.13}$$

This assumption is known as the *local approximation* and has interesting consequences when used in conjunction with the fact that W is small enough to be treated as a perturbation. The first order perturbation expression for the interaction between an electron and a system of ions can be written in terms of the so-called *plane wave matrix elements*, which are the quantities $\langle \mathbf{k}+\mathbf{q} \mid W \mid \mathbf{k} \rangle$ where the wave functions $|\mathbf{k}\rangle$ are the plane waves $\exp(i\mathbf{k}\cdot\mathbf{r})$. These quantities turn out to be independent of $\mathbf{k}$ and, in fact, become the Fourier transform of $W(r)$

$$\langle \mathbf{k}+\mathbf{q} \mid W \mid \mathbf{k} \rangle = \frac{1}{V}\int \exp[-i(\mathbf{k}+\mathbf{q})\cdot\mathbf{r}]W(\mathbf{r})\exp[i\mathbf{k}\cdot\mathbf{r}]\,d\mathbf{r}$$

$$= \frac{1}{V}\int W(r)\exp[-i\mathbf{q}\cdot\mathbf{r})\,d\mathbf{r} = \tilde{W}(q) \qquad (10.14)$$

where V is the volume of the system.

Returning to $W(r)$, we can assume that it is made up of the sum of N individual ion contributions, $w(r)$ say, each of which moves with the ions so that

$$W(\mathbf{r}) = \sum_j w(\mathbf{r}-\mathbf{r}_j) \qquad (10.15)$$

The matrix elements between the plane wave states $|\mathbf{k}\rangle$ and $|\mathbf{k}+\mathbf{q}\rangle$ can now be simplified as follows

$$\langle \mathbf{k}+\mathbf{q}| W |\mathbf{k}\rangle = \tilde{W}(\mathbf{q}) = \frac{1}{V}\int \sum_j w(\mathbf{r}-\mathbf{r}_j)\exp(-i\mathbf{q}\cdot\mathbf{r})\,d\mathbf{r}$$

$$= \frac{1}{V}\sum_j \exp[-i\mathbf{q}\cdot\mathbf{r}_j]\int w(\mathbf{r}-\mathbf{r}_j)\exp[-i\mathbf{q}\cdot(\mathbf{r}-\mathbf{r}_j)]\,d\mathbf{r}$$

$$= S(\mathbf{q})\tilde{w}(\mathbf{q}) \qquad (10.16)$$

where the factorization follows if we make the change of variable $\mathbf{R} = \mathbf{r}-\mathbf{r}_j$ in the integration. The matrix elements are expressed as a product of two factors, the *structure factor* $S(\mathbf{q})$ and the *screened ion form factor*, $\tilde{w}(\mathbf{q})$. The structure factor is given by

$$S(\mathbf{q}) = \frac{1}{N}\sum_j \exp(-i\mathbf{q}\cdot\mathbf{r}_j) \qquad (10.17)$$

and depends only on the ion positions. The screened ion form factor is given by an integral over the volume occupied by one particle

$$\tilde{w}(\mathbf{q}) = \frac{1}{V}\int_{\Omega_0} w(r)\exp(-i\mathbf{q}\cdot\mathbf{r})\,d\mathbf{r} \qquad (10.18)$$

where $\Omega_0 = V/N$ and is the atomic volume. Notice that $\tilde{w}(\mathbf{q})$ is the Fourier transform of a single ion pseudopotential and that it is independent of the ion position. In this sense, we expect $w(r)$ to depend only on the distance r from the ion center so that it is a central, spherically symmetric potential. Hence the form factor will depend only on the magnitude of $\mathbf{q}$. If we did not use the local approximation, eq. 10.13, to obtain this result the form factor would depend on both the momenta, $\mathbf{q}$ and $\mathbf{k}$. Nevertheless, it can be shown that the factorization of the plane wave matrix elements (eq. 10.16) into the structure factor and the form factor is still possible, and this factorization is of central importance in the theory of liquid metals [1].

It is worth mentioning some of the approximations used to calculate the screened ion form factor. Ashcroft [4] has used perhaps the simplest approach, assuming that the ion-electron interaction is unscreened outside some radius r_c and that it is completely screened inside this distance. This approximation states that the unscreened pseudopotential is

$$w^0(r) = \begin{cases} 0 & r < r_c \\ \dfrac{-Ze^2}{r} & r_c \leq r \end{cases} \tag{10.19}$$

It follows that the Fourier transform, or form factor, is given by

$$\tilde{w}^0(q) = -\frac{4\pi Ze^2}{\Omega_0 q^2} \cos(qr_c) \tag{10.20}$$

A similar expression for the form factor has been proposed by Ho [5], who modified the Ashcroft model for the unscreened ion pseudopotential by assuming that the potential is slightly negative inside the core

$$w^0(r) = \begin{cases} -v_0, & r < r_c \\ \dfrac{-Ze^2}{r}, & r_c \leq r \end{cases} \tag{10.21}$$

In this case the unscreened form factor is given by

$$\tilde{w}^0(q) = -\left(\frac{4\pi}{\Omega_0 q^2}\right)\left[\frac{v_0}{q}\sin(qr_c) + (Ze^2 - v_0 r_c)\cos(qr_c)\right] \tag{10.22}$$

Other models have been proposed for the single ion form factor, most notably that of Heine and Abarenkov [6], whose model follows more closely the spirit of pseudopotential theory and incorporates both projection operators and nonlocal character in the model. Their model includes some dependence on the conduction electron energy in the pseudopotential and represents the unscreened potential as

$$w^0(r) = \begin{cases} \sum_l A_l(E)\mathbf{P}_l, & r < r_c \\ -\dfrac{Ze^2}{r}, & r_c \leqslant r \end{cases} \tag{10.23}$$

where $\mathbf{P}_l$ are projection operators using the lth angular momentum component of the wave function. The coefficients $A_l(E)$ are parameters which vary slowly with the energy of the conduction electron, E, and with the ion radius r_c. They were fitted to energy levels of the free ion obtained from spectroscopic measurements [7, 8]. The Heine-Abarenkov potential has been modified and extended by Shaw and other workers [9], whose more elaborate treatment is necessary for those metals where the core energy bands are not well separated from the conduction band energies. However, it appears to be quite adequate for simple metals that have well separated energy bands to simulate the ion field by a small constant local potential.

The screening effect of the conduction electrons outside the core can be added independently to each Fourier component of $w(r)$:

$$\tilde{w}(q) = \frac{\tilde{w}^0(q)}{\varepsilon(q)} \tag{10.24}$$

where $\varepsilon(q)$ is the *static dielectric function* of an electron gas [10]. This assumption is a good approximation in a metal, where the conduction electrons are very mobile. The dielectric function can be expressed as an analytic form, using the *Hartree dielectric function* [2, 10]

$$\varepsilon(q) = 1 + \frac{8\pi k_F m e^2}{h^2 q^2}\left\{1 + \frac{4k_F^2 - q^2}{4k_F q}\log\left|\frac{2k_F + q}{2k_F - q}\right|\right\} \tag{10.25}$$

where m is the electron mass and k_F is the *Fermi wave number.* The dielectric function is continuous everywhere although its first derivative has a logarithmic singularity at the point $q = 2k_F$. This singularity has a major effect on the Fourier transform of $\tilde{w}(q)$, producing the so-called *Friedel oscillations* in the pseudopotential in coordinate space at large distances.

We now wish to obtain a useful expression for the effective pair potential of a simple metal within the framework of pseudopotential theory. At this stage we have available an expression for the screened ion pseudopotential form factor, incorporating the dielectric function and the bare ion form factor, eq. 10.24. To determine the form of the interaction between two such ions we calculate the total energy per ion of a metal, assuming this to be the sum of a direct contribution, due to the Coulomb interactions and an indirect contribution, due to the presence of electrons.

The indirect ion-ion interaction is obtained from the *band structure energy*

$$E_{bs} = \frac{N\Omega_0}{8\pi^3} \int |S(q)|^2 F(q)\, d\mathbf{q} \tag{10.26}$$

where $F(q)$, the *energy-wave number characteristic*, depends on the choice of pseudopotential and the atomic volume but not on the position of the ions. Shaw [11] has considered the functional form of $F(q)$ and shows that for the local pseudopotential approximation

$$F(q) = \frac{\Omega_0 q^2}{8\pi e^2} \left\{ \frac{1}{\varepsilon(q)} - 1 \right\} |\tilde{w}^0(q)|^2 \tag{10.27}$$

Combining eqs. 10.17 and 10.26 we can write

$$\begin{aligned} E_{bs} &= \frac{\Omega_0}{4\pi^3 N} \int \left(\sum_j \exp[-i\mathbf{q}\cdot\mathbf{r}_j] \right) \left(\sum_k \exp[i\mathbf{q}\cdot\mathbf{r}_k] \right) F(q)\, d\mathbf{q} \\ &= \frac{\Omega_0}{4\pi^3} \int F(q)\, d\mathbf{q} + \frac{\Omega_0}{8\pi^3 N} \sum_{j\neq k} \int \exp[-i\mathbf{q}\cdot(\mathbf{r}_j - \mathbf{r}_k)] F(q)\, d\mathbf{q} \end{aligned} \tag{10.28}$$

The first term in this expression is independent of r whereas the second term has the form of a sum over pair interactions. Consequently, we can define the indirect potential between the ions to be

$$\phi_{\text{ind}}(r) = \frac{\Omega_0}{4\pi^3} \int \exp[-i\mathbf{q}\cdot\mathbf{r}] F(q)\, d\mathbf{q} = \frac{\Omega_0}{\pi^2 r} \int_0^\infty q F(q) \sin(qr)\, dq \tag{10.29}$$

so that the indirect contribution to E_{bs} takes the form

$$E_{bs}(\text{indirect}) = \frac{1}{2N} \sum_{j\neq k} \phi_{\text{ind}}(r) \tag{10.30}$$

We can now obtain the total effective atom-atom potential by adding to $\phi_{\text{ind}}(r)$ the direct Coulomb interaction between the ions so that

$$\phi(r) = \phi_{\text{direct}}(r) + \phi_{\text{ind}}(r) = \frac{Z^2 e^2}{r} + \frac{\Omega_0}{\pi^2 r} \int_0^\infty q F(q) \sin(qr)\, dq \tag{10.31}$$

which is the required result. The effective charge Z is that of the pseudoatom rather than that of the atom itself. It is fundamental to the theory of liquid metals that the structure dependent part of the internal energy can be expressed as a sum of two-body interactions.

The foregoing discussion has shown that the effective interatomic pair potential depends on the form of the energy-wave number characteristic $F(q)$, and hence on the unscreened ion form factor $\tilde{w}^0(q)$ and the dielectric function $\varepsilon(q)$. Unless these functions are assumed to have simple analytic forms, the pair potential in eq. 10.31 must be evaluated numerically at each value of r. In practice this does not present many problems despite the apparent long range r^{-1} dependence in $\phi(r)$. In fact, long range contributions from the integral in the second term effectively cancel such contributions from the first leaving an expression which falls off much more rapidly. This net cancellation gives a large positive repulsion at short distances followed by oscillatory behavior, the Friedel oscillations, which behave at large distances as $\cos(2k_F r)/r^3$. The general shape of the effective pair potential between metal atoms is indicated in Fig. 10.1. The detailed nature of Friedel oscillations is not yet fully understood and several studies [12] have suggested that the long range oscillatory interactions may be much weaker than previous studies have indicated, particularly in the alkali metals. Leribaux and Boon [13] have reported some calculations which suggest a reason for this attenuation. They show that in those liquid metals for which the form factor $\tilde{w}(q)$ is

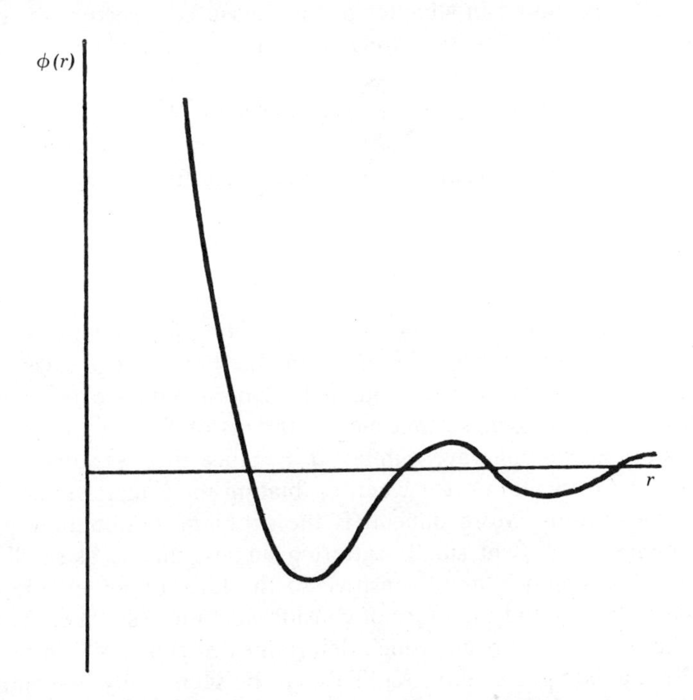

Figure 10.1 General shape of the effective pair potential between metal atoms.

close to zero at $q=2k_F$, the Friedel oscillations have nearly zero amplitude and are replaced by higher order oscillations proportional to $\cos(2k_F r)/r^5$. As we will indicate later, this uncertainty in the effective pair potential at long distances is rather difficult to detect experimentally. In any case, it does not detract from the usefulness of effective pairwise interactions in the study of liquid metals.

10.2 Relation Between Structure and Effective Interionic Potential

The bulk of the experimental results on liquid metals that are of interest to us come from X-ray scattering and neutron diffraction scattering data. We are not directly concerned with other studies of liquid metals such as electron transport, impurities in metals, and phonon spectra. The measured static structure factor $S(q)$ is the ratio of scattered intensity to incident intensity and plays a role quite similar to the lattice-type structure factor referred to in the previous section. Experimental data furnished by scattering experiments lead us directly to the radial distribution function $g(r)$. As noted in Chapter 5, the measured structure factor and the radial distribution function are Fourier transforms of each other

$$g(r)=1+\frac{1}{2\pi^2\rho r}\int_0^\infty q\sin(qr)[S(q)-1]\,dq \tag{10.32}$$

Equation 10.32 can be inverted, to express $S(q)$ in terms of $g(r)$:

$$S(q)=1+\frac{4\pi\rho}{q}\int_0^\infty r\sin(qr)[g(r)-1]\,dr \tag{10.33}$$

and so measurements of $S(q)$ can, in principle, be used to determine $g(r)$. In practice, difficulties arise because of the finite range over which measurements of $S(q)$ may be obtained. Experiments are restricted by maximum beam energies, so that measurements of $S(q)$ cannot be made beyond some maximum wave number $q_{\max}$. Since $S(q)$ approaches unity at large q, the truncation error when evaluating the integral in eq. 10.32 need not be serious. More difficult is the problem of obtaining precise measurements of $S(q)$ at small scattering angles, that is at small wave numbers. This region is quite sensitive to the finer details of the intermolecular potential and therefore of considerable interest. There is a well known thermodynamic result which determines $S(q)$ at $q=0$ in terms of the isothermal compressibility, K_T. This can be seen easily by comparing eq. 10.33 and eq. 3.63. At small q, $\sin(qr)=qr+O(q^3)$ so that eq. 10.33

becomes

$$S(q) = 1 + 4\pi\rho \int_0^\infty r^2[g(r) - 1]\, dr + O(q^2) \tag{10.34}$$

Using the result given in eq. 3.63 we obtain in the limit $q \to 0$

$$S(0) = \rho k T K_T \tag{10.35}$$

where k is the Boltzmann constant.

Theories of $g(r)$ can be used together with $S(q)$ obtained from scattering studies to estimate the interaction $\phi(r)$ operating between the ions in the metal. The principal theories used to estimate the effective pair potential are the Born-Green(BG), Percus-Yevick(PY), and hypernetted-chain(HNC) integral equations [14]. These approximate equations were introduced in Section 3.8 and their application to other liquid systems was discussed in later chapters. Before the BG equation can be used to estimate $\phi(r)$ it must be supplemented by the Kirkwood superposition approximation for the three-particle distribution function (see eq. 3.86). By substituting eq. 3.86 into eq. 3.83 and integrating the latter equation, we obtain after some manipulation the form of the BG equation used in studies of liquid metals:

$$\frac{\phi(r)}{kT} = -\log g(r) - \rho \int E(|\mathbf{r} - \mathbf{r}'|)(g(r') - 1)\, d\mathbf{r}' \tag{10.36}$$

where

$$E(t) = \frac{1}{kT} \int_\infty^t g(r) \frac{d\phi}{dr}\, dr \tag{10.37}$$

We have shown in eq. 3.102 and eq. 3.104 that the HNC and PY approximations can be written in terms of $g(r)$, the direct correlation function, $c(r)$, and the interaction potential. Solving these equations for $\phi(r)$ gives two approximations from which experimental estimates of $\phi(r)$ can be made

$$\frac{\phi(r)}{kT} = g(r) - 1 - c(r) - \log g(r) \tag{10.38}$$

from the HNC approximation and

$$\frac{\phi(r)}{kT} = \log\left[1 - \frac{c(r)}{g(r)}\right] \tag{10.39}$$

from the PY approximation. If we remember that $c(r)$ was defined by the Ornstein-Zernike equation

$$h(\mathbf{r}) = c(\mathbf{r}) + \rho \int c(\mathbf{s}) h(\mathbf{r} - \mathbf{s})\, d\mathbf{s} \tag{10.40}$$

where $h(r)=g(r)-1$, we can Fourier transform to obtain $\tilde{c}(q)=\tilde{h}(q)/[1-\rho\tilde{h}(q)]$ and then use eq. 10.33, giving

$$\tilde{c}(q)=\frac{[S(q)-1]}{S(q)} \tag{10.41}$$

Thus we can obtain both $c(r)$ and $g(r)$ from measurements of $S(q)$ and so estimate $\phi(r)$.

We have seen in earlier chapters that the approximate theories relating $\phi(r)$ and $g(r)$ are most accurate in the low density region. An alternative to using the approximations to estimate $\phi(r)$ is to begin from an assumed form for $\phi(r)$ and compute $S(q)$ using one of the theories, or machine simulations, and so test the usefulness of model pseudopotentials. Both approaches have had varying degrees of success and in the following section, we review the status of those scattering experiments and theoretical calculations for liquid metals that have been used to obtain information about the effective pair potential.

10.3 Results for Liquid Metals

Johnson and March [15] were among the first to suggest that integral equation approximations could be used in the theory of liquid metals. In their initial analysis they used the BG equation together with the superposition approximation to calculate the effective pair potential for liquid sodium from $g(r)$ obtained from neutron scattering data [16]. Subsequently, Johnson, Hutchinson, and March [17] used both the BG and PY theories to compute the effective pair potential for several metals, again using distribution functions obtained from neutron and X-ray scattering results. They found in all cases that their computed potentials for the liquid metals had a short range repulsion, a deep minimum, and was followed by marked long range oscillations. This latter result for the ion-ion potential was in sharp contrast to the corresponding result for the liquid inert gas pair potentials, such as that found for argon [18], which had a long range van der Waals tail. At first sight, it seemed to be a convincing confirmation of the theory of the effective ion-ion potential in terms of the pseudopotential developed in this chapter. However, subsequent analyses [19, 20] questioned whether the unambiguous determination of the pair potential from the measured liquid metal structure factor was possible. Difficulties were caused by the sensitivity of $S(q)$ at small wave numbers to the details of the pair potential. Ashcroft and Lekner [20], examined the structure of liquid metals from a different viewpoint, using the hard sphere potential to compute the structure factor in the PY approximation and comparing the result with experiment. The PY expression for $S(q)$ from the hard sphere potential was obtained analytically by Wertheim and Thiele [21] and contains a single independent parameter, σ, the hard sphere diameter. In particular, the hard sphere model predicts that the structure factor is independent of temperature. Ashcroft and Lekner [20] accounted for the temperature dependence of $S(q)$ by making σ temperature dependent. They were able to show that the structure factors of 17 metals near their melting points were fitted remarkably well by the hard sphere PY expression for $S(q)$.

The hard sphere PY results of Ashcroft and Lekner were extended to liquid sodium-potassium alloys by Enderby and North [22] who compared their results with X-ray data. Again they found that this relatively crude pair potential reproduced the general structural behavior given by the scattering data. Greenfield et al. [23] attempted to improve the agreement with experiment by introducing a second independent parameter. However, their generalization did not significantly improve the overall fit to the experimental data for sodium and potassium.

Blum and Narten [24] used an extension of the PY hard sphere model—the Mean Spherical Model—to obtain $g(r)$ for sodium and aluminum. In their calculations they used several model pair potentials developed from simulation studies [25, 26] (see eqs. 10.42, 10.43, and 10.44) and a somewhat more realistic effective interionic potential reported by Shyu, Singwi, and Tosi [27]. Blum and Narten obtained good agreement with the lengthy machine results from their simpler computational technique. Similarly, Leribaux and Miller [28] reported an optimized iterative solution of the PY equation for the liquid alkali metals near their melting points. Their radial distribution functions agreed favorably both with molecular dynamics calculations [25, 26] and with experimental data [16].

A somewhat different approach was considered by Jones [29] who used the Mansoori-Canfield variational method [30] to calculate the equation of state for liquid metals, using PY hard sphere $g(r)$ for the reference fluid. In this method, the hard sphere diameter is used as a variational parameter to minimize the free energy. Jones reported good agreement with the properties of liquid sodium.

The qualitative conclusion drawn from these results is that most details of the structure of liquid metals are determined by the short range ion core repulsions. The finer details of the potential, such as the Friedel oscillations, produce small quantitative effects which are superimposed on the dominant repulsive interactions. In particular, the long range part of the effective pair potential is most important for low wave number values of $S(q)$, the small angle scattering region which is most difficult to measure experimentally.

All the studies discussed so far have used one or more of the integral equation approaches to connect the pair potential with $S(q)$ and $g(r)$. We turn now to calculations based mainly on Monte Carlo (MC) and molecular dynamics (MD) simulations using effective pair potentials for liquid metals. In principle, these essentially "exact" calculations avoid the errors inherent in integral equation methods that are particularly important at low temperatures and high densities. In practice, machine calculations use interaction potentials truncated beyond a few molecular diameters, and they also contain errors arising from neglecting the long range potential. Fortunately, the errors in $S(q)$ can be minimized by assuming that $g(r)$ satisfies the PY equation for $r > r_c$ where r_c is the cutoff distance. This is usually an excellent approximation.

Paskin and Rahman [25] investigated the dynamics and structure of liquid sodium using long range oscillatory potentials of the form obtained by Johnson, Hutchinson, and March (JHM) [17]. The radial distribution function and self-diffusion coefficient were calculated using an MD simulation with 686 particles and a potential truncated at 8.18 Å. Their model potentials had the form

$$\phi(r) = -A\left(\frac{r_0}{r}\right)^3 \cos\left\{7.812\left[\left(\frac{r}{r_0}\right)+\beta\right]\right\}$$
$$+9050.0 \exp\left[5.0724 - 10.7863\left(\frac{r}{r_0}\right)\right] \text{K} \qquad (10.42)$$

where $r_0 = 3.72$ Å and A and β were taken as adjustable parameters. Using the values $A = 556$ K, $\beta = 0.5954$ chosen to fit the JHM potential, Paskin and Rahman were able to recover the radial distribution function from which the potential was obtained. (The potential was derived using PY and BG equations and the X-ray diffraction data of Orton et al. [31].) At the time, Paskin and Rahman were misled by a misprint in the JHM paper [17] and incorrectly assumed that the potential was obtained from neutron scattering results. For sodium, the two sets of data differed considerably. With modified parameters, $A = 313$ K, $\beta = 0.5689$ they were also able to fit the radial distribution function obtained from the neutron data reasonably well. Subsequently, Paskin [32] demonstrated that the existing data did not necessarily require a potential with long range oscillations. He used a Lennard-Jones 12-6 potential with parameters $\varepsilon = 620$ K, $\sigma = 3.32$ Å, to show that the large differences in the shape of the two potentials gives only small differences in $g(r)$ in the region 5–7 Å. Paskin's comparison of the two radial distribution functions is shown in Fig. 10.2. Apparently, the qualitative features of the short range liquid metal structure depend primarily on the depth and position of the potential minimum and not to any great extent on the potential shape. This suggests that a Law of Corresponding States argument may hold for the liquid metals, and this is shown in Fig. 10.3 where the radial distribution functions for the liquid alkali metals are scaled so that the position of the first maxima coincide [32]. Thus one expects that a combination of any reasonable computational technique and a realistic pair potential will reproduce the general features of $g(r)$ for liquid metals. Paskin noted that even the random packing of hard spheres [33] yields moderate agreement with experiment for the radial distribution function.

Schiff [26] also used computer simulation techniques to examine the sensitivity of $S(q)$ to potential models for liquid sodium and aluminum. He considered four oscillatory two-body effective pair potentials with both hard and soft cores. As an example, one potential for sodium had a

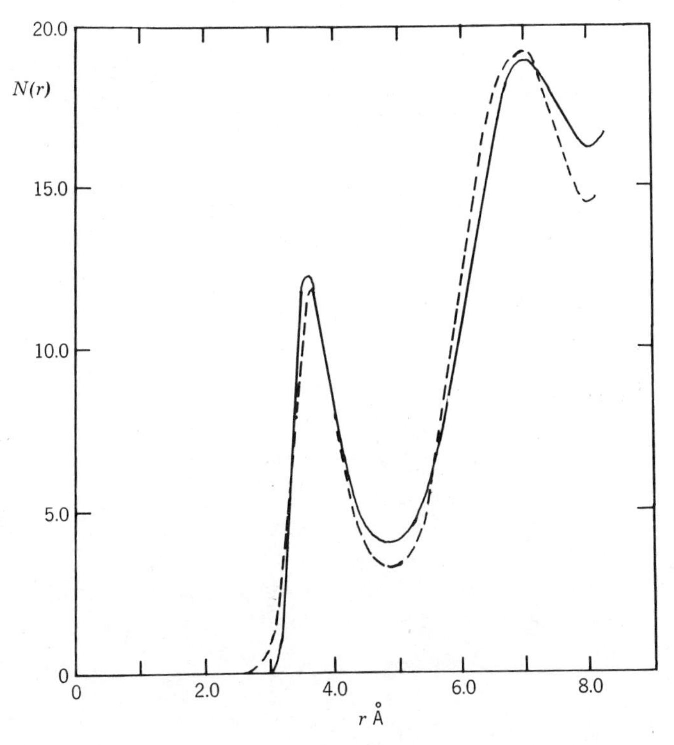

Figure 10.2 Radial distribution function for liquid Na near the melting point, in the form $N(r) = 4\pi r^2 g(r)$. ---- calculated using a long range oscillatory potential; ——— calculated using the LJ potential. (Adapted with permission from *Advan. Phys.*, **16,** 223, 1967.)

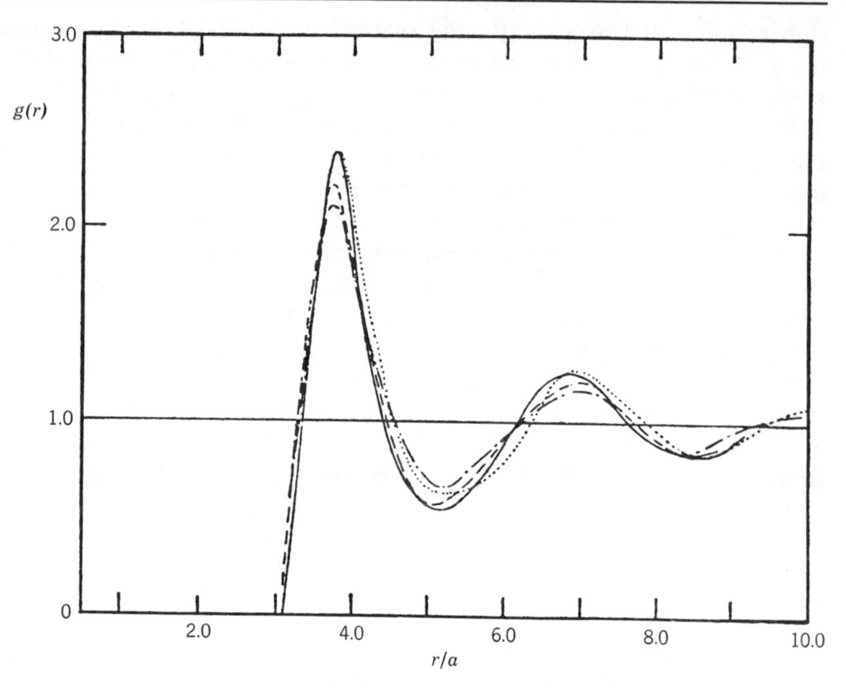

Figure 10.3 Radial distribution functions, of liquid metals and liquid argon near their melting points, scaled so that the positions of the first maxima coincide. · · · argon, $a = 1$ Å; —— cesium, $a = 1.39$ Å; – – – sodium, $a = 1$ Å; - - - - rubidium, $a = 1.3$ Å. (Adapted with permission from *Advan. Phys.*, **16,** 223, 1967.)

soft core of the Born-Mayer type and was written

$$\phi(r) = \left(\frac{A}{r^3} + \frac{B}{r^5}\right) \cos(2k_F r) + \left(\frac{C}{r^4} + \frac{D}{r^6}\right) \sin(2k_F r) + E \exp\left(F - G\frac{r}{r_0}\right) \qquad (10.43)$$

with parameters $A = -0.42$, $B = -0.56$, $C = -2.96$, $D = 1.46$, $E = 15.11$, $F = 5.07$, $G = 10.79$, $r_0 = 1.15$, $2k_F = 5.987$, given in terms of length and energy units $\sigma = 3.24$ Å and $\varepsilon = 599$ K. Another of his potentials, for aluminum in this case, was harder, having an inverse twelfth power repulsive core

$$\phi(r) = \left(\frac{A}{r^3} + \frac{B}{r^5} + \frac{C}{r^7}\right) \cos(2k_F r) + \left(\frac{D}{r^4} + \frac{E}{r^6}\right) \sin(2k_F r) + \frac{F}{r^{12}} \qquad (10.44)$$

with parameters $A = 0.66$, $B = 4.22$, $C = -2.61$, $D = -0.54$, $E = 0.67$, $F = 1.49$, $2k_F = 8.97$, given in terms of length and energy units $\sigma = 2.56$ Å and $\varepsilon = 1198$ K.

Schiff found that the oscillations in the structure factors were more damped for the softer potentials. However, differences between the structure factors calculated using the hard and soft cores was negligible for values of q less than that corresponding to the first maximum and only became prominent at large wave numbers. Even then, a noticeable effect on $S(q)$ was produced only by using unreasonably large oscillations in $\phi(r)$. This shows that it is very difficult to give convincing evidence for the existence of long range oscillations in the effective pair potential using experimental structure data alone.

Subsequently, Fowler [34] used MC calculations, together with the potential that Shyu, Singwi, and Tosi [27] derived from pseudopotential arguments, to compute $g(r)$ and $S(q)$ for sodium. Fowler found excellent agreement with $g(r)$ obtained from X-ray diffraction measurements [35] and from the MD work of Rahman [36]. In addition to computing $S(q)$ by Fourier transformation of $g(r)$, Fowler made a direct computation of $S(q)$ in the small q region, where the errors in the Fourier transformation are most significant, obtaining excellent agreement with experiment. At about the same time Murphy and Klein [37] reported a similar Monte Carlo calculation of $g(r)$ and $S(q)$ for liquid sodium. They used a realistic potential proposed by Duesbury and Taylor [38], and Basinski et al. [39], which was volume dependent but contained no adjustable parameters. In addition to giving good agreement with experimental scattering data, this potential also reproduced lattice constants and phonon frequencies of the solid [40]. Both the effective pair potentials mentioned in this paragraph have much attenuated long range oscillations. The effect of these oscillations on the structure factor is small, and it is unlikely that more refined calculations and measurements of $S(q)$ will provide a definitive test of their presence, or absence, in the effective pair potential for liquid metals.

10.4 Effective Pair Potentials for Molten Salts

Despite the presence of many body forces in ionic salts, investigations have shown that interactions between the ions can be represented by pairwise additive forces. This simplification has permitted the systematic study of the thermodynamic and structural properties of molten salts. Several interactions contribute to the ion-ion potential, the most important of which are the Coulomb interactions between the permanent charges on the ions. Next, there are terms due to interactions between the permanent charge on one ion and the dipole moments induced in other ions by the electrostatic fields arising from these charges. These terms are usually treated approximately, using the rigid ion approximation or a shell approximation [41]. The rigid ion model assumes that the polarizability of the ions is zero, so that the induced interactions can be ignored, and the shell models assume that a shell of polarizable material surrounds the ion nucleus so that the center of this shell can move relative to the center of the ions. Variants of the shell model include the breathing shell model, in which the radius of the polarizable shell can fluctuate. Other important terms in the ion-ion potential are also found in substances such as the inert gases. Thus there is a short range repulsion term which arises from the overlap forces between ions and van der Waals terms describing the long range dispersion forces.

One of the most frequently used effective pair potentials is the Born-Mayer-Huggins potential [42]. If two ions, labeled i and j, are a distance r apart this potential is

$$\phi_{ij}(r) = \frac{z_i z_j e^2}{r} + c_{ij} b \exp\left[\frac{\sigma_{ij} - r}{d}\right] - \frac{C_{ij}}{r^6} - \frac{D_{ij}}{r^8} \tag{10.45}$$

Here z_i, z_j are the ionic charges, C_{ij} and D_{ij} are the coefficients of the dipole-dipole and dipole-quadrupole dispersion terms, and σ_{ij} is the mean of the ionic diameters σ_{ii} and σ_{jj}. The parameter b is the same for all salts, d is a "hardness parameter" characteristic of the particular salt, and c_{ij} are numerical coefficients introduced by Pauling [43]. This potential has been applied successfully to liquid alkali halides and other molten salts. Parameters appearing in the repulsive term in eq. 10.45 have been obtained by Tosi and Fumi [44, 45] from an extensive analysis of the lattice constants, lattice energies, and compressibilities of the solid alkali halides. The exponential form used to describe the short range repulsion is sometimes replaced by an inverse power repulsion and, for example, Pauling [43] used

$$\frac{c_{ij}(b_i + b_j)^m}{r^n} \tag{10.46}$$

where b_i, b_j are constants characteristic of the ions, and m, n are taken to be the same for all salts. Parameters in the interionic potentials useful for the alkali halides were given by Woodcock and Singer [46] and are listed in Table 10.1. As an example of the typical shape for these ion-ion potentials we show in Fig. 10.4 the potassium chloride potentials taken from this table. As might be expected, the like-ion potentials are repulsive everywhere and only the unlike-ion potential has an attractive region.

Table 10.1 Parameters in the Tosi-Fumi and Pauling interionic potentials for alkali halides

Salt	C_{++}	C_{+-} $\times 10^4$ K Å^6	C_{--}	D_{++}	D_{+-} $\times 10^4$ K Å^8	D_{--}	σ_{++}	σ_{--} Å	d	b_+ $\times(10^{12}$ K Å$^n)^{1/m}$	b_-
LiF	0.053	0.58	10.5	0.02	0.43	12.3	1.632	2.358	0.299	10.0	16.7
NaF	1.22	3.26	12.0	0.58	2.75	14.5	2.340	2.358	0.330	13.6	16.7
KF	17.6	14.1	13.5	17.4	15.2	15.9	2.926	2.358	0.338	16.7	16.7
RbF	43.0	22.4	13.7	59.3	28.9	16.7	3.174	2.358	0.328	17.8	16.7
LiCl	0.053	1.5	80.4	0.02	1.7	162.0	1.632	3.170	0.342	10.0	19.9
NaCl	1.22	8.11	84.0	0.58	10.1	169.0	2.340	3.170	0.317	13.6	19.9
KCl	17.6	34.8	90.2	17.4	52.9	181.0	2.926	3.170	0.337	16.7	19.9
RbCl	43.0	57.2	94.2	59.3	97.1	188.0	3.174	3.170	0.318	17.8	19.9
NaBr	1.22	10.1	142.0	0.58	13.7	326.0	2.340	3.432	0.340	13.6	20.8
KBr	17.6	43.5	149.0	17.4	71.7	340.0	2.926	3.432	0.335	16.7	20.8
RbBr	43.0	71.7	156.0	59.3	130.0	355.0	3.174	3.432	0.335	17.8	20.8

For lithium salts the coefficients $c_{++} = 2.0$, $c_{+-} = 1.375$, and $c_{--} = 0.75$; for all other salts $c_{++} = 1.25$, $c_{+-} = 1.0$, and $c_{--} = 0.75$. The parameters $b = 2448$ K and in the inverse power potential $n = 8.28$ and $m/n = 1.5$.

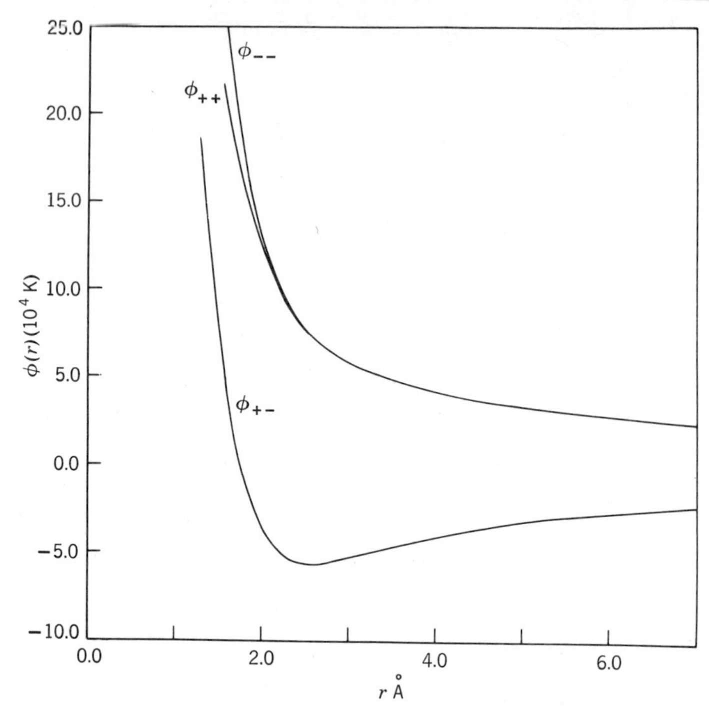

Figure 10.4 Ion-ion potentials for potassium chloride using the Born-Mayer-Huggins form of repulsion.

Recent calculations with effective pair potentials for molten salts have almost always used machine simulation methods. In principle, the methods are similar to those used for studies of the inert gases, although in practice difficulties arise from the very long range Coulombic interactions. A lattice, or liquid, of like-charged ions is inherently unstable and so some care must be taken with these terms. Normally, interactions are summed over all pairs of ions separated by a distance less than $\frac{1}{2}L$, where L is the length of one side of the periodic cube containing the fluid, and a correction term to account for interactions at long range is added later. For molten salts, a significant contribution to the total energy comes from Coulombic interactions outside this distance and consequently these terms must be included at each step in the simulation. One of the most common methods for computing the long range Coulomb energy contributions is the Ewald summation method [47] although other procedures have been tried [48]. This method was developed for lattice sums in crystalline solids and its use in liquid state calculations depends on the periodic image convention described in Section 3.10.

The Ewald method divides the slowly converging sum for the electrostatic energy per unit cell into two series which converge rapidly [41]. We can write the single particle charge density for a system of point charges, with charges z_k at sites $\mathbf{r}_k$

$$\rho(\mathbf{r}) = \sum_k z_k\, \delta(\mathbf{r}-\mathbf{r}_k) \tag{10.47}$$

where $\delta(\mathbf{r}-\mathbf{r}_k)$ is the Dirac delta function. This sum can be divided into two parts by adding and subtracting a normalized Gaussian charge distribution so that

$$\rho(\mathbf{r}) = \rho_1(\mathbf{r}) + \rho_2(\mathbf{r}) \tag{10.48}$$

where

$$\rho_1(\mathbf{r}) = \sum_k [z_k\, \delta(\mathbf{r}-\mathbf{r}_k) - \alpha^3 \pi^{-3/2} \exp(-\alpha^2 |\mathbf{r}-\mathbf{r}_k|^2)]$$

$$\rho_2(\mathbf{r}) = \sum_k z_k [\alpha^3 \pi^{-3/2} \exp(-\alpha^2 |\mathbf{r}-\mathbf{r}_k|^2) - L^3] \tag{10.49}$$

The second term in $\rho_2(\mathbf{r})$ represents a continuous charge distribution of opposite sign to the Gaussian charge distribution given in its first term. This continuous distribution gives zero net contribution to the charge distribution for a system of equal numbers of positive and negative ions. The identical (except for sign) Gaussian terms in $\rho_1(\mathbf{r})$ and $\rho_2(\mathbf{r})$ are normalized, with half width α^{-1}, a parameter that is later chosen to optimize convergence of the lattice sums. The potential energy contribution corresponding to $\rho_1(\mathbf{r})$ is given by

$$\phi_1 = e^2 \sum_i z_i \sum_{j>i} z_j \frac{\operatorname{erfc}(\alpha r_{ij})}{r_{ij}} - \frac{\alpha e^2}{\sqrt{\pi}} \sum_i z_i^2 \tag{10.50}$$

where r_{ij} is the distance between ions i and j, erfc is the complementary error function

$$\operatorname{erfc}(x) = \frac{2}{\sqrt{\pi}} \int_x^\infty \exp(-u^2)\, du \tag{10.51}$$

and where the summations are taken over all ions in the cell. Similarly, the potential energy corresponding to the charge distribution $\rho_2(\mathbf{r})$ can be written

$$\phi_2 = \frac{e^2}{\pi L} \sum_i z_i \sum_j z_j \sum_{\mathbf{n}}{}' \frac{\exp(-\pi^2 |\mathbf{n}|^2/\alpha^2 L^2)}{|\mathbf{n}|^2} \cos \frac{2\pi \mathbf{n}\cdot\mathbf{r}_{ij}}{L} \tag{10.52}$$

The sum over $\mathbf{n}$ is taken over reciprocal lattice vectors of the simple cubic cell [49] but excluding the zero vector and including only one of each pair $\mathbf{n}' = -\mathbf{n}$. The triple summation in ϕ_2 is time consuming, although Adams and McDonald [50] have shown how to rearrange the expression into two double summations which can be computed more quickly. Returning to the problem of evaluating the interaction potential, the rapid convergence of ϕ_2 is apparent from the presence of the negative exponential factor multiplying each term. The rapid convergence of ϕ_1 arises from the fact that the Gaussian and point charge distributions cancel at large distances. Note that the size of the parameter α has opposite effects on the convergence of ϕ_1 and ϕ_2, because of the way it appears in the exponentials in the two cases. Thus if α is small, ϕ_1 converges slowly and ϕ_2 converges rapidly whereas the converse holds if α is large. Since the application of the Ewald method involves truncation errors from two infinite series, a compromise must be made between accuracy and expenditure of computer time. A typical value of α that gives good convergence is $\alpha \approx 5/L$ [46, 50].

10.5 Results for Molten Salts

Early work on liquid ionic salts and their mixtures concentrated on studies of their thermodynamic properties using the Law of Corresponding States [51, 52]. The basic assumption was that the effective pair potentials in a given class of molten salts differed only by a scaling factor. Given this premise, it was possible to obtain good qualitative agreement with experiment for a large class of ionic salts using dimensional arguments alone.

One of the first exploratory Monte Carlo calculations of the structural properties of liquid ionic salts was undertaken by Woodcock and Singer [46]. They used the Born-Mayer-Huggins type potential, eq. 10.45, with parameters appropriate to potassium chloride. Using 216 particles (108 cations, 108 anions) in their simulation, Woodcock and Singer calculated $g_{ij}(r)$ together with several thermodynamic quantities, including the internal energy, pressure, molar heat capacities, and compressibility at several temperatures. Their calculations gave good agreement with experiment as well as new information about the structure of the liquid salt. Following this, Woodcock [53] reported molecular dynamics calculations for other liquid alkali chlorides.

Prompted by experimental results on the Raman spectrum of the fused salt solution, Sundheim and Woodcock [54] investigated the liquid K_2MgCl_4 using a molecular dynamics computation in an effort to analyze the spectrum in terms of ion cluster structures. The potential was taken as the sum of pairwise additive interactions, similar to those discussed earlier but omitting the van der Waals dispersion terms. The radial distribution functions were obtained for each of the ion pairs. Sundheim and Woodcock analyzed the magnitude and shape of the nearest neighbor peaks in $g_{ij}(r)$ and detected symmetry arrangements which were favored in the molten salt. Their approach, using purely ionic potentials, provided a valuable tool for investigating structural details of such liquids. Rahman et al. [55] undertook a similar investigation of the liquid BeF_2 and of its mixtures with LiF, again in an attempt to describe the structure in terms of central ionic forces. Their distribution functions yielded average nearest neighbor distances and coordination numbers in good agreement with results derived from X-ray diffraction studies [56].

Following earlier investigations of pure salts, [46, 57] Larsen et al. [58] used Monte Carlo calculations to explore the thermodynamic properties of mixtures of molten salts. Using a NaCl-KCl mixture, and a Born-Mayer-Huggins form for the ion-ion potential, they concluded that detailed information about the mixing process could be obtained from Monte Carlo calculations provided potential parameters for the salt mixture were known.

Adams and McDonald [50] have made a Monte Carlo study of several alkali halides at room temperature and at temperatures near their melting points. They used two different forms of the repulsive interaction, eqs. 10.45 and 10.46, using the parameters of Tosi and Fumi (see Table 10.1). They found that the Born-Mayer-Huggins form of the potential was superior to the Pauling form in all respects, although there was evidence to suggest that both models were not sufficiently flexible to represent the experimental data at the higher temperatures. It may be that potential models for high temperature calculations must include the polarizability of the ions.

The discussion in this section is by no means an extensive summary of the research done on molten salts but rather is an attempt to show the direction in which recent studies have developed. It appears that one can obtain reasonable quantitative agreement with experiment by assuming that central potentials act between the ion pairs. Furthermore, machine simulations with these potentials provide valuable evidence from which the structural details of molten salts may be determined.

References

1. J. E. Enderby, 1972, Chapter 14 in *Liquid Metals, Chemistry and Physics*, S. Z. Beer, Ed., Dekker, New York.
2. I. M. Torrens, 1972, *Interatomic Potentials*, Academic, New York.

3. P. A. M. Dirac, 1947, *The Principles of Quantum Mechanics*, 3rd ed., Oxford University Press.
4. N. W. Ashcroft, 1966, *Phys. Lett.*, **23,** 48; 1968, *J. Phys., C (London)*, **1,** 232.
5. P. S. Ho, 1968, *Phys. Rev.*, **169,** 523.
6. V. Heine and I. V. Abarenkov, 1964, *Phil. Mag.*, **9,** 451.
7. A. O. E. Animalu and V. Heine, 1965, *Phil. Mag.*, **12,** 1249.
8. I. V. Abarenkov and V. Heine, 1966, *Phil. Mag.*, **12,** 529.
9. R. W. Shaw and W. A. Harrison, 1967, *Phys. Rev.*, **163,** 604; R. W. Shaw, 1968, *ibid.*, **174,** 769.
10. J. Lindhard, 1954, *Kgl. Dan. Vidensk. Selsk. Mat.-Fys. Medd.*, **28,** 8.
11. R. W. Shaw, 1969, *J. Phys., C (London)*, **2,** 2335.
12. N. W. Ashcroft, 1972, in *Interatomic Potentials and Simulation of Lattice defects*, P. C. Gehlen, J. R. Beeler Jr., and R. I. Jaffee, Eds., p. 91, Plenum, New York.
13. H. R. Leribaux and M. H. Boon, 1975, *Phys. Rev.*, B**11,** 2412.
14. R. O. Watts, 1973, in *Specialist Periodical Reports of the Chemical Society; Statistical Mechanics*, K. Singer, Ed., Vol. 1. Chemical Society, London.
15. M. D. Johnson and N. H. March, 1963, *Phys. Lett.*, **3,** 313.
16. N. S. Gingrich and L. Heaton, 1961, *J. Chem. Phys.*, **34,** 873.
17. M. D. Johnson, P. Hutchinson, and N. H. March, 1964, *Proc. Roy. Soc. (London), Ser. A*, **282,** 283.
18. J. A. Barker, R. A. Fisher, and R. O. Watts, 1971, *Mol. Phys.*, **21,** 657.
19. D. M. North, J. E. Enderby, and P. A. Egelstaff, 1968, *J. Phys., C (London)*, **1,** 1075; P. C. Gehlen and J. E. Enderby, 1969, *J. Chem. Phys.*, **51,** 547.
20. N. W. Ashcroft and J. Leckner, 1966, *Phys. Rev.*, **145,** 83.
21. M. S. Wertheim, 1963, *Phys. Rev. Lett.*, **10,** 321; E. Thiele, 1963, *J. Chem. Phys.*, **39,** 474.
22. J. E. Enderby and D. M. North, 1968, *Phys. Chem. Liquids*, **1,** 1.
23. A. J. Greenfield, N. Wiser, M. R. Leenstra, and W. van der Lugt, 1972, *Physica*, **59,** 571.
24. L. Blum and A. H. Narten, 1972, *J. Chem. Phys.*, **56,** 5197.
25. A. Paskin and A. Rahman, 1966, *Phys. Rev. Lett.*, **16,** 300.
26. D. Schiff, 1969, *Phys. Rev.*, **186,** 151.
27. W. M. Shyu, K. S. Singwi, and M. P. Tosi, 1971, *Phys. Rev.*, B**3,** 237.
28. H. R. Leribaux and L. F. Miller, 1974, *J. Chem. Phys.*, **61,** 3327.
29. H. D. Jones, 1971, *J. Chem. Phys.*, **55,** 2640.
30. G. A. Mansoori and F. B. Canfield, 1969, *J. Chem. Phys.*, **51,** 4958.
31. B. R. Orton, B. A. Shaw, and G. I. Williams, 1960, *Acta Met.*, **8,** 177.
32. A. Paskin, 1967, *Advan. Phys.*, **16,** 223.
33. J. D. Bernal, 1964, *Proc. Roy. Soc. (London), Ser. A*, **280,** 299; G. D. Scott, 1964, *Nature (London)*, **201,** 382.
34. R. H. Fowler, 1973, *J. Chem. Phys.*, **59,** 3435.
35. A. J. Greenfield, J. Wellendorf, and N. Wiser, 1971, *Phys. Rev.*, A**4,** 1607.
36. A. Rahman, 1971, in *Interatomic Potentials and Simulation of Lattice Defects*, P. C. Gehlen, J. R. Beeler Jr., and R. I. Jaffee, Eds., p. 233, Plenum, New York.
37. R. D. Murphy and M. L. Klein, 1973, *Phys. Rev.*, A**8,** 2640.
38. M. S. Duesbery and R. Taylor, 1969, *Phys. Lett.*, **30**A, 496.
39. Z. S. Basinski, M. S. Deusbery, A. P. Pogany, R. Taylor, and Y. P. Varshni, 1970, *Can. J. Phys.*, **48,** 1480.
40. R. R. Glyde and R. Taylor, 1972, *Phys. Rev.*, B**5,** 1206; D. J. W. Geldart, R. Taylor, and Y. P. Varshni, 1970, *Can. J. Phys.*, **48,** 183.

41. W. Cochran, 1971, "Lattice Dynamics of Ionic and Covalent Crystals," in *CRC Critical Reviews in Solid State Sciences.*
42. M. L. Huggins and J. E. Mayer, 1933, *J. Chem. Phys.*, **1,** 643.
43. L. Pauling, 1928, *J. Am. Chem. Soc.*, **50,** 1036.
44. M. P. Tosi and F. G. Fumi, 1964, *J. Phys. Chem. Solids*, **25,** 45.
45. F. G. Fumi and M. P. Tosi, 1964, *J. Phys. Chem. Solids*, **25,** 31.
46. L. V. Woodcock and K. Singer, 1971, *Trans. Faraday Soc.*, **67,** 12.
47. P. P. Ewald, 1921, *Ann. Phys.* (*Leipzig*), **64,** 253.
48. H. M. Evjen, 1932, *Phys. Rev.*, **39,** 675.
49. C. Kittel, 1966, *Introduction to Solid State Physics*, 3rd ed., Wiley, New York.
50. D. J. Adams and I. R. McDonald, 1974, *J. Phys., C* (*London*), **7,** 2761.
51. H. Reiss, S. W. Mayer, and J. L. Katz, 1961, *J. Chem. Phys.*, **35,** 820.
52. M. Blander, 1967, *Adv. Chem. Phys.*, **11,** 83.
53. L. V. Woodcock, 1971, *Chem. Phys. Lett.*, **10,** 257.
54. B. R. Sundheim and L. V. Woodcock, 1972, *Chem. Phys. Lett.*, **15,** 191.
55. A. Rahman, R. H. Fowler, and A. H. Narten, 1972, *J. Chem. Phys.*, **57,** 3010.
56. W. R. Busing, 1972, *J. Chem. Phys.*, **57,** 3008.
57. J. Krog-Moe, T. Østvold, and T. Førland, 1969, *Acta Chem. Scand.*, **23,** 2421.
58. B. Larsen, T. Førland, and K. Singer, 1973, *Mol. Phys.*, **26,** 1521.

Author Index

The listing indicates pages on which an author's work is referred to, although his name is not necessarily cited in the text. Numbers in *italics* indicate the pages on which the complete references are given.

Subject Index